Natural Gas: A Commercial Perspective

Andrej Pustišek · Michael Karasz

Natural Gas: A Commercial Perspective

Springer

Andrej Pustišek
University of Applied Sciences, Stuttgart
Karlsruhe Institute of Technology (KIT), Karlsruhe
2Pi-Energy GmbH, Stuttgart
Germany

Michael Karasz
The Energy House GmbH
Leipzig
Germany

ISBN 978-3-319-53248-6 ISBN 978-3-319-53249-3 (eBook)
DOI 10.1007/978-3-319-53249-3

Library of Congress Control Number: 2017930165

Printed on acid-free paper

This Springer imprint is published by Springer Nature
The registered company is Springer International Publishing AG
The registered company address is: Gewerbestrasse 11, 6330 Cham, Switzerland

Contents

About the Authors

Andrej Pustišek teaches energy economics at the University of Applied Science, Stuttgart, Germany and at the University of Karlsruhe (KIT), Germany, owns a consulting companys providing services to international natural gas and energy companies, *inter alia*, in arbitration proceedings, and lectures international courses on natural gas. He studied Geophysics and Economy in Vienna, Cologne and Münster. He holds a Ph.D. in Geophysics and a Ph.D. in Economy. In 1990, he started in international natural gas purchasing of Wintershall, Kassel, Germany. As of 1993 he headed different departments of WINGAS, Kassel, Germany (a Gazprom and BASF joint venture). From 2002 to 2012, he worked for E.ON Ruhrgas, Essen, Germany; first as Senior Vice President International Natural Gas Sales and from 2008 as Senior Vice President Portfolio Management.

Michael Karasz is managing director of THE ENERGY HOUSE GmbH a company he founded in 2011. He holds master degrees in Geology and Business Administration from the University of Cologne, Germany. His long-lasting involvement in the natural gas business started in the early 90's at WINGAS, Kassel, Germany (a Gazprom and BASF joint venture). As of 1998 he held different senior management positions (*inter alia*, natural gas purchasing and portfolio management) both, in Germany and Central & Eastern Europe with E.ON, EEX, RWE, and Bayerngas. In these positions he successfully negotiated and concluded complex long-term purchase, transportation, storage, and sales agreements. He has extensive expertise in preparing and successfully executing international arbitration proceedings.

List of Figures

List of Tables

1 Prologue

Odorless and colorless, 'conventional' but clean, easy-to-handle, available for more than the future century, relatively simple and uniform in composition, natural gas can be used for multiple purposes. It is a source of primary energy for heating, industrial processes, electricity production or transportation, a feedstock, or can even be converted to liquids. As such, natural gas has been long regarded as the (conventional) fuel of the future. Growth rates of natural gas' shares in primary energy consumption have been impressing and significant—worldwide. Infrastructure, such as storages, pipelines, and LNG facilities, has been constructed successfully and helped fulfilling demand requirements while safeguarding security of the supply and flexibility of deliveries.

On a worldwide scale, the demand for natural gas keeps increasing. Yet, consumption in European markets, with a few notable exceptions, either stagnates or decreases and might continue to do so. Obviously, in these markets, natural gas is considered (relatively) expensive, still CO_2 polluting, at least partly 'captured' in mono- or oligopolistic market structures and less secure than alternative fuels, as it is predominantly network based and available from remote sources only—more and more from outside the European Union. In short, at the time of writing this book, market participants in Europe mostly perceive that the political support for natural gas, granted throughout the past decades, diminishes. This, however, does not render natural gas insignificant or even irrelevant. In most parts of the world, natural gas is well established as one of the most important, with regard to market share, primary fuels. This applies, despite decrease or stagnation, also to Europe.

Given the importance of natural gas as a primary source of energy, it remains surprising why only a few authors are engaged in discussing the challenging, interesting, and dynamic subject of the natural gas business. Therefore, it is the authors' intention to deliver a contribution to this subject by making the reader familiar with, piquing his curiosity in more details about, and commanding admiration of the up-to-date achievements of the natural gas industry.

To this end, a textbook is provided covering the commercial and economic topics of the natural gas business and markets and discussing interdependencies with other fuels and between international natural gas markets, as well as the impact of technical issues on (economic or commercial) results.

With such objectives and restrictions, the book may serve to provide novices in the natural gas business with first ideas. Whether the novice is a bachelor student, a trainee in a company or (governmental) institution, or an (otherwise) experienced manager accepting a challenge in the natural gas business, is irrelevant. The intention is that all readers get an understanding of the basic concepts of the natural gas industry. To support this understanding, the general description will be illustrated by 'real-world examples'. These examples are—probably needless to say, as they are written by European authors with

A. Pustišek and M. Karasz, *Natural Gas: A Commercial Perspective*,
DOI 10.1007/978-3-319-53249-3_1

European backgrounds—inevitably, European biased.

In addition, we endeavor to provide ideas, background information, and several detailed discussions of advanced topics, which we hope, might be useful for the practitioners' day-to-day business. Nonetheless, the selection of the topics discussed in this book is unavoidably subjective, as, of course, there are no logical (or, in any other way, objective) rules for their selection.

It is equally important to note what this book does not attempt to accomplish. This list is (necessarily) longer than the one above. Among the important topics, which may be expected in a book about natural gas, are: the description of the technical background or details, be it for exploration, production, transportation, storage, distribution, or utilization. So, the descriptions of these subjects provided in this book just illustrate in highly simplified manner the basics considered necessary for an understanding of the commercial perspective of the industry. Hence, the treatment of these topics merely serves the purpose to provide an overview. The interested reader is referred to the further reading (see Appendix A). The same applies to the description of (general or specific) economic or commercial, mathematical, and geological, physical or chemical basics and details. In the section on portfolio management, the main principles are described but not discussed in detail. This applies to methods used for portfolio optimization and risk management. The reason is simple; any textbook-like approach will necessarily over-simplify and, consequently, lack practical relevance,[1] or, if, e.g., portfolio optimization were covered in depth, the section would by far exceed the intended size of the entire book.

Furthermore, national details of sales, transportation, and storage, including the contracts and regulations on various levels of the natural gas industry, will be discussed only when used as examples—it is not the intention of this book to provide details of individual markets, neither is it intended to serve as a database. The intention of providing selected market data is to support the reader's understanding of markets. As we have experienced rapid and remarkable changes of objectives, tasks, strategies, and even names of companies, institutions, and associations, in the general text we try, as far as reasonably possible, to refrain from referring to specific ones. Finally, we will refrain from discussing or describing organizational issues. Natural gas companies worldwide are not uniformly organized, neither the processes nor the structures. Moreover, organization helps companies coping with the challenges of the business but an understanding of the organization is not considered essential for an understanding of the business.

Since we are convinced that only by combining experience with knowledge or, in other words, practice with theory, will the reader gain a (profound) understanding of the subject, exactly this amalgamation of alleged antipodes is attempted in this book. Successful or not—this judgement shall be left to the reader.

It has become a habit to express sincere thanks to a (sometimes) long list of friends and colleagues at the end of a prologue. We would like to restrict ours to a short list. When thanking in particular (in alphabetical order) Silke Baumann, Darya Bomberger, Georg Fuchs, Harald Granser, Harald Kraft, and Heiko Lohmann for their support and help preparing the manuscript, this does not mean that all valuable, challenging, and intellectually thrilling comments and advices from others are neglected or forgotten. By expressing our general wholehearted anonymous thanks, we are at least sure not to have omitted anyone.

[1]Therefore, we will also refrain from formulating any approach in mathematical form.

2 Introduction

2.1 Concept

The guiding principle for structuring this book is the natural gas value chain—exactly in this order, i.e., 'natural gas' and then (its) 'value chain'.

Explanation of commercially relevant elements of what is commonly referred to as the 'natural gas business' is believed to be achieved best by choosing a 'mid-stream bias', i.e., by focusing on the stages of transportation, storage, and sales.[1] Implicitly or explicitly, such discussion will involve all 'levels' of the natural gas industry, i.e., all distinguishable groups of enterprises active in bringing natural gas from the source to the sink, i.e., to the end user's burner tip.[2]

The detailed discussion of sales, transportation, and storage is preceded by brief discussions of fundamentals, specific for the natural gas industry. First, an overview of worldwide natural gas market is provided by depicting selected figures, and then non-economic and non-commercial fundamentals are discussed. This comprises the description of the chemical and physical properties of natural gas (in more-or-less technical terms) and environmental effects, as well as—to facilitate the general understanding but not being a focus of this book—sources and sinks, i.e., the initial and final stages of the value chain, namely, 'exploration and production' and 'final usage'. However, the description of these stages will have to remain superficial. In the next preliminary treatment, i.e., the economic and commercial fundamentals, the 'product' and the market structures are described. This delineates the framework for the subsequent description of the selected stages of sales, transportation, and storage. As all entrepreneurial activities are reflected in contracts, contractual elements often are used to structure and explain the subject. Basic elements of portfolio management, i.e., the main ideas and objectives, are used to demonstrate the interdependence of the stages, strategic and tactic options for market participants, and the importance of risk identification.

To facilitate reading, units and a glossary of terms are included, as well as a list of literature recommended for further reading that covers primarily basics, which could not and should not be mentioned or discussed in this book, *inter alia*, the economic and technical background knowledge required in the natural gas business. Throughout the book, case studies and examples will be included wherever appropriate.

In short[3]: Natural gas, a hydrocarbon, formed over millions of years, is trapped in subsurface reservoirs worldwide. Various, however, predominantly seismic, methods of exploration are applied in order to 'find', i.e., discover, the

[1]See Fig. 2.1. The separation of these three stages of the value chain is, by default, consistent with the requirements of unbundling, see Sect. 5.3.

[2]These levels are represented by producers, importers, regional utilities, regional distributors, local distributors, and end user for the commodity sales. For transportation and for storage the equivalent transport system operators and storage system operators exist. See also the market structure in Sect. 5.4.

[3]Partly depicted in Fig. 2.1.

A. Pustišek and M. Karasz, *Natural Gas: A Commercial Perspective*,
DOI 10.1007/978-3-319-53249-3_2

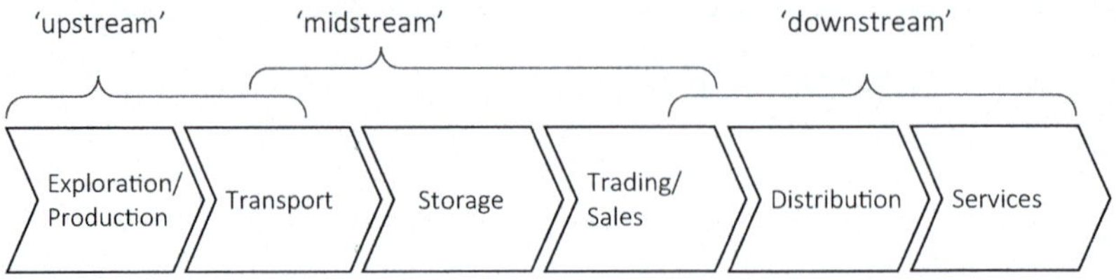

Fig. 2.1 Simplified natural gas value chain

reservoirs and, subsequently, to start production. The natural gas produced, i.e., 'brought to the surface', is processed, eventually cooled to be transported as LNG, and transported, in liquid or gaseous state, from the production site to the region or country, where it is consumed, i.e., (primarily) combusted to accomplish the energetic requirements of households or industrial end users, to produce electricity, or used as a feedstock. As demand for quantities at each point in time typically does not comply with production, natural gas is stored on its way from source to sink—often in vicinity of consumption sites. Additional services may be offered to the end user—primarily to increase demand and reduce problems of utilization.

Market participants acting at each level[4] could cover several, but not necessarily all, stages of the value chain. So, even if, e.g., a regional utility (or regional distributor) is neither active in exploration and production nor in sales to end users, it might offer sales, transportation, storage, and other services, i.e., act along a selected stage or stages of the value chain. However, products, prices, and costs incurred can best be distinguished for the stages of the value chain. Moreover, the stages of the value chain are, contrary to some 'levels' of the industry, universally present. Hence, the description of commercial perspectives for the industry will be structured 'along the stages of the value chain', not for different levels of the industry. For sales, transportation, and storage, the products and pricing principles are explained in detail. In addition, some details of technical basics of transportation and storage as well as costs will be given.

Any commercial transaction is based on contracts. Yet, despite being essentially a legal topic, commercially relevant contractual issues, be it for sales, transportation or storage, are an integral part of any explication of the natural gas business. The business and even its technical opportunities and restrictions are reflected in contracts and, therefore, such contracts are both instructive and relevant. Consequently, the main commercially relevant contractual elements, in addition to products and prices, will be included in each of the sections covering one element of the value chain. By default, this will also be consistent with the regulatory requirements concerning, in particular, unbundling.

2.2 Historic Outline

While the history of oil can be traced through a broadly internationally perspective, the natural gas industry was developed differently in different regions. Nothing like a 'worldwide history of the natural gas industry' exists. Therefore, the history of the industry will be outlined by describing briefly the development of the natural gas industry in selected regions.[5]

2.2.1 North America

The first natural gas company of the world was Fredonia Gas Light Company founded in 1858.

[4]For an explanation and illustration of the 'levels of the industry', see Sect. 5.5 and Fig. 5.4.

[5]The interested reader is referred to various national publications. An overview, broader and deeper than the one presented here, is given by Colombo et al. (2016).

The first natural gas well was drilled more than 30 years earlier in 1821. Yet, until the middle of the Nineteenth century, 'town' gas dominated the market. This was nearly exclusively used for lighting. Heating, electricity production, and industrial usages were developed during the second half of the Nineteenth century only. In the 1890s, natural gas 'pipelines' began to emerge from Texas and Oklahoma. In the 1950s, 93% of world's natural gas was produced in the US. Continuous growth caused natural gas to account for 44% of primary energy consumption in 1965. Autarky in the US ended at the beginning of the 1970s, when natural gas imports from Canada began. In later years, increasing natural gas consumption in the US was also supplied by LNG imports. LNG-receiving terminals were constructed primarily in the Northeast and the South of the US. The shale-gas boom, starting by the middle of the first decade of the new millennium, made most of the receiving terminals otiose. As several companies in the US are preparing for exports, liquefaction plants have been constructed or are under construction. First exports of LNG were executed in early 2016.

2.2.2 Europe

Europe relied for decades on coal gas that was abundantly available. Ruhrgas, later for decades the most important player in the German natural gas industry, was founded in 1926 as 'Aktiengesellschaft für Kohleverwertung' (stock company for coal utilization). The era of natural gas started in Europe not before the Groningen field in the Netherlands was discovered and production started in 1965. In the 1970s, imports from the Soviet Union (giving rise to intense political discussions at the time) and Algeria, as well as Norway started, while town gas production gradually declined, effectively ceasing in the 1980s. Starting at the end of the Twentieth century indigenous production in the European Union decreased and imports increased. Diversification of supplies, either by LNG or via pipelines, is, therefore, one of the political priorities of the European Union.[6]

2.2.3 Russia

The natural gas industry in the Soviet Union started effectively in the 1960s only. The Ministry of the Gas Industry was created in 1965 and promoted the increase of natural gas consumption. In the 1960s, exports began, first to Eastern Europe and by the end of the decade with the first exports to Western Europe. By the end of the 1980s, Gazprom was designated the state-owned monopoly for natural gas.

How Brotherhood Changed the Political Landscape

It is common knowledge that, today, Russia earns a considerable share of its export income with the export of natural gas. This success story started on 3 December 1964 when an intergovernmental agreement between the Soviet Union and Czechoslovakia was signed in Moscow. Both countries were, since 1949, members of the Council for Mutual Economic Assistance, or COMECON, an organization that focused mainly on building trustful commercial relationships between the participating countries. One of these commercial relationships was the delivery of crude oil and natural gas from the Soviet Union into the COMECON countries, as did the December 1964 agreement. The agreement foresaw the building of the Brotherhood pipeline (братство or bratstvo) from Daschawa in today's Ukraine to Bratislava (former Czechoslovakia), today's capital of Slovakia, and subsequent deliveries of natural gas. The pipeline went into operation in June 1967, and the natural gas was delivered to a power plant and chemical production facilities.

Geographically, Bratislava is not far from Vienna, and so Austria became a few years later the first Western state that imported natural gas from the Soviet Union.

[6] For the history of regulation in Europe, see Sect. 5.3.

The basis for these deliveries, which started in 1971, was a contract between the Soviet Union and the Austrian company ÖMV (today OMV) signed in June 1968. The contract included a so-called barter component, i.e., Western companies (in this case, Austria's VÖEST and Germany's Mannesmann and Thyssen) delivered pipes, and, in return, the Soviet Union delivered natural gas. This barter idea became the blueprint for other natural gas sales of the Soviet Union to, e.g., Germany and Italy, and is the nucleus of Russia's strong position today as an exporter of natural gas to Central and Western Europe.

2.2.4 Japan

The Japanese archipelago is not connected by natural gas pipelines to the Asian mainland, and natural gas was not used until the 1960s. Then, in the early 1960s, the interest of Japanese companies to import LNG was triggered as a consequence of the enactment of environmental regulations. The first LNG import to Japan started in 1969. Further contracts were signed in the 1970s, *inter alia*, for imports from Brunei and Indonesia, principally to cope with an increased demand for natural gas in the agglomerations of Tokyo and Osaka. Increased utilization of natural gas for electricity generation necessitated further imports from Malaysia and Australia. Japanese companies were also pivotal for the development of the Ras reservoir offshore Qatar. Starting at the beginning of the new millennium, Japanese companies also participated in or initiated upstream developments in Russia, Australia, Mozambique, and Papua New Guinea and endeavored to import natural gas from the US.

The successful development of the natural gas industry in Japan has helped other industries to develop and attain leading global positions, amongst them shipbuilders, machinery manufacturing, and thermal-power system construction.

2.2.5 South-East Africa[7]

In Mozambique and Tanzania, natural gas was found already in the 1960s. However, natural gas production started only decades later. Only beginning in 2009 did major oil companies, like Anadarko, Eni, Statoil, (former) BG Group, and Ophir Energy, make several huge natural gas discoveries. As the total reserves are still not finally appraised, estimation for Mozambique's recoverable natural gas resources range from 2.8 to 5.1 tcm, and Tanzania's at about 1.1 tcm. Companies are planning to build LNG trains, both in Mozambique and in Tanzania. It is envisaged to construct up to four LNG trains in Mozambique within the next years. Even further extensions are discussed for the mid-term future.

Tanzania proposed to the operators the construction of an offshore LNG facility with two liquefaction trains. Yet, exploration in the area is still ongoing.

Already today, natural gas is used to generate electricity in Tanzania, and is exported by pipeline from Mozambique to South Africa. Originally considered as a by-product of oil, natural gas is gaining importance in the region. However, marketability requires investments in further exploration and infrastructure—for processing and transportation, eventually also for storage, but general infrastructure, like roads, harbors and airports, too.[8]

2.2.6 Global LNG

The first (US) patent for LNG handling/shipping was awarded in 1917, but it was not before the early 1940s that an LNG peak-shaving plant was constructed in Cleveland, Ohio. The first experimental LNG cargo reached Canvey Island in the UK in 1959, and only five years later, i.e., 1964,

[7]See Copinschi and Smedley (2016), p. 198 et seq.

[8]This has been recognized: For example, airports and roads in Mozambique have been constructed with the help of Chinese investors, in preparation for future exports of oil and natural gas to China.

the state-owned natural gas-producing companies of Algeria and later Libya supplied France, the UK, Italy, and Spain with LNG. Japanese imports started at the end of the decade.[9] The first imports to South Korea took place in 1974, to Taiwan in 1990. Australia started exporting LNG in 1991 and Trinidad and Tobago in 1999.

However, only in the mid-2000s did LNG become what the industry observers dub as 'one of the hottest topics of the natural gas industry'.

One of the main 'game changers' was the perceived change in the US supply and demand balance. Deliveries dedicated for the US were detoured, as the 'shale-gas revolution' started and solved perceived US-supply constraints. US import projects were cancelled. The world's LNG market globalized. However, the main geographic regions, i.e., the Atlantic Basin, involving trade in Europe, northern and western Africa, and the US Eastern and Gulf coasts and the Pacific Basin, involving trade in South Asia, India, Russia, and Alaska, both complemented by Middle-Eastern LNG-exporting countries between these regions, remained and still characterize the global LNG market. Yet, implication for price alignment between these markets started to be evident only by 2015.

2.3 Preliminary Overview—Development from Scratch

As in most of the network industries, both natural gas networks and consequently the markets, are characterized by economies of scale, sometimes economies of scope and also direct, as well as indirect, network effects.[10] Furthermore, other economic, commercial, legal, and technical restrictions apply. Therefore, when starting from scratch or intending to extend a natural gas business in a region or country, several requirements have to be fulfilled in order to ensure the economic viability, technical operability, and legal compatibility. The essential requirements are briefly described.

First and foremost, natural gas has to be available, i.e., sufficient reserves of natural gas should be present and ready for development and production. The closer these reserves are located to the market, the easier and cheaper the development of a natural gas market will be. The availability of a potential downstream market thereby fosters upstream developments. Contracts for the sale of natural gas can be used by producers to finance their investments in production facilities.

Sweden

The Swedish natural gas market is less developed than are others in Europe. With a population of nearly 10 million people, Sweden's natural gas consumption did not exceed 20 TWh/a during recent years. Overall, only 3.5% of the total Swedish primary energy requirements are covered by natural gas. Yet, in the southwest of Sweden, where a natural gas infrastructure has been established, this figure increases to approximately 20% of the primary energy consumption.[11]

There are several reasons for this (practically) insignificant market penetration. Except for some small quantities of biogas (and biomethane), Sweden has to import all its natural gas. The only pipeline connection to Sweden exists via Denmark.[12] Physical diversification of supplies is not a feasible option. At Skallen, one lined rock-cavern storage facility with a (very limited) capacity of 10 million Nm^3 was constructed. In addition, Sweden's population density of 24 inhabitants/km^2 is considerably lower than the average for the

[9] See Sect. 2.2.4.

[10] Network effects are often equated to economies of scale. This view is not shared. While economies of scale (as well as economies of scope) refer to the supply side, (other) network effects result from demand side effects, e.g., interoperability.

[11] See Energimarknadsinspektionen (2012), p. 1.

[12] South of Stockholm, at Nynäshamn, an LNG facility is in operation since 2011. The supplies received are used in the area of Stockholm only. A connection to the natural gas transportation system in the southwest of Sweden does not exist.

European Union, with approximately 120 inhabitants/km^2 or, to choose an extreme example, the Netherlands with more than 500 inhabitants/km^2.[13] This implies that (potential) investors cannot profit from noticeable economies of scale (or sometimes also referred to as 'economies of density'). The end users are sparsely distributed, and the marginal costs of connecting the next end user to the grid are high, as it (at least on average) requires considerable pipeline extension.

In summary, the market participants and investors claim that the following factors are hindering a sustainable and economically viable extension of the Swedish natural gas market: Supplies are not diversified and only hardly diversifiable; local geology thwarts plans to construct significant storages; the specific transportation costs, i.e., costs per quantity, are high, as long distances have to be bridged; and, finally, end users in most parts of Sweden have not had the chance to get acquainted with natural gas.

If a natural gas infrastructure, consisting primarily of transportation, LNG, and storage facilities does not exist, and sales of natural gas are intended, such infrastructure has to be constructed. In order to be economically viable, primarily the following prerequisites should be ideally fulfilled:

- Potential demand/end users has/have to exist or can reasonably be expected to develop. Ideally, demand/end users will:
  - be 'large', ensuring 'sufficient' revenues and reduction of specific costs;
  - be diversified with regard to price (structure), quantity and flexibility requirements, and segments, in order to reduce risks,[14] in particular:
    - credit risk,[15]
    - price risk,[16]
    - volume risk,[17]
    - flexibility requirement (by equilibration of deliveries to end users with different requirements[18]);
  - be located in clusters, i.e., in case a new region has to be developed, economies of scale will be more pronounced in clusters, as the marginal costs for connecting additional end users will decrease.
- Infrastructure can be developed at (relatively) low costs.

  This implies that a transportation and distribution system can be constructed efficiently, i.e., that a sufficiently large number of end users can be connected to the system. Therefore, locally clustered demand/end users will decrease marginal costs for connecting additional end users.

  In addition, infrastructure to safeguard security of supply, as well as short-, mid- and long-term flexibility will have to be constructed. As the 'tools' used to provide such services are predominantly either underground storages located in the vicinity of demand or (flexible) production, the (geologic) conditions to develop storage facilities

[13] See World Bank (2016).

[14] Excursus: In this book, the following distinction will be used: While 'uncertainty' describes a state or situation (in the future) for which neither the values of (a) variable(s) nor its/their probability distribution(s) is/are known, 'risk' refers to a situation (in the future) where the probability distribution(s) of (a) variable(s) is/are known but not the value(s). (Often risk is solely referring to the probability of loss).

[15] Credit risk refers to the risk that a debtor defaults, i.e., will not pay back a loan.

[16] Price risk is caused by the probability of any 'adverse movement' in the (market) price of an asset, commodity, or any other good.

[17] Volume risk, having been defined in the traditional natural gas market as the risk of incompatible sales and supplies, i.e., over- or undersupply, (by definition) does not exist in liquid markets. Here any quantity can be purchased or sold, yet at different prices, i.e., in modern markets, volume risks are converted to price risks.

[18] Also referred to as 'foisonnement' (using metaphorical French).

have to be favorable or the production has to be in or near the region of consumption.

- The investors have to have sufficient capital at their disposal to bear all costs for infrastructure construction, operation, and maintenance.
- A legal (or regulatory) framework, including technical standards, has to be established providing investment security and fostering further market development.
- Supplies are available at competitive prices, either from indigenous production or from international sources. Ideally such supplies will:
  - on an annual basis, match demand,
  - be diversified in order to:
    - ensure optimal utilization of the network,
    - increase security of supply by reducing the impact in case one source is interrupted,
    - enable the customer to optimize its portfolio, and
  - fulfill potential regulatory requirements.

If, on the other hand, infrastructure exists already and non-discriminatory third-party access to infrastructure is granted (or its own infrastructure is available) and it is intended to enter a market competitively, then competitive sales products, as well as competitive supplies, low costs of infrastructure utilization, and a legal framework encouraging such market entry, are essential.

WINGAS

While today entry into most mature natural gas markets is facilitated by non-discriminatory rules of third party access to transportation and storage capacities, this was not the case in Germany in the early 1990s. To supply its mother company BASF, in 1990, Wintershall made efforts to purchase natural gas directly from North Sea producers and in parallel asked incumbent Ruhrgas for access to its transportation capacities. Such access was not granted. Natural gas from producers in the Norwegian North Sea was also not available. Instead, Wintershall, supported by BASF, and Gazprom agreed to co-operate. To this end, joint ventures, one of them WINGAS, were founded, and pipelines across Germany and Germany's largest storage at Rehden were constructed. Due to non-existing third party access to pipelines, some of the pipelines were laid parallel to existing pipelines. At the same time, the sales organization was established, and major long-term delivery contracts with Gazprom's export affiliate, Gazexport, were concluded. Investment-based market entry was achieved in a mature market, with already existing, but at that time not accessible, infrastructure.

References

Colombo S, Harrak M El, Sartori N (2016) The future of natural gas–markets and geopolitics. In: Colombo S, El Harrak M, Sartori N, (eds) Lenthe/European Energy Review, Hof van Twente

Copinschi P, Smedley M (2016) Sub-saharan Africa: a future global gas player? In Colombo S, El Harrak M, Sartori N (eds) The future of natural gas. Market and geopolitics. Hof van Twente, pp 181–207

Energimarknadsinspektionen (2012) An overview of the Swedish natural gas market

World Bank (2016) Population density. Available at http://data.worldbank.org/indicator/EN.POP.DNST?year_high_desc=false

3 Market Data

Summary

- The world's natural gas consumption in 2015 was 131 EJ. Natural gas accounted for approximately 24% of the world's primary energy requirements.
- While the share of natural gas in the primary energy consumption of Western Europe and the US is approximately 30%, it is more than 50% in Russia and 11% in the Asia-Pacific region.
- The world's largest natural gas reserves are located in Iran, Russia, and Qatar.
- In recent years, the largest natural gas production was reported to be in the US, followed by Russia, Iran, and Qatar.
- In absolute terms, the natural gas consumption in the US is the world's largest, followed by Russia, China, and Iran. European natural gas consumption decreased during recent years.
- The total storage capacity is largest in the US, followed by Russia, the Ukraine, and Germany.
- The largest LNG-liquefaction capacities have been constructed in Qatar. Considerable investments have also been made in Australia and Indonesia.
- In Japan, the US, and South Korea, the LNG-regasification capacities are the world's largest. Yet, in the US, after self-sufficiency in natural gas supply has been achieved during recent years, their regasification capacities are mainly idle.
- The prices for natural gas traded at hubs in Europe align, i.e., the differences are small and the correlation coefficients high.
- In Europe, the correlation coefficients between crude-oil prices and natural gas prices traded at hubs increased in recent years after having been lower in the preceding decade.

3.1 Primary Energy Consumption

The world's energy consumption in 2015 was at 550.5 EJ.

The various primary fuels' shares are depicted in Table 3.1. Behind oil and coal, natural gas is the third most important primary energy.

The natural gas share in primary energy consumption was largest in Trinidad & Tobago (91.4%). Table 3.2 shows the shares of natural gas in primary energy consumption for various world regions and selected countries.

A. Pustišek and M. Karasz, *Natural Gas: A Commercial Perspective*,
DOI 10.1007/978-3-319-53249-3_3

Table 3.1 Primary fuels' shares in 2015 world energy consumption

	Oil	Coal	Natural gas	Hydro electric	Nuclear energy	Renewables (other than hydro)	Total
Consumption [EJ]	181.3	160.8	131.3	37.4	24.4	15.3	550.5
Share (%)	32.93	29.21	23.85	6.79	4.43	2.78	100.0

See British Petroleum (2016), p. 41

Table 3.2 Natural gas share in 2015 in primary energy consumption in selected regions and countries

Country	Share of natural gas in primary energy consumption [%]
Trinidad & Tobago	91.4
Qatar	78.9
Bangladesh	78.6
Algeria	64.3
Russia	52.8
Total Middle East	49.9
Italy	36.5
UK	32.1
Total Europe & Eurasia	31.9
Total North America	31.5
US	31.3
Total Africa	28.0
OECD countries	26.5
Total World	23.8
Total South & Central America	22.5
European Union	22.2
Non-OECD countries	21.9
Germany	20.9
Brazil	12.6
Total Asia Pacific	11.5
China	5.9
South Africa	3.6
Sweden	1.5

See British Petroleum (2016), p. 41

3.2 Reserves

Excursus

Though similar, reserves and resources are not the same. In a nutshell, reserves are those resources that are geologically well-known and, with today's technology, economically producible. Of course, many definitions for resources and reserves exist in different countries[1] but

[1]Compare with the Society of Petroleum Engineers (SPE) (2005) which provides an overview based on different definitions.

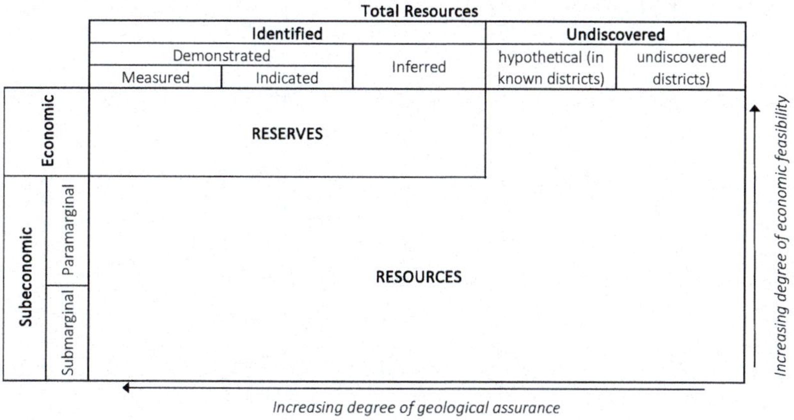

Fig. 3.1 McKelvey diagram (Kleppe and McKelvey 1976, p. A2)

nearly all of them are based on the so-called McKelvey box (see Fig. 3.1). Named after Vincent E. McKelvey, former director of the US Geological Survey, this diagram classifies reserves and resources according to the degree of geological certainty (horizontal axis) and the degree of economic feasibility (vertical axis), i.e., only those resources that are geologically well-known and economic to produce are considered as reserves. This implies that the same volume of natural gas in the ground might change its status from resource to reserve and vice versa. For example, by advances in drilling technology, shale-gas resources in North America became reserves in the early 2000s.

At the end of 2015, the total proven reserves in the world were estimated to be 186.9 tcm.[2] The 25 largest reserve-holding countries are described in Table 3.3.

None of them is member of the European Union.[3] In total, only 0.7% of the world reserves are located in the European Union. Due to the relatively large North American, Norwegian, and Australian natural gas reserves, the OECD countries together hold 10.5% of the world reserves. The remaining is located in non-OECD countries.[4]

[2]See British Petroleum (2016), p. 20.

[3]The largest reserve holder within the EU is The Netherlands ranked 26 on the list of largest reserve-holding countries.

[4]See British Petroleum (2016), p. 20.

3.3 Production

Worldwide natural gas production reached 3538.6 bcm[5] in 2015. Again, the largest share was produced in non-OECD countries with 63.2% (OECD share: 36.8% and share of EU: 3.4%).[6] Obviously, OECD and EU produce a higher share than their share of worldwide reserves. Therefore, the R/P ratio[7] is in the EU at 10.8 years, in all OECD countries at 15.1 years, and in non-OECD countries at 74.5 years. The R/P ratio for the world is at 52.8 years. The development of worldwide natural gas production since 1970 is depicted in Fig. 3.2.

Global natural gas production has more than tripled since 1970. The 25 largest natural gas producing countries of 2015 are enumerated in Table 3.4.

Whereas some countries show steep production increases (e.g., US and Qatar), others are in the declining production phase (e.g., the UK and the Netherlands).

[5]In this section volumes will be quoted in [bcm] = [Gm^3] = [billion m^3] in accordance with use in the industry.

[6]See British Petroleum (2016), p. 22.

[7]The R/P ratio, the so-called reserves-to-production ratio, is calculated by dividing the reserves by the annual production. Assuming constant production for the future and no further changes to reserves, the R/P ratio shows the length of time (in years) that the reserves would last.

Table 3.3 Largest reserve-holding countries

Rank	Country	Reserves [tcm]
1	Iran	34.0
2	Russia	32.3
3	Qatar	24.5
4	Turkmenistan	17.5
5	US	10.4
6	Saudi Arabia	8.3
7	UAE	6.1
8	Venezuela	5.6
9	Nigeria	5.1
10	Algeria	4.5
11	China	3.8
12	Iraq	3.7
13	Australia	3.5
14	Indonesia	2.8
15	Canada	2.0
16	Norway	1.9
17	Egypt	1.8
18	Kuwait	1.8
19	Libya	1.5
20	India	1.5
21	Malaysia	1.2
22	Azerbaijan	1.1
23	Uzbekistan	1.1
24	Kazakhstan	0.9
25	Oman	0.7

See British Petroleum (2016), p. 20

3.4 Consumption

The 25 largest natural gas consuming countries of 2015 are enumerated in Table 3.5.

Worldwide natural gas consumption reached 3468.6 bcm in 2015. The largest share was consumed in non-OECD countries with 53.5% (OECD share: 46.5% and share of EU: 11.5%).[8] Obviously, OECD and EU consume at a higher share than their share in worldwide reserves or production.

While natural gas consumption in most countries increased during the last years and is expected to increase in the near- and mid-term future, in Europe a decline was observed,[9] and most of the forecasts expect natural gas consumption to stagnate or further decrease in the future.

3.5 Trade and Prices

Tables 3.4 and 3.5 show that production and consumption of natural gas in many cases do not coincide geographically. (One notable exception being the US, which, however, together with

[8]See British Petroleum (2016), p. 23.

[9]See, e.g., the consumption for Italy, the UK, and Germany in Table 3.5.

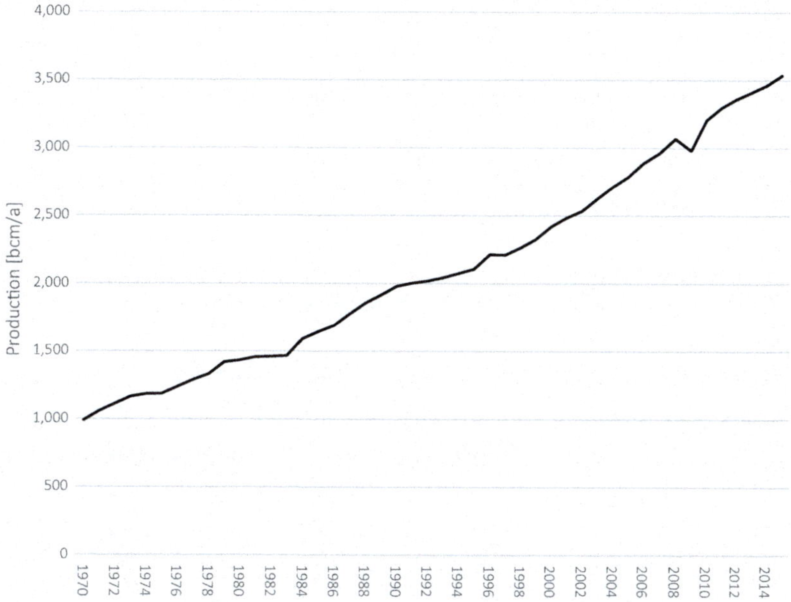

Fig. 3.2 Development of worldwide natural gas production since 1970 [See British Petroleum (2016)]

Table 3.4 Largest natural gas producing countries since 1970

Rank	Country	Production [bcm]						
		1970	1980	1990	2000	2005	2010	2015
1	US	595.1	549.4	504.3	543.2	511.1	603.6	767.3
2	Russia	179.1[a]	393.8[a]	590.0	528.5	580.1	588.9	573.3
3	Iran	3.7	4.8	26.2	59.6	102.3	152.4	192.5
4	Qatar	1.0	4.7	6.3	24.9	45.8	131.2	181.4
5	Canada	56.7	74.8	108.6	182.2	187.1	159.9	163.5
6	China	3.0	14.7	15.8	28.1	51.0	99.1	138.0
7	Norway	–	25.1	25.5	49.8	85.8	107.3	117.2
8	Saudi Arabia	1.6	9.7	33.5	49.8	71.2	87.7	106.4
9	Algeria	2.5	14.7	49.4	87.8	88.2	80.4	83.0
10	Indonesia	1.2	18.5	43.9	69.6	75.1	85.7	75.0
11	Turkmenistan	b	b	79.5	42.5	57.0	42.4	72.4
12	Malaysia	–	2.5	17.2	46.6	63.8	60.9	68.2
13	Australia	1.7	11.1	19.7	32.1	39.2	52.6	67.1
14	Uzbekistan	b	b	36.9	51.1	54.0	54.4	57.7
15	UAE	0.8	7.5	20.1	38.4	47.8	51.3	55.8
16	Mexico	11.2	25.7	27.1	38.3	52.2	57.6	53.2
17	Nigeria	0.1	1.7	4.0	11.8	25.0	37.3	50.1
18	Egypt	0.1	2.2	8.1	21.0	42.5	61.3	45.6
19	Netherlands	26.7	76.4	61.0	58.0	62.5	70.5	43.0
20	Pakistan	3.5	7.2	12.2	21.5	39.1	42.3	41.9
21	Thailand	–	–	6.5	20.2	23.7	36.2	39.8

(continued)

Table 3.4 (continued)

Rank	Country	Production [bcm]						
		1970	1980	1990	2000	2005	2010	2015
22	UK	10.5	34.8	45.5	108.4	88.2	57.1	39.7
23	Trinidad & Tobago	1.8	2.7	5.2	15.5	33.0	44.8	39.6
24	Argentina	6.0	8.4	17.8	37.4	45.6	40.1	36.5
25	Oman	–	0.9	3.3	10.9	22.1	29.3	34.9

[a]Figures for Soviet Union
[b]Figures included in Russian figure. See British Petroleum (2016)

Table 3.5 Largest natural gas consuming countries since 1970

Rank	Country	Consumption [bcm]						
		1970	1980	1990	2000	2005	2010	2015
1	US	598.6	562.9	542.9	660.7	623.4	682.1	778.0
2	Russia	181.0[a]	360.3[a]	407.6	360.4	394.0	414.1	391.5
3	China[c]	3.0	14.7	15.8	28.3	50.9	115.0	200.6
4	Iran	2.7	4.8	24.1	62.9	102.7	152.9	191.2
5	Japan	3.4	24.1	48.1	72.3	78.6	94.5	113.4
6	Saudi Arabia	1.6	9.7	33.5	49.8	71.2	87.7	106.4
7	Canada	36.4	52.2	66.9	92.7	97.8	95.0	102.5
8	Mexico	10.2	23.0	27.5	40.8	60.9	72.5	83.2
9	Germany	15.3	58.3	60.9	79.2	86.3	84.1	74.6
10	UAE	0.8	4.9	16.9	31.4	42.1	60.8	69.1
11	UK	11.3	44.8	52.4	96.8	94.9	94.2	68.3
12	Italy	13.6	25.4	43.4	64.9	79.1	75.6	61.4
13	Thailand	–	–	6.5	21.9	32.5	45.1	52.9
14	India	0.7	1.2	12.0	26.4	35.7	61.5	50.6
15	Uzbekistan	b	b	35.7	45.7	42.7	40.8	50.3
16	Egypt	0.1	2.2	8.1	20.0	31.6	45.1	47.8
17	Argentina	6.0	11.5	20.3	33.2	40.4	43.2	47.5
18	Qatar	1.0	4.7	6.3	10.9	18.7	32.1	45.2
19	South Korea	–	–	3.0	18.9	30.4	43.0	43.6
20	Turkey	–	–	3.4	14.6	26.9	39.0	43.6
21	Pakistan	3.5	7.2	12.2	21.5	39.1	42.3	43.4
22	Brazil	0.1	1.0	3.1	9.4	19.6	26.8	40.9
23	Malaysia	<0.05	2.5	7.6	26.6	34.9	34.5	39.8
24	Indonesia	1.2	7.0	16.9	32.5	35.9	43.4	39.7
25	France	9.1	24.0	28.1	39.7	45.6	47.3	39.1

[a]Figures for Soviet Union
[b]Figures included in Russian figure
[c]Including Hong Kong. See British Petroleum (2016)

Canada, has to be regarded as a continent rather than a country.) Therefore, international (and, via LNG global) trade flows from high-production countries to high-consumption countries are quite common in the natural gas business. These trade flows are depicted in Fig. 3.3.

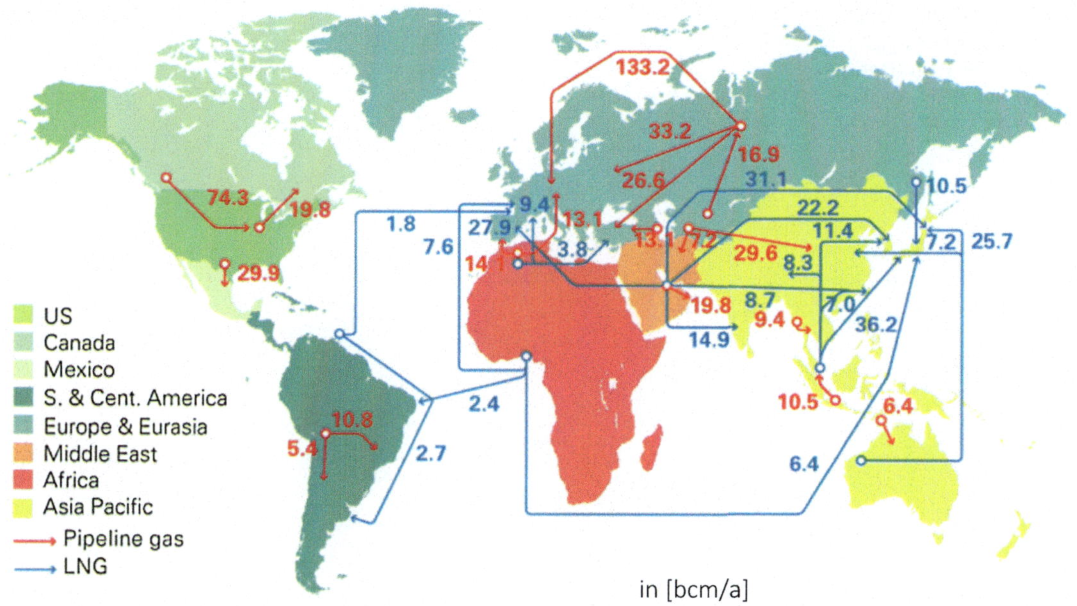

Fig. 3.3 Major trade movements of natural gas incl. LNG in 2015 [British Petroleum 2016, p. 29 (reproduced with permission of British Petroleum)]

The largest net exporters, i.e., countries with high indigenous production and relatively low consumption, are Canada, Nigeria, Norway, Russia, Turkmenistan, Qatar, Malaysia, and Australia.

Countries with relatively low indigenous production and high consumption are the largest net importers.[10] Among these are Mexico, the UK, Spain, France, Italy, Germany, Turkey, China, South Korea, and Japan.

As natural gas has a much lower energy density than oil, the costs for transport and storage are relatively higher. Given the (gaseous) state, natural gas transportation is restricted to pipeline transport or liquefaction. For decades, this restriction caused the lack of any global natural gas markets being established. Regional markets dominated and price differences between these continued to exist for longer periods than otherwise.

Despite increasing LNG trade fostering 'globalization' of the natural gas industry, prices in different countries did not align worldwide until the end of 2015. In Fig. 3.4, some benchmark prices for natural gas traded in different regions are shown. Only at the beginning of 2016, mainly as a result of abundant supply and advances in technology, did worldwide natural gas prices start to align. The 'shale-gas boom' started in the US in the middle of the first decade of the twenty first century. The financial crisis began in 2008. Natural gas prices worldwide plummeted in 2008 and 2009. In subsequent years, they recovered in Europe and Japan. However, in the US, the oversupply situation caused prices to stay low and to develop uncorrelated with prices in other regions and uncorrelated with the prices of alternative fuels. That is, while US natural gas and crude-oil prices were, at least in the long-term, fairly well correlated until the end of 2008, the correlation coefficient decreased as of 2009 but increased (again) in 2014.[11] Notably in the period

[10]Leaving out transit quantities that distort many statistical data.

[11]The decrease of a correlation coefficient is often referred to as 'decoupling'. However, as this term is also used to describe any increase of the difference of two time series irrespective of the correlation coefficient, the term is interpreted ambiguously and shall not be used subsequently.

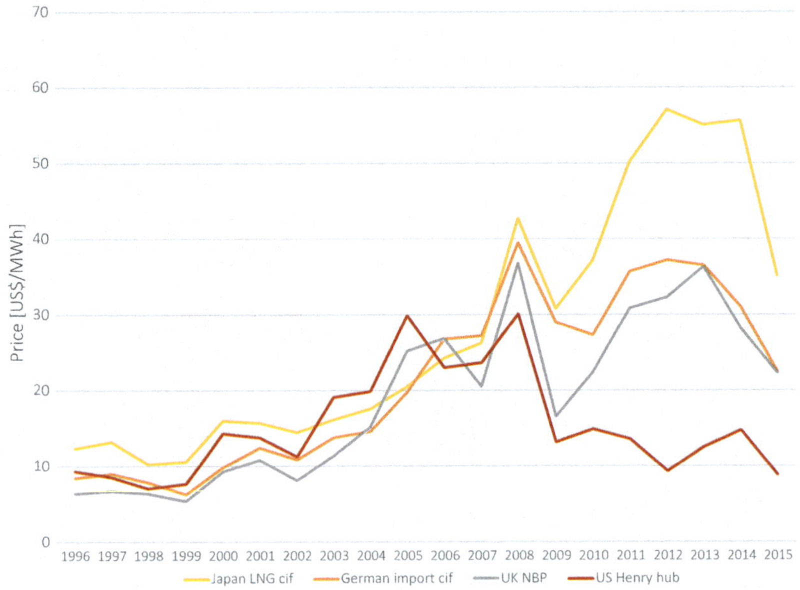

Fig. 3.4 Development of worldwide natural gas prices since 1996 (British Petroleum 2016. Average prices for calendar years)

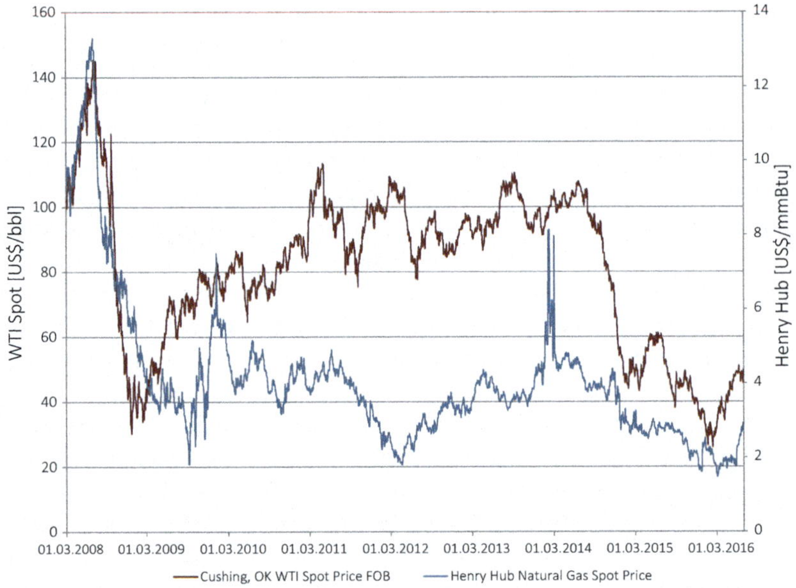

Fig. 3.5 US: crude oil (WTI) versus natural gas (Henry Hub) prices compared [WTI spot price (US$/bbl); natural gas at Henry Hub spot price (US$/mmBtu). Daily data from 1 March 2008 until 30 June 2016 (trading days). Data provided by 'Montel Nachrichtenservice' (www.montel.de). For a further comparison of long-term inter-fuel prices, see also Sect. 5.1.2]

2012–2013 (calendar years), the correlation coefficient between crude oil and natural gas in the US was even negative. This is depicted in Fig. 3.5 and Table 3.6.[12]

In Europe, prices developed differently than in the US. The correlation coefficient between crude oil (Brent) and natural gas (NBP) prices did not change considerably. It was in the period 2008–2016 and in the period 2014–2016 slightly over 60%. Thus, the correlation coefficients do not reveal a considerable similarity of the time series. This changed in the period 2015–2016, when the correlation coefficient between crude oil (Brent)

[12]It has to be noted, to use a well-known phrase, that a high correlation coefficient does not imply causation. Further, two highly correlated time series have a (relatively) constant difference, not necessarily a difference of zero.

Table 3.6 Correlation coefficients between US crude oil (WTI) versus US natural gas (Henry Hub) for different periods

Period	Correlation coefficient [%]
2008–2016	51
2011–2014	8
2012–2013	−10
2014–2016	90

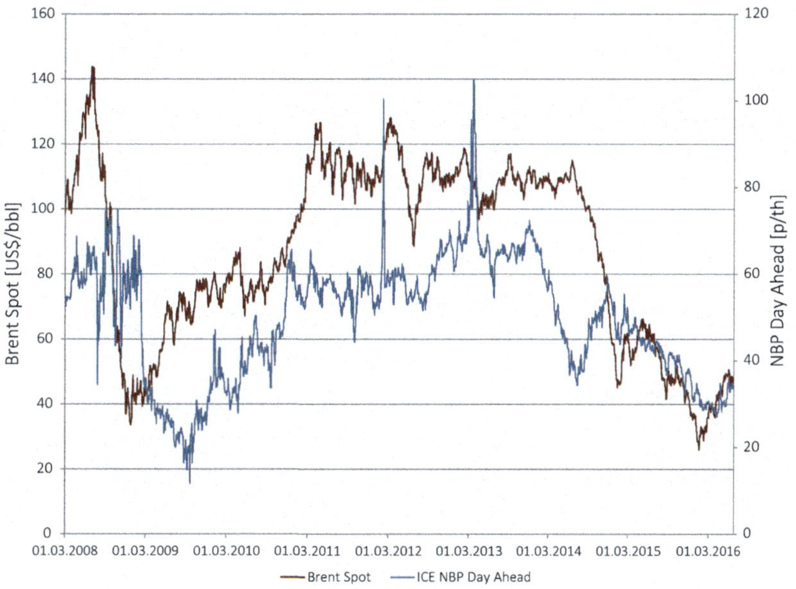

Fig. 3.6 Europe: crude oil (Brent) versus natural gas (NBP) prices compared [Brent spot price (US$/bbl); natural gas at NBP spot price (day-ahead) (pence/therm). Daily data from 1 March 2008 until 30 June 2016 (trading days). Data provided by 'Montel Nachrichtenservice' (www.montel.de). For a further comparison of long-term inter-fuel prices, see also Sect. 5.1.2]

Table 3.7 Correlation coefficients between Europe crude oil (Brent) versus natural gas (NBP) prices compared

Period	Correlation coefficient (%)
2008–2016	65
2014–2016	62
2015–2016	79

and natural gas (NBP) increased to 79%, as shown in Fig. 3.6 and Table 3.7.

A comparison of US natural gas prices and European natural gas prices shows that, despite the 'globalization' of the natural gas industry having been initiated approximately in 2005 as a consequence of international LNG deliveries, only during recent years, has the correlation coefficient started to increase, i.e., prices have started to align. This is shown in Fig. 3.7 and Table 3.8.

The prices at European hubs are aligned. A comparison of natural gas day-ahead prices at three main European natural gas hubs, NBP, TTF, and NCG, shows significant correlation coefficients between all these prices. This applies even if prices at NBP are quoted (and analyzed) in £. Minor distortions are caused by exchange-rate fluctuations. Yet, not only are the correlation coefficients between prices quoted at different hubs high, but also the differences between such prices are low or non-existent. This implies that the (intrinsic) value of transportation capacity decreases.[13] This is shown in Fig. 3.8 and Table 3.9.

[13]For a detailed discussion, see Sect. 5.5.4.2.3.2.

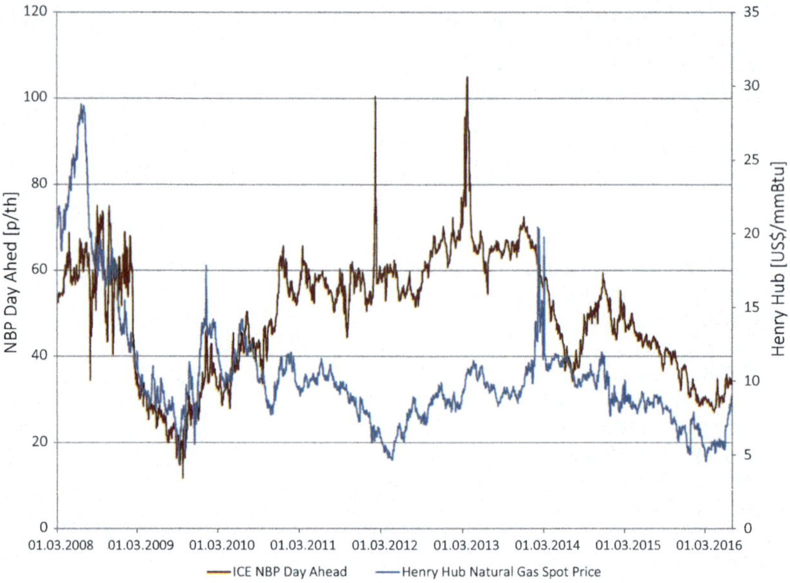

Fig. 3.7 US (Henry Hub) versus Europe (NBP) natural gas prices compared (in original units) [Natural gas at Henry Hub spot price (US$/mmBtu); natural gas at NBP spot price (day-ahead) (pence/therm). Daily data from 1 March 2008 until 30 June 2016 (trading days). Data provided by 'Montel Nachrichtenservice' (www.montel.de)]

Table 3.8 Correlation coefficients US (Henry Hub) versus Europe (NBP) natural gas prices compared

Period	Correlation coefficient for	
	Data in original units [%]	Data converted to [€/MWh] [%]
2008–2016	27	27
2011–2014	−13	−15
2012–2013	55	48
2014–2016	58	67
2015–2016	84	84

The correlation coefficient for data presented in Fig. 3.7 is shown under the heading 'Data in original units'. In addition, the data are converted using the daily exchange rates for [US$/€] and [€/£] to [€/MWh]. As shown in Table 3.8, the correlation coefficients do not differ considerably

3.6 Physical Capacities

Pipelines, LNG facilities, and underground storages are the essential parts of any natural gas infrastructure. While an overview of pipeline capacities (even at selected) delivery points would go beyond the purpose of this book, a depiction of the European high-pressure pipeline system, main locations of underground storages, and LNG facilities is intended to provide an initial idea of the countries with developed natural gas markets. However, in the particular case of underground storages, the reverse conclusion is not necessarily true, i.e., developed natural gas markets can also exist with only limited underground storage capacities.

3.6.1 Transportation

Taken as an example, the European natural gas pipeline system evolved continuously throughout the recent decades.[14] Currently, it connects the sources in the North Sea, Russia, and Northern Africa with all the main centers of utilization. Yet, in most areas, distinct directions of flow exist. For example, no natural gas is transported from the

[14]Similar developments could be observed in other regions as, e.g., the US, Russia, or Japan.

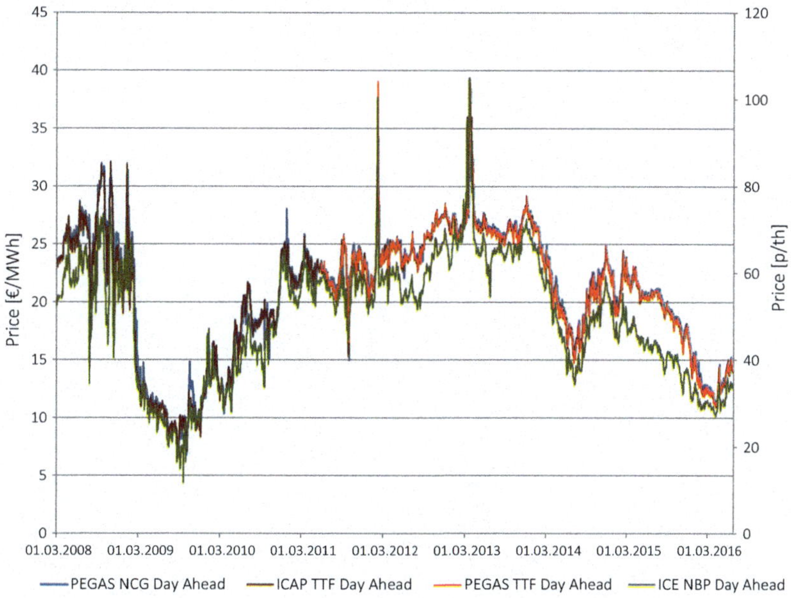

Fig. 3.8 Europe [NBP (right axis), TTF, NCG (left axis)] natural gas prices compared [PEGAS TTF spot prices (day-ahead) (€/MWh); ICAP TTF spot prices (day-ahead) (€/MWh); PEGAS NCG spot prices (day-ahead) (€/MWh); ICE NBP spot prices (day-ahead) [pence/therm]. Daily data from 1 March 2008 until 30 June 2016 (trading days). Data provided by 'Montel Nachrichtenservice' (www.montel.de)]

Table 3.9 Correlation coefficients between European day-ahead quotations at hubs (NBP, UK; TTF, NL; NCG, DE)

Correlation coefficients	2008–2016		
	NCG [%]	TTF [%]	NBP [%]
NCG	100		
TTF	99.2	100	
NBP	96.5	96.9	100
Correlation coefficients	2014–2016		
	NCG [%]	TTF [%]	NBP [%]
NCG	100		
TTF	99.7	100	
NBP	95.6	95.4	100

south via the Italian grid to Central and Northern Europe or, to Central Europe, where Russian natural gas is (physically) dominant. The European high-pressure pipeline grid is depicted in Fig. 3.9.

3.6.2 Underground Storages

The 20 countries with the largest working gas capacities in underground storages are enumerated in Table 3.10. Storage capacity availability is essential for providing security of supply and flexibility[15] and, thus, the development of a natural gas industry in a region or country. Yet, the development of underground storages is restricted by geological conditions.

[15]See Sects. 5.4, 6.3.1.1.4 and 8.2.

3.6.3 LNG

3.6.3.1 Liquefaction Plants

At the end of 2015, 90 liquefaction trains were in operation worldwide.[16] Out of these, 41 were located in the Pacific basin, 29 in the Atlantic basin and 20 in the Middle East. Their total

[16]All numbers subsequently cited do not include trains not in operation (for different reasons) in Angola, Egypt, Libya, and Yemen.

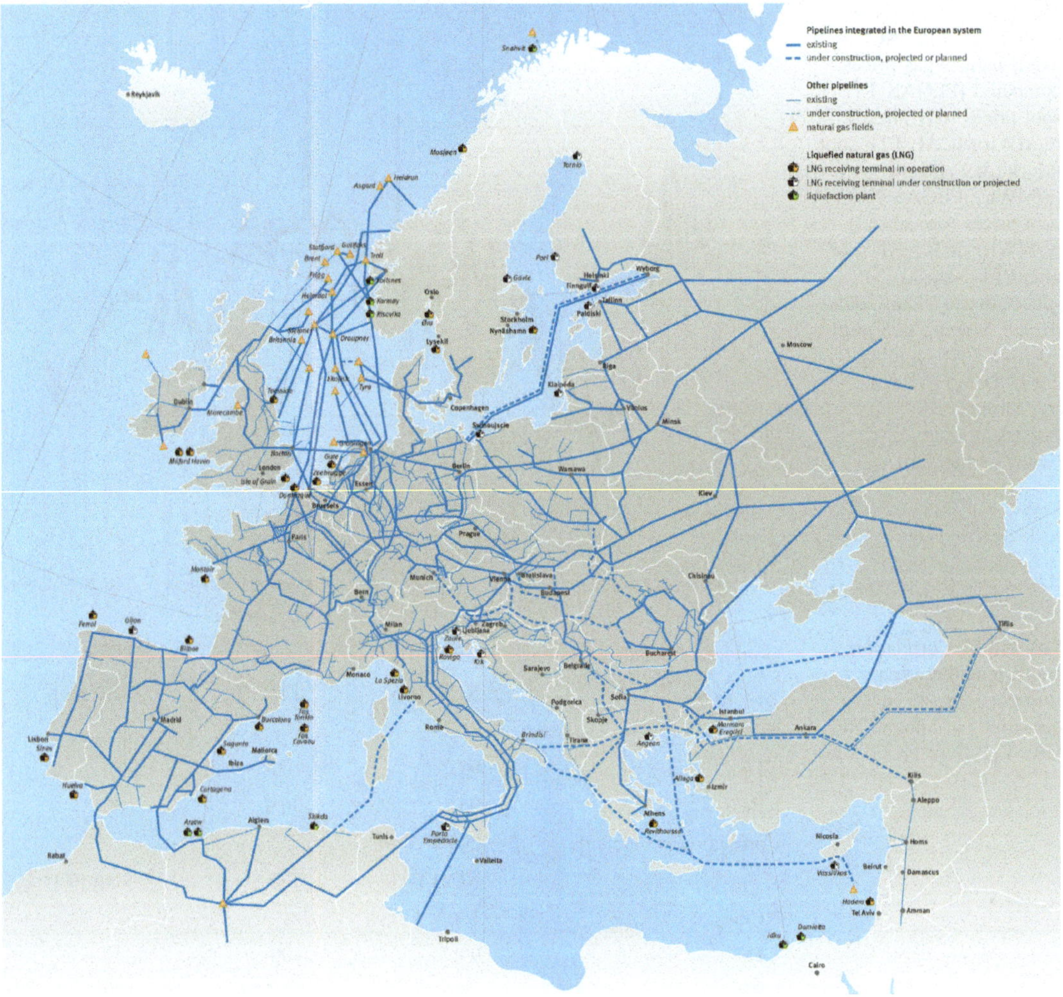

Fig. 3.9 European high-pressure pipeline grid 2015 (Picture provided by © Eurogas 2015)

nominal liquefaction capacity was at 288.5 mmtpa,[17] thereof in the Pacific basin 121.2 mmtpa, in the Atlantic basin 73.8 mmtpa, and in the Middle East 93.5 mmtpa.[18]

The ten countries with the largest available LNG liquefaction capacities at the end of 2015 are enumerated in Table 3.11.

3.6.3.2 LNG Tanker

At the end of 2015, 449 LNG tankers existed. Approximately 63% of these vessels were younger than 10 years. Their overall capacity was at 64.6 million m^3 of LNG. Most of the vessels were relatively large: 412 of them had a capacity greater than 90,000 m^3 of LNG, and 67 had a capacity of more than 170,000 m^3 of LNG. The largest vessels[19] can load more than 265,000 m^3 of LNG.

[17]Unit used in this form worldwide: 1 mmtpa = 1,000,000 tonnes/year = 10^6 t/a. Note: in several publications 'mtpa' instead of 'mmtpa' is used as the abbreviation for 10^6 t/a.

[18]All data from Groupe International des Importateurs de Gaz Naturel Liquéfié (GIIGNL) (2016), p. 22 et seq.

[19]These are the vessels: Aamira, Al Dafna, Al Mafyar, Al Mayeda, Bu Samra, Mekaines, Mozah, Rasheeda, Shagra, Umm Slal, and Zarga. All of them were built by Samsung Heavy Industries Co., Ltd., and all are sailing under the

Table 3.10 Countries with largest working gas capacities

Rank	Country	Working gas capacity [bcm]	Number of facilities
1	US	128.1	419
2	Russia	70.4	23
3	Ukraine	32.2	13
4	Germany	24.6	51
5	Canada	20.6	61
6	Italy	17.1	11
7	Netherlands	12.8	5
8	France	12.8	16
9	Austria	8.2	9
10	China	6.6	21
11	Hungary	6.5	6
12	Uzbekistan	6.2	3
13	Iran	6.0	2
14	UK	5.3	8
15	Kazakhstan	4.7	3
16	Azerbaijan	4.2	3
17	Czech Republic	3.5	8
18	Spain	3.4	4
19	Slovakia	3.3	3
20	Romania	3.1	8

Anonymus (2015), p. 405

Table 3.11 LNG liquefaction capacities at the end of 2015 (excl. the plants in Angola, Egypt, Libya, and Yemen that had stopped for different reasons)

Rank	Country	Nominal liquefaction capacity [mmtpa]
1	Qatar	77.0
2	Australia	36.6
3	Indonesia	31.9
4	Algeria	28.5
5	Malaysia	24.2
6	Nigeria	21.8
7	Trinidad & Tobago	15.5
8	Oman	10.7
9	Russia	9.6
10	Brunei	7.1

All data from Groupe International des Importateurs de Gaz Naturel Liquéfié (GIIGNL) (2016), p. 22 et seq

3.6.3.3 Regasification Plants

At the end of 2015, 117 regasification terminals (onshore as well as offshore) worldwide were in operation.[20] More than half the terminals (64) were located in Asia, mainly in Japan. In North and South America, in total 24 terminals

(Footnote 19 continued)
flag of the Marshall Islands for Shell International Trading & Shipping Co. Ltd. as manager.

[20] All numbers are exclusive of the mothballed Spanish regasification terminal El Musel near Gijón.

Table 3.12 LNG regasification capacities at the end of 2015

Rank	Country	Nominal send-out capacity [bcm/a]
1	Japan	271.1
2	US	178.6
3	South Korea	152.9
4	Spain	60.1
5	China	55.7
6	UK	53.6
7	India	28.1
8	Mexico	23.3
9	France	21.3
10	Brazil	15.8
11	Italy	15.1
12	Taiwan	13.7
13	Egypt	13.0
14	Turkey	12.2
15	Netherlands	12.0

All data from Groupe International des Importateurs de Gaz Naturel Liquéfié (GIIGNL) (2016), p. 29 et seq. Data exclusive of the mothballed Spanish El Musel plant

were in operation, in Europe (incl. Turkey) 23, and in the Middle East the remaining six terminals. Their total nominal send-out capacity was at 1041.9 bcm/a, thereof in Asia 553.4 bcm/a, in the Americas 249.6 bcm/a, in Europe 199.8 bcm/a, and in the Middle East 39.1 bcm/a. The total storage capacities at these plants was approximately 55.7 million m^3 LNG.[21]

The 15 countries with the largest available LNG regasification capacities at the end of 2015 are enumerated in Table 3.12.

References

Anonymus (2015) Untertage-Gasspeicherung in Deutschland. Erdöl Erdgas Kohle 131(11):398–406

British Petroleum (2016) BP statistical review of world energy

Groupe International des Importateurs de Gaz Naturel Liquéfié (GIIGNL) (2016) The LNG Industry: 2016 Edition

Kleppe TS, McKelvey VE (1976) Principles of the mineral resource classification system of the U.S. Bureau of Mines and U.S. Geological Survey, Washington, D.C.

Society of Petroleum Engineers (SPE) (2005) Comparison of selected reserves and resource classifications and associated definitions

[21] All data from Groupe International des Importateurs de Gaz Naturel Liquéfié (GIIGNL) (2016), p. 29 et seq.

Non-economic and Non-commercial Fundamentals

4

Summary

- Natural gas is a natural product, consisting mainly of methane (CH_4), whereby the state depends on temperature, pressure, specific volume, and its chemical composition.
- Compressed natural gas (CNG) is produced from natural gas by compression to approximately 200 bar.
- Liquefied natural gas (LNG) is natural gas that is cooled down to approximately −161.5 °C where methane liquefies.
- Propane and butane are also referred to as liquefied petroleum gas (LPG). Together with ethane, LPGs, pentanes, and higher hydrocarbons are usually referred to as natural gas liquids (NGL).
- The most important parameters describing the combustion properties of natural gas are (gross) calorific value and Wobbe index.
- All gas-fired equipment and applications are designed (and sometimes fine-tuned) for a particular natural gas quality. Gas specifications are necessary to define the minimum quality requirements that natural gas has to fulfil.
- Production and use of natural gas has an impact on the environment. These environmental effects might be categorized according to the part of the ecosystem influenced. Climate effects, air pollution, water pollution, and land use do occur. In general, natural gas combusts cleaner than other fossil fuels.
- The source material for natural gas genesis is organic matter, e.g., marine microorganisms like algae.
- Natural gas fields are commercially recoverable accumulations of natural gas in the Earth's crust.
- Conventional natural gas fields are not found in source rocks, but in reservoir rocks. A source rock is a sedimentary rock in which petroleum and natural gas are formed. A reservoir rock is generally any porous and permeable rock containing hydrocarbons. Reservoir rocks are predominantly sedimentary (sandstones and carbonates).
- Other than conventional reservoirs, which allow for economic production without massive stimulation treatments, unconventional sources show usually quite low permeabilities and can therefore only be produced at economic flow rates with the assistance of large stimulation treatments and special recovery technologies.

A. Pustišek and M. Karasz, *Natural Gas: A Commercial Perspective*,
DOI 10.1007/978-3-319-53249-3_4

- Unconventional sources are tight gas sands, shale gas, and coal-bed methane.
- Anthropogenic production of 'natural' gas delivers biogas, biomethane, and coal gas.
- Natural gas offers a wide range of applications.
- In the residential and commercial segment, natural gas is mainly used for cooking, heating, and cooling.
- In the industrial segment, natural gas might be used for heating or as chemical feedstock.
- Technologies to produce electricity from natural gas in centralized power plants are steam turbines, combustion turbines, and combined cycles.
- Furthermore, natural gas is used in the transportation sector.

4.1 Technical Properties

As the name suggests, natural gas is a natural product whereby the state of the product depends on temperature, pressure, specific volume, and its (chemical) composition. The chemical and physical properties of natural gases have to be taken into account when trading natural gas on an international (or for LNG on a global) scale.[1]

4.1.1 Reference Conditions

Under 'normal circumstances' natural gas is in gaseous state. As such, natural gas is compressible. Depending on temperature and pressure conditions, a given volume of natural gas contains, to put it simply, a certain number of molecules of methane, ethane, and the other constituents of natural gas and, hence, a specific energy content. The volume of natural gas increases proportionally to its temperature, and it decreases with pressure. Therefore, when specifying parameters or measuring (and by measuring also valuing) natural gas volumes, it is necessary to consider metering conditions. These metering-reference conditions are determined by the specific temperature and pressure conditions.

It is important to distinguish these metering reference conditions from so-called standard reference conditions. Standard reference conditions define the conditions at which natural gas is valued. This is necessary, as natural gas is typically not priced per volume but per energy content.[2] Of course, definitions are always arbitrarily—so are the definitions of the standard reference conditions. In theory, parties to a contract could define their own standards to be applied for the contract in question. However, in practice, certain standards have been developed for the industry in different parts of the world.

The most common standard reference conditions in Western Europe are the so-called 'norm conditions' at a temperature of 0 °C (i.e., 273.15 K) and a pressure of 1.01325 bar. One cubic meter [m^3] of natural gas at these conditions is also referred to as a norm cubic meter [Nm^3].[3] Other commonly used standards are shown in Table 4.1.

The European Association for the Streamlining of Energy Exchange-gas (EASEE-gas) aims to develop common business practices in order to support the creation of an efficient and effective European natural gas market. As different reference conditions and measurement units are regarded to encumber such common market, EASEE-gas recommends using[4]:

[1]See Sects. 4.1.4 and 6.3.6.

[2]See Sect. 6.3.2.

[3]If not stated explicitly otherwise, in this book, the norm conditions 0 °C and 1.01325 bar are used. For further information on units and conversions, see Appendix B.

In addition to differing temperatures and pressures, cubic meters [m^3] are not used everywhere as a base geometrical volume. Mainly, Anglo-Saxon countries use cubic feet [ft^3] instead.

[4]See European Association for the Streamlining of Energy Exchange—Gas (2003), p. 3; formally ratified by EASEE-gas Executive Committee on 18 September 2014.

Table 4.1 Reference conditions for the definition of units

Name	Pressure [kPa]	Pressure [unit used]	Temperature [K]	Temperature [unit used]	Used in, e.g.
Norm	1.01325×10^5		273.15	0 °C	Germany, Denmark, Austria
Standard (European)	1.01325×10^5		288.15	15 °C	Italy, UK
Standard (US)	1.01325×10^5	1 atm = 14.696 psi	288.71	60 °F	US, Middle East, Africa, Latin America
Basic	1.01325×10^5		298.15	20 °C	Russia, Eastern Europe[a]

[a]This standard is also called 'Normal Temperature and Pressure' (NTP)

- pressure: [bar], whereby the system operators shall indicate whether it refers to absolute pressure (bar a) or gauge pressure (bar g);
- temperature: [°C]; and
- volume: [m^3 (@ 0 °C, 1.01325 bar)].

To determine the norm volume [or 'norm cubic meters', (V_{nr})] from measured volume [or measured 'cubic meters' (V_r)], the following equation is applied[5]:

$$V_{nr} = V_r * \frac{(p_{Amb} + p_E - \chi p_S)T_n}{p_n T} * \frac{1}{\sigma_{Deviation}}$$

with:

T_n	273.15 K
p_n	1.01325 bar
T	ambient temperature (temperature) [K]
$\sigma_{Deviation}$	gas law deviation coefficient (non-dimensional)
p_{Amb}	ambient barometric pressure (pressure) [bar]
P_E	measured pressure in pipeline (pressure) [bar]
P_S	water vapor partial pressure (pressure) [bar]
χ	relative gas humidity (non-dimensional)
V_{nr}	norm volume (volume) [m^3]
V_r	measured volume (volume) [m^3]

[5]For further reading, see, e.g.: Cerbe et al. (2004).

Metering and Reference Conditions

Company A delivers to company B a volume V_r of 20,000 m^3/h. The measured pressure P_E is 10 bar and the temperature 12 °C. The relative gas humidity shall be at 20% and the ambient barometric pressure is $p_{Amb} = 980$ mbar. The water-vapor partial pressure P_S = 0.014028 bar and the gas law deviation coefficient $\sigma_{Deviation}$ is calculated[6] to be 0.9756.

Using the above equation shows that 20,000 m^3 measured under these conditions correspond to 212,746 Nm^3.

The equation can also be used as a 'rule of thumb' to calculate the volume at different temperatures. To do this, it is assumed that:

$$\frac{(p_{Amb} + p_E - \chi p_S)}{p_n} \frac{1}{\sigma_{Deviation}} = 1$$

Then, the equation is simplified to:

$$V_{nr} = V_r \frac{273.15}{T}$$

i.e.:

$$\frac{V_{nr}}{V_r} = \frac{273.15}{T}$$

or

[6]By using the approximation formula $\sigma_{Deviation} = 1 - \frac{p_{Amb} - p_E}{450\,\text{bar}}$.

Table 4.2 Typical compositions of natural gas used in Europe (@0 °C, 1.01325 bar)[a]

Component		Unit	Russian H-gas	North Sea H-gas	Danish H-gas	Dutch L-gas	German L-gas
Methane	CH_4	mol%	96.96	88.71	90.07	83.64	86.46
Nitrogen	N_2	mol%	0.86	0.82	0.28	10.21	10.24
Carbon dioxide	CO_2	mol%	0.18	1.94	0.60	1.68	2.08
Ethane	C_2H_6	mol%	1.37	6.93	5.68	3.56	1.06
Propane	C_3H_8	mol%	0.45	1.25	2.19	0.61	0.11
Butanes	C_4H_{10}	mol%	0.15	0.28	0.90	0.19	0.03
Pentanes	C_5H_{12}	mol%	0.02	0.05	0.22	0.04	0.01
Hexanes and higher	C_{6+}	mol%	0.01	0.02	0.06	0.07	0.01
Oxygen	O_2	mol%	<0.001	<0.001	<0.001	<0.001	<0.001
Total sulphur	S	mg/m³	<3	<5	<3	<3	<3

[a]Deutscher Verband der Gas- und Wasserwirtschaft (DVGW) (2013), p. 27

$$\frac{T}{273.15} V_{nr} = V_r$$

Russian [m³]
100 million m³ (at 20 °C, 'Russian' natural gas) are 93,177,554 Nm³, since:

$$V_{nr} = 100{,}000{,}000 * \frac{273.15}{293.15} = 93{,}177{,}554.$$

4.1.2 (Chemical) Composition

Other than so-called biomethane (or other sorts of biogases and synthetic gases[7]), natural gas is not anthropogenic. Natural gas is formed from organic matter within sediments, which in the past have been exposed to high pressure and temperature conditions.[8] Therefore, it is a mixture of different chemical components, which reflect the geological conditions under which it was formed.

Typically, natural gas consists predominantly of methane (CH_4). Other components are higher hydrocarbons (ethane, propane, butane, etc.) but also nitrogen, oxygen, carbon dioxide, and hydrogen sulphide. An overview of typical compositions of natural gases used in Europe is shown in Table 4.2.

Table 4.3 gives an overview of typical LNG compositions whereby LNG compositions might change during storage, transport, and handling of LNG, since small portions of LNG will boil-off. As it is economically not feasible to keep LNG at all stages of the value chain at a temperature below −161.5 °C, some components with very low boiling temperatures like methane or nitrogen will boil off. Since they boil off faster than other components, this effect leads to a small change in the chemical composition of LNG while it is stored or transported.[9]

As natural gases are typically blended from many different fields, the composition of natural gas produced from individual fields might differ considerably. For example, natural gas produced from the Lacq field in France might contain only 70% methane but very high contents of carbon dioxide and hydrogen sulphide. Other natural gases, such as the natural gas produced from the Altmark field in Germany, show very high nitrogen contents of more than 60% and correspondently low methane content. Some gases also contain noble gases like helium and argon.

[7]For details, see Sect. 4.2.3.

[8]For details, see Sect. 4.2.1.1.

[9]This effect is also named ageing or weathering of LNG.

Table 4.3 Worldwide average LNG compositions (@15 °C, 1.01325 bar)[a]

Origin	Nitrogen	Methane	Ethane	Propane	Butanes and higher
	N_2 (%)	CH_4 (%)	C_2H_6 (%)	C_3H_8 (%)	C_{4+} (%)
Algeria—Arzew	0.56	87.98	9.00	1.99	0.47
Algeria—Bethioua 1	1.20	87.59	8.39	2.12	0.70
Algeria—Bethioua 2	0.92	91.39	7.17	0.52	–
Algeria—Skikda	1.02	91.19	7.02	0.66	0.11
Egypt—Damietta	0.08	97.70	1.80	0.22	0.20
Egypt—Idku	–	97.20	2.30	0.30	0.20
Libya	0.69	81.57	13.38	3.67	0.69
Nigeria	0.08	91.28	4.62	2.62	1.40
Abu Dhabi	0.29	84.77	13.22	1.63	0.09
Oman	0.35	87.99	7.27	2.92	1.57
Qatar	0.36	90.10	6.23	2.32	0.99
Trinidad	0.03	96.82	2.74	0.31	0.10
US—Alaska	0.17	99.73	0.08	0.01	–
Australia—NWS	0.09	87.39	8.33	3.35	0.84
Brunei	0.05	90.61	4.97	2.89	1.48
Indonesia—Arun	0.06	91.16	6.01	1.84	0.93
Indonesia—Badak	0.02	89.76	5.06	3.54	1.62
Malaysia	0.16	91.15	4.96	2.79	0.94

[a]British Petroleum (2011), p. 83

Other natural gases might contain undesirable ingredients of, e.g., mercury, lead, thallium, or cadmium. These ingredients need to be removed from the natural gas, as they are toxic and harmful to the safe operation. They might block the equipment. For instance:

- hydrogen sulphide (H_2S) is harmful for different reasons. Firstly, it is toxic and therefore a potential risk to human health; secondly, if combusted, sulphur dioxide (SO_2) is formed, which is environmentally harmful and, thirdly, it is operationally harmful as it might lead to corrosion;
- the total sulphur content usually includes (apart from hydrogen sulphide) carbonyl sulphur and mercaptans, as well as elementary sulphur. Further to its corrosiveness, these sulphur components might lead to the plugging and therefore blocking of pipes;
- nitrogen (N_2) and carbon dioxide (CO_2) are inert gases, i.e., they do not provide any heat content to the natural gas stream. The higher the content of these inert gases, the lower the value of the natural gas since more volume has to be transported in order to achieve a certain heat content. In addition to this economic effect, carbon dioxide is harmful as it reacts with free water to form an acid, which again is corrosive; and
- most natural gases also contain water (H_2O). Because of its corrosiveness (especially in connection with sulphur or carbon dioxide) and since it might form hydrates (e.g., with hydrocarbons) which again might block pipes, it is removed from natural gas as much as possible.

These impurities are usually removed as early as possible from the natural gas, i.e., near the production site, either to meet product specifications or for economic reasons. Helium and argon, for example, are of great economic interest.

Dependent on pressure and temperature conditions, the hydrocarbons mentioned in Tables 4.2 and 4.3 are either in the solid, gaseous, or liquid state. Under normal conditions,[10] methane, ethane, propane, and butanes are in the gaseous state, whereas pentanes and higher are liquid. Therefore, pentanes and higher are also referred to as condensates.

As propane, n-butane and i-butane[11] can be liquefied without refrigeration through moderate pressurization, they are also named Liquefied Petroleum Gases (LPG). Sometimes they are referred to as 'bottle gas', in accordance with their preferred form of use in bottles in industry or homes.

Pentanes and higher[12] hydrocarbons are also referred to as natural gasoline, whereby they usually do not contain hydrocarbon molecules heavier than C_{10}.[13]

Together with ethane, LPGs and pentanes and higher hydrocarbons are usually referred to as natural gas liquids (NGL). These hydrocarbons are condensed from natural gas at the moment they are exposed to normal conditions. For storage, shipment, and consumption, NGLs are usually kept in liquid state.

This terminology is illustrated in Fig. 4.1.

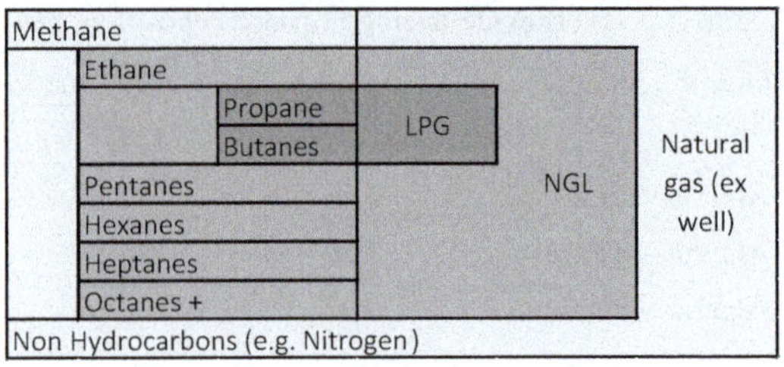

Fig. 4.1 Natural gas terminology

So-called compressed natural gas (CNG) and liquefied natural gas (LNG) have the same chemical composition as (gaseous) natural gas. However, whereas CNG is produced from natural gas by compression up to approximately 200 bar, LNG is natural gas that is cooled down to approximately −161.5 °C at which temperature methane liquefies. The energy density of LNG is approximately 600 times higher than that of natural gas under normal conditions. For CNG, the density factor depends on the pressure.

Treatment of natural gas has to meet two main objectives:

- adjustment of the natural gas composition to standard quality requirements, either for pipeline transport or—in the case of LNG—for liquefaction; and
- availability in economically sufficient quantities for the recovery of valuable by-products, e.g., noble gases or NGLs.

This treatment either might take place directly at the well (in case the natural cannot be transported to a processing plant without causing pipeline hazards) or in centralized processing plants.

4.1.3 Physical Properties

The chemical composition of natural gases determines their physical properties. To describe the physical properties of natural gas, it is first necessary to define exactly the respective parameters. The most common parameters in practice are used to describe the so-called combustion properties. In addition, parameters like molecular mass, molar volume, density and relative density, viscosity, and compressibility are necessary for a complete description of the physical properties of natural gases. All parameters defined, jointly describe the natural gas quality.

[10] See Sect. 4.1.1.

[11] n-butane (or normal butane) and i-butane (or iso-butane) are isomers of butane, i.e., they have the same chemical composition as C_4H_{10}, but a different arrangement of molecules, which leads to slightly different chemical and physical properties, e.g., the boiling points (at atmospheric pressure) differ between i-butane (−12.1 °C) and n-butane (−0.5 °C).

[12] That is, hexane (C_6H_{14}), heptane (C_7H_{16}), octane (C_8H_{18}), nonane (C_9H_{20}) and decane ($C_{10}H_{22}$).

[13] This C_{10} refers to only one part of the hydrocarbon molecule.

4.1.3.1 Combustion Properties

Chemically, the combustion equals oxidization (of natural gas). This process is an exothermic reaction, i.e., 'energy is released' as it occurs. For pure methane and oxygen, the chemical reaction is:

$$CH_4 + 2O_2 \rightarrow CO_2 + 2H_2O$$

However, methane (CH_4) is not the only constituent of natural gas and the combustion usually does not take place in pure oxygen (O_2) but in air, which has an oxygen content of approximately 21%. Therefore, in practice, the combustion reaction is more complex and oxides of nitrogen (NO_x) and oxides of sulphur (SO_x) are formed—environmentally harmful, undesirable by-products of combustion.[14]

The most important parameters describing the combustion properties are calorific value and Wobbe index. Others include maximum flame speed, the volume of combustion air required, the maximum flame temperature, the lift index, and the soot index.

4.1.3.1.1 Calorific Value

As described, the combustion of natural gas is an exothermic reaction. The amount of heat, i.e., energy, released by complete combustion of (a defined volume of) natural gas is called the calorific value. Sometimes it is also referred to as the 'heating value'. When determining the gross calorific value,[15] GCV, all water formed during the combustion process is condensed to the liquid state. As the additional energy for this condensation is considered, the gross calorific value is higher than the net calorific value[16] where the water formed during combustion remains in the vapor state.[17]

Calorific values are expressed in energy per volume, e.g., [MJ/m^3] or [kWh/m^3] or [Btu/ft^3]. In parts of Eastern Europe, also [kilocalories = kcal] are used as an energy unit.[18]

Again, when measuring a calorific value, it is crucial to define reference conditions (for pressure and temperature) for the combustion process. Typical definitions for gross calorific value in Europe and North America are:

- 'gross calorific value', GCV, shall mean that the amount of heat, i.e., energy, produced by the complete combustion of one Nm^3 of natural gas at 25 °C and an absolute pressure of 1.01325 bar with excess air at the same temperature and pressure as the natural gas when the products of combustion are cooled to 25 °C and when the water formed by combustion is condensed to the liquid state and the products of combustion contain the same total mass of water vapor as the natural gas and air before combustion; or
- 'total heating value', GCV, shall mean that amount of heat, i.e., energy, produced by the combustion, at constant pressure, of the quantity of the gas that would occupy a volume of 1 ft^3 at a temperature of 60 °F if saturated with water vapor and under a pressure equivalent to that of 30″ of mercury at 32 °F and under standard gravitational force in air of the same temperature and pressure as the gas, when the products of combustion are cooled to the initial temperature of gas and air when the water formed by combustion is condensed to the liquid state.

Important parameters in such definition are:

- the volume of natural gas burned,[19]
- the combustion temperature, and
- the pressure.

[14]Compared with other fuels, the combustion of natural gas is relatively clean. See also Sect. 4.1.5.

[15]Sometimes gross calorific value is also referred to as total calorific value or gross heating value.

[16]Sometimes net calorific value is also referred to as net heating value.

[17]The difference and relationship between gross and net calorific value depends on the natural gas composition and the condensation enthalpy of water.

[18]Measuring energy in 'kilocalories' may not be unambiguous. Whereas MJ (and, by a constant conversion factor of 3.6 MJ/kWh, effectively also kWh) is a SI unit solely depending on the international conventions for length (meter), mass (kilogram), and time (second), the definition of a kilocalorie itself depends on conditions regarding pressure and temperature. For further details, see Appendix B.

[19]See Sect. 4.1.1.

Table 4.4 Calorific values for different gases

		GCV_n		NCV_n		NCV_n/GCV_n
		MJ/Nm^3	kWh/Nm^3	MJ/Nm^3	kWh/Nm^3	
Methane	CH_4	39.83	11.06	35.89	9.97	0.9012
Ethane	C_2H_6	70.33	19.54	64.38	17.88	0.9154
Propane	C_3H_8	101.14	28.10	93.12	25.87	0.9207
n-Butane	C_4H_{10}	134.12	37.25	123.86	34.40	0.9235

EASEE-gas recommends using the so-called standard conditions, i.e., one Nm^3 with a combustion reference temperature of 25 °C and a pressure of 1.01325 bar (see Footnote 4). The gross calorific value under such conditions is abbreviated by GCV_n; the net calorific value under such conditions is abbreviated by NCV_n. Whereas these norm conditions are used in many parts of Continental Europe, the UK defines 15 °C as reference temperatures for metering and combustion, and the US uses 60 °F (i.e., approximately 15.6 °C) for both.[20]

Table 4.4 shows gross and net calorific values for various gases.

Calorific values for natural gases can be calculated from the calorific values of its components and their molar composition, i.e., a natural gas which consists of 100 mol% methane has a gross calorific value of $GCV_n = 11.06\ kWh/Nm^3$, whereas a natural gas having only 98 mol% methane and 2 mol% ethane has $GCV_n = 11.23\ kWh/Nm^3$ ($0.98 * 11.06\ kWh/Nm^3 + 0.02 * 19.54\ kWh/Nm^3$).

4.1.3.1.2 Wobbe Index

The gross Wobbe[21] index (or Wobbe number) is defined by:

$$W_{G,n} = \frac{GCV_n}{\sqrt{\rho_n}}$$

with $\rho_n = \frac{\rho_{Gas}}{\rho_{Air}}$; whereby ρ_n is the relative density[22] of the natural gas, i.e., the ratio of its density, ρ_{Gas}, to density of dry air, ρ_{Air}. As the relative density is dimensionless, the unit of the Wobbe index is identical to the unit of the calorific value.

The net Wobbe index is defined analogously using NCV_n instead of GCV_n, i.e.:

$$W_{N,n} = \frac{NCV_n}{\sqrt{\rho_n}}$$

Natural gases, with different chemical compositions and, thus, probably different calorific values, produce the same amount of heat per time unit if their Wobbe indices are equal. Therefore, most natural gas specifications stipulate Wobbe index ranges instead (or at least in addition to) calorific value ranges.

Table 4.5 shows gross and net Wobbe indices for various gases.

4.1.3.2 Other Properties

In order to avoid liquid phases of water or hydrocarbons in a given natural gas, the gas temperature should not fall below the so-called water dew point, or, the hydrocarbon dew point, respectively.

Both dew points depend on the chemical composition of the natural gas and on the pressure. The higher the water content, the higher the water dew point and, analogically, the higher the pressure, the higher the water dew point.

The methane number[23] is a parameter for the knock resistance of a gaseous fuel. Minimum requirements for the methane number have to be fulfilled to achieve safe operation of a given engine. Methane has by definition a methane number of 100, whereas hydrogen has a methane number of zero. The methane number of natural

[20] These temperatures are not to be confused with the reference conditions.

[21] Named after the Italian engineer Goffredo Wobbe, who developed the index in the 1920s.

[22] Sometimes also referred to as specific gravity.

[23] The methane number is the analogue to the octane number for gasoline.

Table 4.5 Wobbe indices for different gases

		$W_{G,n}$		$W_{N,n}$		ρ_n
		MJ/Nm^3	kWh/Nm^3	MJ/Nm^3	kWh/Nm^3	
Methane	CH_4	53.47	14.85	48.18	13.38	0.555
Ethane	C_2H_6	68.70	19.08	62.89	17.47	1.048
Propane	C_3H_8	81.13	22.54	74.70	20.75	1.554
n-Butane	C_4H_{10}	92.66	25.74	85.57	23.77	2.095

gas depends on its chemical composition whereby higher hydrocarbons reduce the methane number and high contents of inert gases like carbon dioxide and nitrogen increase the methane number.

4.1.4 Specifications and Interchangeability of Natural Gases

All gas-fired equipment and applications are designed (and sometimes fine-tuned) for a particular natural gas quality. Optimal conditions for these applications (including measurement equipment) are achieved when the natural gas quality, i.e., the composition, the combustion parameters, and all other factors described, remain constant over time.

Due to weather conditions, the outside temperature and pressure change frequently. This implies a change of the reference conditions.[24] In addition, usually, grids receive natural gases from several sources possessing differing qualities. Moreover, the off-take from different sources changes over time. In case the system includes storage facilities, withdrawal from or injection into these leads to further variability. Further, users of natural gas use different quantities of natural gas in different periods (e.g., dependent on weather conditions). The combination of all these factors necessarily leads to zones within a pipeline system, where the prevailing natural gas quality is not constant over time. Realizing (and accepting) such variability led to the definition of the specification of natural gas quality by either system operators (transport, storage, liquefaction, regasification) and governments or regulatory bodies.

These specifications define ranges, i.e., minimum and maximum values a certain parameter might have when natural gas is brought into a system. These ranges are defined in a way that natural gas applications are able to work without materially affecting their operational performance, their safety, and their efficiency. In any case, as long as natural gases within a defined specification are combusted, problems with environmentally harmful emissions and poor combustion quality or risks of damaging combustion equipment can be avoided and a safe and reliable operation of the natural gas infrastructure is achieved.

4.1.4.1 Specifications

Different countries define different specifications for natural gas quality. That is, neither the parameters themselves nor their values are harmonized internationally.[25] Clearly, this might lead to situations where natural gas within the specification of one country or system does not comply with the specification of a neighboring country or system, not because the natural gas's quality changes but because different specifications define different ranges. Such non-harmonization might cause additional costs for market participants (if they decide to bring the natural gas quality within specification), or at least expose them to contractual risks (e.g., if a trader accepts the risk to deliver natural gas out of specification).[26]

[24] See Sect. 4.1.1.

[25] Note that the subsequent description refers to the specifications of natural gas quality parameters, i.e., the minimum and maximum values permitted, in contrast to the actual parameter value at a given point in time.

[26] See Drasdo et al. (2013) and the example in Sect. 6.3.6.

Table 4.6 Quality specification according to EASEE-gas

Parameter	Unit	Min	Max
Wobbe index	kWh/m^3	13.60	15.81
Relative density	m^3/m^3	0.555	0.700
Total sulphur	mg/m^3	–	30
Hydrogen sulphide and carbonyl sulphide (as sulphur)	mg/m^3	–	5
Mercaptanes (as sulphur)	mg/m^3	–	6
Oxygen	mol%	–	0.001
Carbon dioxide	mol%	–	2.5
Water dewpoint	°C @ 70 bar(a)	–	−8
Hydrocarbon dewpoint	°C @ 1–70 bar(a)	–	−2

Harmonization[27] of natural gas specifications is not only supposed to reduce contractual risks but also to enhance safety and reliability of natural gas infrastructure. In order to streamline interoperability at cross border points in Europe, EASEE-gas attempted to harmonize quality specifications. EASEE-gas proposed to use the parameters and ranges shown in Table 4.6.[28]

Despite these efforts, there are still other natural gas specifications in existence. One example is Germany where the specification is defined in the Code of Practice DVGW G 260(A).[29] The Wobbe index (for high-calorific gas) in German distribution systems must to be in a range from 13.6 to 15.7 kWh/m^3, i.e., narrower than proposed by EASEE-gas. In line with this, the relative density has a wider allowed range from 0.55 to 0.75. Other members of the European Union, let alone non-European countries, define other parameters to specify natural gas and other ranges of these or other parameters. In some countries, not the physical parameters but the chemical composition is used to specify natural gas.[30]

Different Wobbe Indices for Central Europe

Already by comparing only a few countries in Central Europe, differences in natural gas quality specifications can be observed. As shown in Fig. 4.2, the ranges of Wobbe Indices accepted in different countries are not compatible. For example, while natural gas with a Wobbe index of $W_{G,n} = 15.7\ kWh/m^3$ is permitted in Germany, the upper limit accepted in the Net4Gas system in the Czech Republic is $W_{G,n} = 15.24\ kWh/m^3$.

A further specification issue arises from odorization. As natural gas is odorless, so-called odorants are added for safety reasons to enhance the detection of leaks. In the various national grids in Europe, different odorants are used. In addition, some countries (e.g., France) have chosen to odorize natural gas already in the high-pressure grid, i.e., when imported, while in others (e.g., Germany) natural gas is odorized in distribution grids only. Obviously, this might hinder the free flow of natural gas in high-pressure grids, as odorized natural gas should not flow into a 'non-odorized' pipeline system.

Another peculiarity of Germany and its neighbors The Netherlands, Belgium, and France is the existence of different natural gas grids for different gas qualities. Locally produced low-calorific gas (L-gas) or Groningen gas is transported in different

[27] In the sense of 'unification'.

[28] European Association for the Streamlining of Energy Exchange—Gas (2005), p. 4.

[29] See Deutscher Verband der Gas- und Wasserwirtschaft (DVGW) (2013).

[30] For example, Bulgartransgaz (2016).

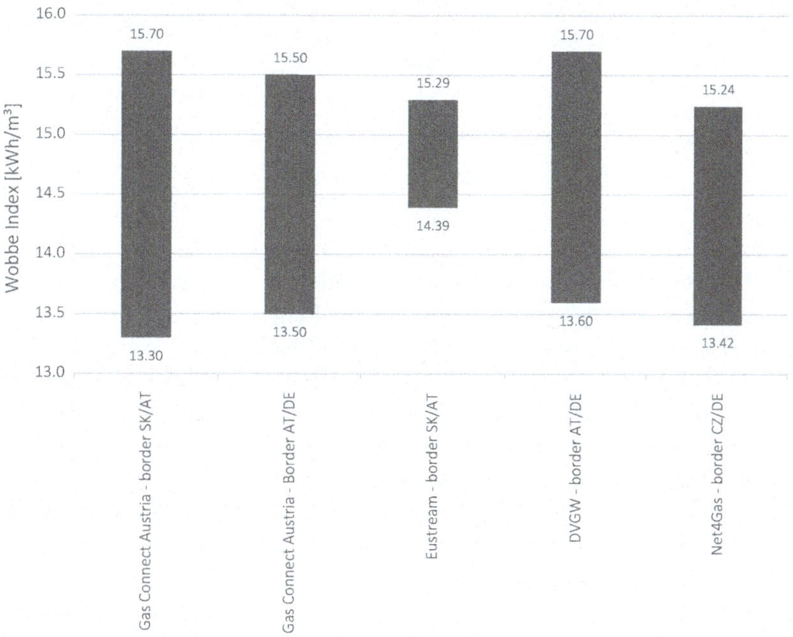

Fig. 4.2 Wobbe Index specifications applicable in DE, AT, CZ, and SK [based on: Eustream (2016), Net4Gas (2010), Gas Connect Austria (2013), and Deutscher Verband der Gas- und Wasserwirtschaft (DVGW) (2013). The values for the Czech and Slovak system, originally expressed at 'basic conditions' (see, e.g., Eustream (2016), p. 6), have been recalculated to norm conditions]

networks than high-calorific gas (H-gas). In addition, different storage facilities are used to store L-gas or H-gas. As the name suggests, L-Gas has a lower calorific value and Wobbe index than H-gas.

Physical and Chemical Properties Matter in the Natural Gas Industry

In October 2015, the local utility, i.e., the 'Stadtwerke' of Schneverdingen-Neuenkirchen converted its network from L-Gas to H-Gas. Schneverdingen-Neuenkirchen is a small town in Lower Saxony, Germany. Here 6000 households and two industrial sites are supplied with natural gas. Because of the different calorific value of L-Gas and H-Gas, the nozzles of every single burner had to be replaced and the technical equipment of industrial sites had to be adjusted to the new gas quality. It was more a logistic than a technological challenge. Within six weeks after L-Gas is replaced by H-Gas, primarily the modern condensing boilers must be replaced. The conversion in Schneverdingen-Neuenkirchen was the kick off for a fifteen year-long German project to convert almost all L-Gas regions to H-Gas because of the dwindling L-Gas sources in Germany and the Netherlands.

Yet, it was not the first project of this kind: Before the 'natural gas era' started, a number of cities and areas in Germany were supplied with town gas produced from coal. Town gas has an even lower calorific value than L-Gas. When natural gas was replaced by town gas, the same adjustment as in Schneverdingen-Neuenkirchen took place. In the former German Democratic Republic, town gas constituted a substantial share of the gas supply—despite supplies of Russian natural gas. After the reunification, the conversion to natural gas was one of the tasks for the natural gas industry. In West Germany, conversion from L-Gas to H-Gas in some areas of Hesse and North Rhine-Westphalia had been initiated already in the 1980s, after Russian gas was added to the supply portfolio. The major German gas company Ruhrgas had a specialized affiliate to carry out that task.

Table 4.7 Carbon dioxide emissions coefficients for different fuels

	CO_2 emissions coefficient $[kg_{CO_2}/GJ]$
Natural gas	55.9
Hard coal	93.6
Lignite	110.9
Fuel oil	81.3
Gasoil	74.0

Despite that the problems of harmonization of natural gas quality specifications are predominantly a European issue, their global impact has to be recognized. As long as there essentially exists in each natural gas market worldwide different quality specifications, or even different parameters are used to specify natural gas quality, trade, interchangeability, and security of supply are, at least, put at (contractual) risk or reduced, i.e., in some cases, prevented.

4.1.4.2 Gas Quality Management

Producers of natural gas or system operators have various possibilities to influence the quality of a natural gas stream either by:

- blending, i.e., the commingling of natural gases of different qualities;
- derichment processes, i.e., reducing the Wobbe index and calorific value either by ballasting (injecting nitrogen, carbon dioxide, or air) or by removing higher hydrocarbons; or
- enrichment processes, i.e., increasing the Wobbe index and calorific value either by injecting LPG or by removing inert gases like nitrogen and carbon dioxide.

If these physical management tools are not available or too expansive to perform, the only option left is to consider a softening and widening of allowed ranges of the natural gas specification. Of course, such decisions should take into account the effects such widening might have on the safe and efficient operation of natural gas equipment.

4.1.5 Environmental Effects

As any other human activity, the production and use of natural gas has a (potentially negative) impact on the environment. These environmental effects might be categorized according to the part of the ecosystem influenced. Climate effects, air pollution, water pollution, and land use do occur. In addition to this, low levels of radioactivity might be emitted, and earthquakes might be induced.

4.1.5.1 Climate Effects

Today, it is generally assumed that the man-made emission of so-called greenhouse gases causes an increase of temperatures around the globe. As this might have negative environmental, and with it, social impacts, most governments around the world are in agreement that the emissions of such greenhouse gases must be reduced. Carbon dioxide (CO_2) is at the center of these endeavors.

As already described in Sect. 4.1.3.1, carbon dioxide is emitted during the combustion of natural gas. However, the carbon-dioxide emissions coefficients are smaller than for competing fossil fuels. The coefficients are enumerated in Table 4.7.[31]

That is, when producing one kWh of electricity with natural gas, less carbon dioxide is emitted than producing one kWh with coal or oil. *Ceteris paribus*, switching electricity production from, e.g., coal to natural gas or the use of natural gas in vehicles replacing, e.g., diesel, would (relatively) reduce emissions and, therefore, could be regarded as relative, yet not absolute, climate protection.

However, methane itself is also a climate-wrecking greenhouse gas. The global-warming potential of methane over a 100-year period is 21 times higher than for carbon dioxide.[32] Therefore, for reasons of climate protection, it is pivotal to

[31]Länderarbeitskreis Energiebilanzen (2016). The values in Table 4.7 are for orientation only, as the exact values depend on the fuel's chemical composition.

[32]See Solomon et al. (2007), p. 33.

Table 4.8 Pollutants produced from combustion of various fossil fuels

Relative to natural gas	Natural gas	Oil	Coal
Carbon monoxide	1.0	0.8	5.2
Nitrogen oxides	1.0	4.9	5.0
Sulphur dioxide	1.0	1870.0	4318.3
Particulates	1.0	12.0	392.0

avoid natural gas releases as far as possible. By well leaks, pipeline leaks, or venting,[33] methane might be released into atmosphere.[34]

Further, there exist 'natural' methane releases like the unfreezing of permafrost soil. There and in oceanic deposits, methane has been trapped in a crystal (ice-like) structure together with water. These structures are called methane clathrates or methane hydrates. In addition to their potential use as a source of natural gas in competition with conventional and unconventional gas sources,[35] methane clathrates are regarded as a risk factor to the world climate because they are part of a vicious circle. On the one hand, temperature increases might release more methane clathrates by unfreezing permafrost, and, on the other hand, such release might be regarded as cause for further anthropogenic temperature increases.

4.1.5.2 Air Pollution

Not only in regard to carbon dioxide (CO_2) is natural gas cleaner than other fossil fuels. When combusted, natural gas emits less carbon monoxide (CO), nitrogen oxides (NO_x), sulphur dioxide (SO_2), and particulates than other fossil fuels.

The coefficients for these pollutants caused by combustion of the respective fuels are enumerated in Table 4.8.[36]

In contrast to natural gas, coal and fuel oil also release particles of ash and mercury when burned. In areas with low air quality, polluted with, e.g., smog or acid rain, switching electricity or heat production from liquid and solid fossil fuels to natural gas might help to improve air quality.

4.1.5.3 Water Use and Pollution

Nearly all industrial activities require the use of water, and, so does the production of natural gas. Especially for the production of unconventional reserves, huge amounts of water are needed. To a very large degree, the fluids used for hydraulic fracturing consist of water, which is taken either from groundwater or surface water. Particularly in arid and water-scarce regions, this might lead to or intensify an already strained water-supply situation.

Further to the complications from the consumption of clean water, the disposal of wastewater might pose challenges when producing natural gas. If chemicals were used during production, it might be—depending on their composition—harmful to the environment to release the sewage water into rivers or other surface waters.

If at all, groundwater pollution is usually linked to (deficiencies of) the production of natural gas. Improperly constructed wells, as well as surface leaks, might lead to contamination of groundwater with natural gas, (salty) reservoir water, or other fluids used during production. In case of surface spills, also surface water might be contaminated. The only way to mitigate these environmental risks is a reasonable and prudent management and the use of preferably non-hazardous fluids during production. Beyond that, some countries do not allow natural gas production activities in water protection areas.

[33]See also Sect. 4.3.5.

[34]Other than by the production or use of natural gas methane is also emitted as a result of coal production, the decay of organic waste and agriculture.

[35]However, exploitation of these clathrates is not yet economically viable.

[36]See Energy Information Administration (1999), p. 58.

4.1.5.4 Land Use and Sea-Floor Use

As any human activity uses space and land, so does the use of natural gas. The whole value chain from exploration and production to transportation and storage needs land and thereby influences local ecosystems.[37] This becomes most obvious for natural gas production sites, liquefaction and regasification plants, compressor stations, and underground storage sites. In order to prepare them, operators need to build roads, pipelines, and the aboveground facilities themselves. As in any other building activity, this disturbs local ecosystems and has negative impact on flora and fauna. Furthermore, building activity obviously causes and intensifies erosion. In principle, the same is true during the building phase of natural gas pipelines whereby the pipeline itself and the trenches are covered after the building phase, so that they are nearly invisible during operation.[38]

Once the sites or pipelines are not used any longer, e.g., because natural gas production has become uneconomical, the facilities are decommissioned and abandoned. Depending on the environmental legislation, the areas used must be returned to its original state.

In case of offshore activities, e.g., platforms, subsea completions, and offshore pipelines use the sea floor. As a rule, these offshore facilities are less noticeable; however, in case of an offshore emergency, e.g., a spill, the impact on the environment might be more dramatic than onshore as marine ecosystems are usually more sensitive.

4.1.5.5 Radioactivity

Radioactive elements such as uranium, thorium, or potassium are naturally present in low concentrations in the Earth's crust. Mainly, the decay products of these naturally occurring radioactive materials, e.g., radon and lead,[39] are brought to the surface by producing natural gas or crude oil.[40] Natural gas, being extracted from the subsoil, therefore, may contain radioactive elements. These elements are contained primarily in the material that is co-produced with natural gas, e.g., salty waters, especially when they contain barium sulphate. Jointly with other impurities, these salts are usually removed from the natural gas already near the production site. Therefore, natural gas transported in mid- and downstream pipelines is not radioactive. Even at production sites, the level of radioactivity is usually low and not harmful, comparable to a transoceanic flight.[41]

4.1.5.6 Earthquakes

Natural gas production might induce earthquakes. When extracting natural gas, pressure in the natural gas reservoir changes and might cause a compaction of the reservoir rock that again might cause (induced) earthquakes. Further to this effect, hydraulic fracturing[42] is regarded as an additional risk factor for inducing earthquakes. However, most of these man-made, i.e., induced, earthquakes are very low in intensity and not perceivable at the Earth's surface without specialized equipment.

Groningen Earthquakes

The exception proves the rule. Whereas natural gas production induced earthquakes are usually not noticeable, a different situation exists in the Groningen area. The vast Groningen field in the northeast of the Netherlands is among the 10 largest natural gas fields in the world. The reserves were estimated at 2.8 tcm, and, together with its proximity to potential markets the Groningen field became key to the development of the European natural gas markets after it was discovered in the 1950s. After fifty

[37]Examples are birds attracted by gas flares or whales becoming disoriented due to offshore seismic exploration activities.

[38]Except for forest aisles, since safety considerations prohibit the planting of trees above pipelines.

[39]Mainly the isotopes ^{210}Pb, ^{224}Ra, ^{226}Ra and ^{228}Ra are mobilized during oil and gas production. Compare also with World Nuclear Association (2016).

[40]Not only production of oil and natural gas, but also all mining processes, bring radioactive material to the surface. Of particular importance for the energy industry are the radioactive materials in coal ash produced in power plants.

[41]See Bundesverband Erdgas, Erdöl und Geoenergie e. V. (2016).

[42]See Sect. 4.2.2.2.

years of production, seismic activity increased in the area, and, on August 16, 2012, an earthquake occurred at Huizinge in the Groningen province. The earthquake had a Richter magnitude of 3.6, still a minor earthquake, but perceptible by people. As the Groningen area is not an area of high (natural) seismic activity, an intense political discussion started in the Netherlands about the pros and cons of natural gas production in the Groningen area.

As the Dutch Safety Board stated in 2015[43]:

> *The Dutch state and Dutch society have derived significant benefits from gas extraction from the province of Groningen. Gas revenues are considerable. Natural gas has been one of the cornerstones of the Dutch economy for more than half a century. There is no other product from which the Netherlands has earned as much as from natural gas.*
>
> *Gas extraction has also drawbacks. It was known early on that the extraction of gas would cause subsidence in Groningen and later on the first earthquakes caused by gas extraction occurred in the area. All the parties concerned — including numerous citizens — had long viewed these earthquakes as a cause of damage that had to be compensated. The earthquakes were not deemed to be a safety concern for the citizens of Groningen. Studies suggested that the intensity of future earthquakes would be minimal and even though the maximum intensity had been adjusted upward on a number of occasions, the parties responsible for gas extraction ruled out the idea that the earthquakes would lead to personal accidents. The earthquake in the village of Huizinge in 2012 dispelled this optimistic belief. Since the supervisory authority's warning in early 2013 many people viewed the earthquakes not just as a damage issue but also as a threat to the safety of the citizens of Groningen. The reassuring certainty that the intensity of the earthquakes would not exceed a certain value had vanished.*

As a result of these discussions, production caps were introduced by the Dutch government. On July 24, 2016, it decided that under normal circumstances only 24 bcm/a were allowed to be produced, roughly half the production figures the field achieved only a few years earlier.[44]

4.2 Sources

The most common source for natural gas are underground fields or reservoirs. These are commercially recoverable accumulations of natural gas in the Earth's crust. In addition to these natural sources, anthropogenic sources exist.

4.2.1 Natural Sources

In a reservoir, natural gas can occur as:

- non-associated gas (i.e., natural gas without any, or only limited, content of light-liquid components); or
- associated gas (i.e., natural gas either dissolved in crude oil (solution gas) or as a gas cap in contact with gas-saturated crude oil).

Usually, natural gas from a gas cap is not extracted until most of the underlying crude oil has been produced because liquid components offer a higher commercial value, and the gas cap is needed in order to keep the pressure within the reservoir at a high level. With regard to solution gas, both, natural gas and liquid components are produced at the same time. Again, the commercially dominant components are liquids and, therefore, gas production occurs subsequent to liquids production.

Even in non-associated gas fields, the natural gas produced always contains not only methane but also higher alkanes and other components.[45] Natural gas with a low proportion of higher alkanes is called dry gas. Natural gas with a high

[43]Dutch Safety Board (2015), p. 5.

[44]See Reuters (2016).

[45]See Sect. 4.1.2.

proportion of higher alkanes is referred to as wet gas, since higher alkanes inevitably condense while producing natural gas and thereby change the pressure and temperature conditions. While wet gas is always created along with oil, many dry-gas deposits originate in coal deposits.

Natural gas with a high content of H_2S and CO_2 is also known as sour gas, and that with a low content of H_2S and CO_2 is known as sweet gas.

For hydrocarbon accumulations to be present underground, the following independent features have to exist or to have once existed:

- source rock, rich in organic carbon to be converted to hydrocarbons;
- sufficient heat over long periods (necessary to convert organic carbon into hydrocarbons);
- migration pathways to enable the expelled hydrocarbons to migrate from the source rock to a suitable reservoir rock;
- reservoir rock, with sufficient porosity and permeability;
- efficient seal of impermeable rocks above and against the reservoir rock; and
- closed structure, a geometric disposition of the reservoir and the seal to arrest the upward migration of the hydrocarbons.

4.2.1.1 Formation

Almost without exception, oil and gas, as peat, lignite, and coal are of organic—mostly plant—origin.[46] Highly concentrated organic material occurs for example in swamps, lagoons, and shallow seas. To form hydrocarbons, this organic material needed to have been deposited in an anaerobic, i.e., oxygen-free environment.

The source material for the oil (and natural gas) genesis is organic matter, e.g., marine microorganisms like algae. If these organisms after their demise reach deep, oxygen-deficient marine strata, they avoid being recycled in the biological cycle and instead form together with fine rock material a so-called sapropelic mud. Due to the then emerging diagenesis and supported by anaerobic bacterial fermentation, methane, carbon dioxide and kerogen are formed. Kerogen consists of large hydrocarbon molecules. Increasing temperature leads to its decomposition into smaller hydrocarbons. By further deposition of material above the sapropelic mud, pressure and temperature continue to rise and the so-called catagenesis (between 100 and 150 °C) sets in, during which oil is formed from kerogen and, by further depression, more and more natural gas is generated. Above 150 ° C, metagenesis starts. At these temperatures only natural gas is formed. The rocks in which the hydrocarbons are located immediately after their formation are also referred to as source rocks.

Coal is formed in tropical and subtropical climates under anaerobic conditions by decomposition of higher land plants, i.e., plants at a higher evolutionary stage. Caused by multiple sea-level fluctuations, this material has been overlaid by other sediments, whereby—similar to petroleum formation—pressure and temperature conditions altered in a way that the geochemical process of coalification began. Depending on the pressure and temperature reached, different carbons such as peat, lignite, bituminous coal, and anthracite were formed. As part of this process, methane is formed, so that it is possible to refer to a coal seam as a natural gas source rock.[47]

4.2.1.2 Migration

Conventional natural gas fields are not found in source rocks, but in reservoir rocks. Through further compaction, but also increasing temperature and dehydration, the hydrocarbons together with water leave the source rock (also referred to as primary migration) and 'migrate' in permeable layers or along geological faults upwards until they either exit or are caught in a reservoir (so-called secondary migration). Sometimes they leak out at the surface[48] (tertiary migration). The general upward direction is caused by the fact that oil and gas are less dense than water.

4.2.1.3 Conventional Natural Gas Reservoirs

In a reservoir, the hydrocarbons are caught in a trap-like structure. In a reservoir rock, i.e., a rock with high porosity and high permeability

[46]A few natural gas fields are volcanic in origin.

[47]See Sect. 4.2.1.4.3.

[48]The asphalt lakes in Trinidad are one example.

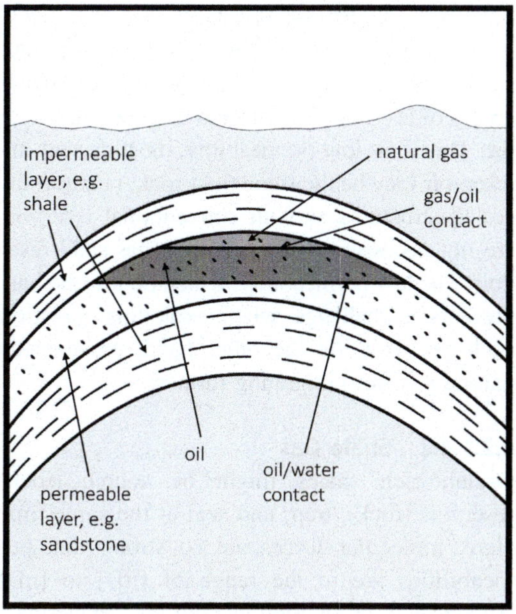

Fig. 4.3 Anticlinal reservoir type

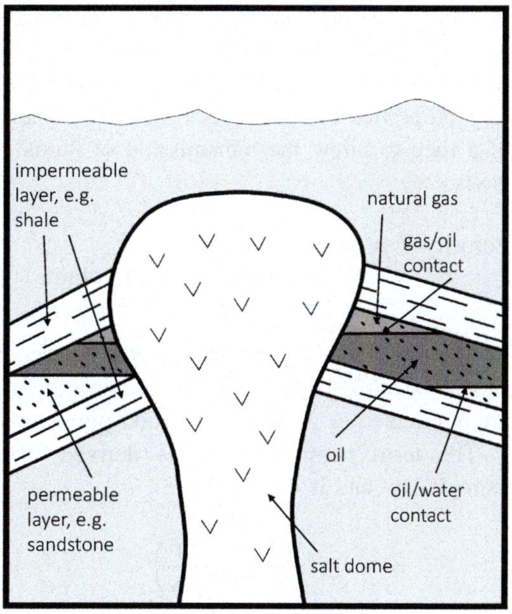

Fig. 4.4 Salt diapiric reservoir type

(frequently sandstone, limestone, or dolomite), hydrocarbons may accumulate, provided that the reservoir rock is sealed at the top and to the side by impermeable rocks (e.g., clays, rock salt). Due to their different densities, natural gas, oil, and water separate in the reservoir rock. Therefore, in a deposit, natural gas is found at the top, in the middle (if any) oil, and (salty) water at the bottom.

The following major types of petroleum traps are widespread[49]:

- anticlinal type,
- (salt-)diapiric type,
- transgression type, and
- fault type

These types are depicted in Figs. 4.3, 4.4 and Fig. 4.5.

Good deposits are characterized by a reservoir rock of high porosity and high permeability.

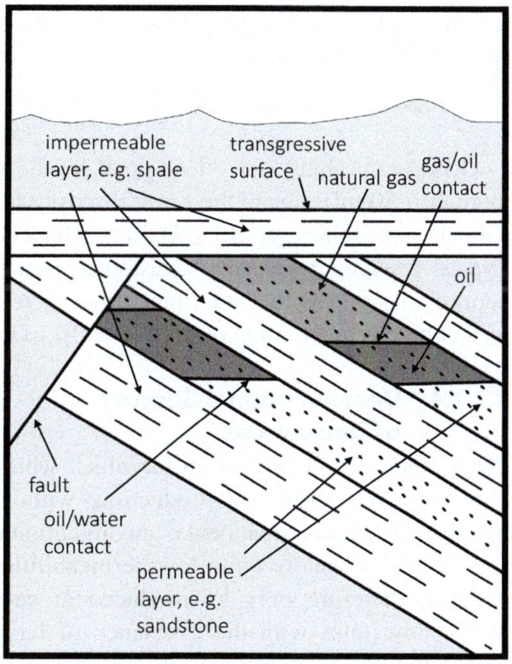

Fig. 4.5 Transgression and fault reservoir type

[49]Many other types like stratigraphic traps formed by variations in lithology of reservoir rocks shall not be discussed here.

Porosity (measured in %) hereby describes the available cavity in the rock; permeability, η, (measured in darcy [D][50] or [m²]) describes the flow properties within the rock, i.e., the capacity of a rock to allow the transmission of fluids or gases.

$$\text{Porosity}\,\Phi = ((\text{total volume} - \text{grain volume})/\text{total volume})\,[\%]$$

Minimum porosity for natural gas reservoirs should be at 7–10%; porosities above 15–20% are referred to as good to very good porosities.

The term permeability η is derived from Darcy's law and is defined by:

$$\lambda = \frac{\eta * A}{\nu} * \left(\frac{\partial p}{\partial x}\right)$$

with:

λ	flow rate $[m^3/s]$,
η	permeability $[m^2]$,
ν	dynamic viscosity [Pa s = Ns/m^2],
$\frac{\partial p}{\partial x}$	fluid potential gradient [Pa/m = N/m^3],
A	cross sectional area of the formation $[m^2]$,

Good permeability for oil occur at levels of more than 50 mD; due to the lower viscosity for natural gas, values of 0.1 mD are sufficient. Typical reservoir rocks like sandstones show a permeability of 1 D to 1 mD; limestones have a typical permeability value of 10 mD to 10 μD.

4.2.1.4 Unconventional Sources of Natural Gas

Other than conventional reservoirs, which allow for an economic production without massive stimulation treatments, unconventional sources show usually quite low permeabilities and can therefore only be produced at economic flow rates with the assistance of large stimulation treatments and special recovery technologies.

[50]One darcy is equal to 9.86923 × 10^{-13} m². The unit darcy is named after Henry Philibert Gaspard Darcy, a French engineer.

4.2.1.4.1 Tight Gas Sands

As the name suggests, tight gas sands are sandstones that are tight, i.e., show a very low permeability of less than 0.1 mD and that contain natural gas. Due to its low-permeability, the trap, seal, and reservoir may be the same tight rock, i.e., trap-like sealed structures as with conventional reservoirs are not necessary to form a tight gas-sand reservoir. The low permeability, which on the one hand presents a challenge when exploiting gas from such reservoir, on the other hand, prevents the natural gas from migrating further.

4.2.1.4.2 Shale Gas

Organic-rich shales might be source (rock), reservoir (rock), trap, and seal at the same time. Here, molecular forces are so strong that permeabilities are in the range of [μD] to [nD]. Advances in production technology[51] have made natural gas production from shale reservoirs a success story in North America, which has had many repercussions in worldwide gas markets.

In the US, the so-called 'shale-gas boom' started by the middle of the first decade of the 21st century. Natural gas production from unconventional sources has increased dramatically. This is shown in Fig. 4.6.

US Shale Gas Revolution

From the industry's perspective, the increased production of shale gas in the US has been a success story. Simultaneously, concerns have been raised about the environmental impact. Particularly, fracing, necessitating the injection of water containing up to 2% chemicals considered toxic, potentially disseminating into the groundwater and the wastewater's disposal, as well as the discussion about potentially induced earthquakes, has triggered broad public discussions.

The population's concerns have led to restrictions on shale gas production—in Europe but also in the US.

[51]See also Sect. 4.2.2.2.

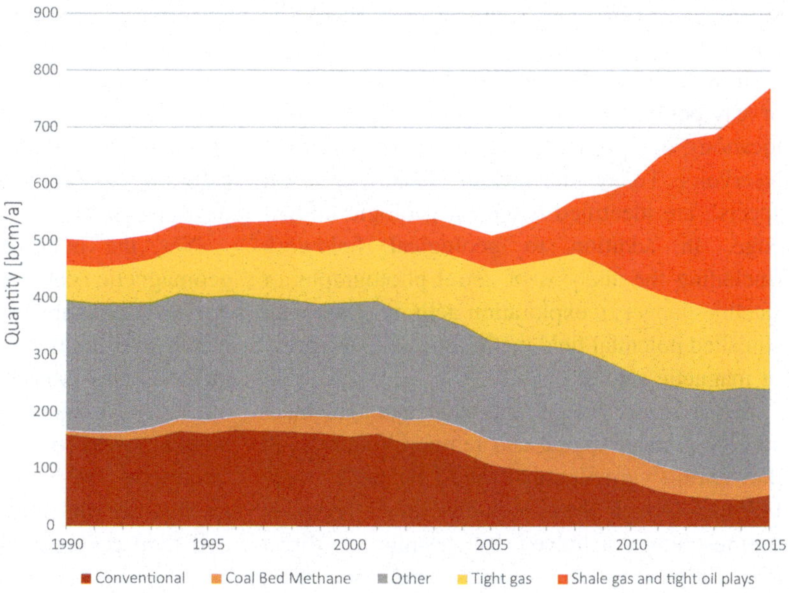

Fig. 4.6 Development of US natural gas production (Energy Information Administration 2016a)

4.2.1.4.3 Coal-Bed Methane

Coal-bed methane is generated during the formation of coal.[52] Adsorption forces keep the methane at the surface of the coal, which makes production of coal-bed methane challenging.[53] As with shale gas, coal seams are source (rock), reservoir (rock), trap, and seal at the same time. When producing coal-bed methane, large amounts of water are produced as well, especially during the early phases of production. With a decreasing amount of water in the coal-seam gas, production rates increase. Mostly, coal-bed methane reservoirs are shallower than conventional gas reservoirs, which makes it cheaper to drill. However, coal-bed wells usually produce at lower production rates than conventional reservoirs. As with other unconventional gas reservoirs, coal bed-methane also is produced by using technologies like horizontal drilling and hydraulic fracturing.

[52]See also Sect. 4.2.1.1.

[53]In coal mines, coal-bed methane poses a threat to security, as it is responsible for fire damps.

4.2.2 Excursus: Upstream—Exploration and Production

4.2.2.1 Exploration

In the prevailing context, exploration, being generally the investigation, search, or examination of an area with the purpose of discovery of resources, shall be understood restrictively as searching for natural gas deposits in the subsurface or structures in which natural gas deposits can reasonably be expected. Exploration is an inverse problem. Given a set of data, in the case of natural gas exploration mostly measured at the Earth's surface, the causal structures and parameters, located in the subsurface, have to be delineated. The result is given (observed or measured), the cause is sought.[54]

Exploration regularly focuses on areas with expansive sediment filling, as it is in these areas most probable to find reservoir rocks[55] with trap

[54]In so-called forward problems, the cause is known and the result is calculated.

[55]A reservoir rock is generally any porous and permeable rock containing hydrocarbons. Reservoir rocks are predominantly sedimentary (sandstones and carbonates). Yet, also fractured igneous and metamorphic rocks are known to be reservoir rocks.

structures. Furthermore, source rocks[56] should be present.

In the past, surface features such as tar seeps or gas pockmarks provided initial signs of the location of primarily shallow hydrocarbon reservoirs. Today, these areas, known as 'prospects', are identified based on geologic knowledge. In addition to geological fieldwork (including the analysis of aerial photographs or satellite imagery), exploration activity starts with so-called potential field methods, i.e., gravimetric or magnetic studies.

Gravimetric and geomagnetic methods measure very small spatial and temporal changes in the Earth's gravitational and magnetic force fields. Although these methods do not provide the resolution achieved by seismic or direct observations, they are well-suited for quick, non-invasive, and relatively cheap analysis of large areas. They are used in the first phases of exploration, as well as for optimizing the location of wells.

To put it simply, (geo-)magnetic methods measure changes in, less the absolute value of, the Earth's magnetic field. The physical parameter of interest is the magnetic polarization, respectively the variation thereof. The polarization depends on the magnetic susceptibility and the (permanent) magnetization of the subsurface. Measurements are performed on land, on sea, in wells, and airborne by using so-called magnetometers. By analyzing and interpreting the data acquired, conclusions about physical parameters and the structure of a sedimentary basin can be drawn.

Minute variations of the Earth's gravitational field, being often less than a few parts per million of the field's absolute value, caused by lateral variations in densities of the subsurface,[57] i.e., by a changing subsurface structure, are measured using gravimetric methods. Measurements of the spatial differences of the Earth's acceleration are performed using so-called gravimeters on land, on sea, or airborne. Observed variations of the gravitational field, i.e., so-called 'anomalies', are then calculated and depicted as differences between a 'theoretical' gravitational field assuming a homogeneous planet Earth and the measured field.

Yet, the interpretation of gravimetric and geomagnetic data is subject to measurement errors, data analysis inaccuracies and, predominantly, the inherent ambiguity of potential field methods: any particular potential field measured on, e.g., a plane, can be caused by an infinite number of various structures in the subsurface. Therefore, any meaningful interpretation of gravimetric and magnetic methods will rely on collateral geologic and geophysical information.

Applying gravimetric and geomagnetic methods requires planning, data acquisition, data processing, and interpretation. The decisions taken during the planning phase, which include, e.g., selection of the methods, delineation of the area, but also the required accuracy, will have a strong impact on the costs of surveys. Once the data are acquired, a process which may take months, they will be processed in order to reduce any effects from insignificant sources, before models of the subsurface are calculated during the phase of interpretation.

In the next step, the key geophysical methods, namely seismic methods and predominantly reflection seismic, are applied to locate a deposit with higher accuracy, i.e., to map geologic structure and stratigraphic features. Reflection seismicity uses elastic waves generated by vibration, air guns, or blasting. In essence, micro-earthquakes are induced to get information about the subsurface. The waves generated are (partly) reflected at subsurface-layer boundaries and then recorded at the surface using 'geophones' (on land) and 'hydrophones' (on water).[58] Layer boundaries are either lithologic (different rocks) or tectonic (faults). The speed at which the waves travel is related to the physical

[56]A source rock is a sedimentary rock in which petroleum is formed.

[57]Only where the density of the Earth varies laterally can gravity variations be measured and interpreted as changes in the subsurface's structure.

[58]Geophones and hydrophones are essentially devices to convert any ground movement, i.e., elastic waves, into an electric signal.

parameters of the subsurface, *inter alia*, the density and elastic parameters. Hence, each wave's 'travel time', or 'run-time', depends on these parameters. In addition to run-times and their differences, the frequencies and amplitudes of the waves are recorded in order to draw conclusions about the structure of the subsurface.

As for potential field methods, the process of getting an 'image' of the subsurface by using seismic methods can be described in four steps: planning, data acquisition, data processing, and data interpretation.

Seismic-data acquisition is conducted 'in the field', i.e., in an area, onshore or offshore, delineated by geologic and potential field methods as worth further exploration. Depending on the area, seismic-data acquisition can take months to years. The seismic acquisition spreads, i.e., the arrangements of shots and receivers in seismic surveys, has to be planned carefully before starting the survey in line with the expected structures. These measurements could be carried out along:

- individual lines, '2D seismic';
- along parallel and intersecting lines, '3D seismic'; or
- performed at different points in time, e.g., during the production phase of a field to determine how production effects the deposit, '4D seismic'.

Yet, increased accuracy implies more data and increased acquisition and processing costs.

Identification and enhancement of desired signals and isolation from undesired ones, so-called 'noise', i.e., an increase of the signal to noise ratio, is performed during the data-processing phase. This includes applying corrections to the data, rearranging, filtering, transforming, and migrating[59] the wave field. Given the plethora of data, processing operations require high-end computers and sophisticated software.

Once data processing has been accomplished, the objective of data interpretation is to derive a (simple and plausible) geologic model compatible with all data observed—if available, including gravimetric and magnetic, and other geologic data, as well as data from wells (petrophysical logs) in the area.

The seismic maps are used to identify structures that either would repay more detailed seismic surveying or would warrant wildcat drilling.

Despite all efforts (and expenses), such a model will never be unambiguous, and only (or, at best) structures in the subsurface or stratigraphic features can be identified. There is no single non-invasive method that can be used to identify directly the presence of hydrocarbons, or, all the more, the volumes, parameters, or composition of hydrocarbons in the underground. However, methods exist that deliver an indication of the presence of hydrocarbons, especially of natural gas. Such seismic methods are called DHI (Direct Hydrocarbon Indicators)—a widely-used seismic technique is AVO (Amplitude Versus Offset).

Yet, this is the best information available. The data are used to identify structures that either would repay more detailed seismic surveying or would warrant drilling of a first hole, a so-called wildcat.[60] After a well has been drilled, but before it is cased, additional information is gained by lowering measuring devices down the borehole. Information of physical parameters of the rocks surrounding the borehole are transmitted to the surface and used to support the interpretation.

As drilling is usually costlier than the applications of geophysical methods described, the

[59]The term 'migration' is used by geophysicists to designate an inversion operation in seismic data processing. Here seismic data are rearranged in order to be able to plot reflections and diffractions at their true locations, which may have been biased by variable velocities and dipping horizons. This does not correspond to the 'geologic' meaning referred to in Sect. 4.2.1.2.

[60]Being an exploratory, i.e., not a production, well drilled in areas where quantities supplying an economically viable production of natural gas (or oil) have not yet been found. Sometimes 'exploratory wells', being wells in areas already explored using geophysical methods, are distinguished from 'wildcats', being then defined as wells in unexplored areas.

success of such drilling is vital for the economic viability of the project. The success is measured often using the so-called 'dry-hole' percentage.[61]

During the drilling process, the drilling fluid with the upwardly transported drill cuttings, along with associated liquids (water, possibly hydrocarbons), is analyzed with chemical, lithological, and micro-paleontological methods. Moreover, drilling core samples are regularly drawn for further investigation and to derive a continuously improving 'image' of the subsurface.

In addition, (partially while drilling) geophysical well logs are carried out. Examples include measurement of electrical resistivity, self-potential, or other geo-electric measurements, radiometric methods, temperature measurements, or, again, seismic methods. By using various methods, these processes try to acquire information about the rocks' physical and chemical parameters, *inter alia*, permeability and porosity, or to detect hydrocarbons directly. If there are positive indications for a discovery, the well is usually tested (by measuring the flow for a short period) to evaluate the productivity.

After discovery of oil and/or natural gas, but before production starts, the appraisal and development phase sets in. First appraisal wells are drilled to reduce existing uncertainty about size, exact location, or better delineation of the field, and the parameters of the reservoir and the reserves, as, e.g., permeability, porosity, and flow rate. Then development wells are drilled to support and initiate the maximization of production, observing predetermined restrictions. Development wells include both first production wells and gas-and-water injection wells used to enhance recovery of the reserves.

[61]A 'dry hole' is any well drilled but failing to serve its purpose of, e.g., producing natural gas, oil, or any other resource in economically viable quantities. The 'dry hole' percentage could be defined as the ratio between the numbers of dry holes to total wells drilled, or the ratio of well depth drilled in dry holes to the total depth drilled. Example: While, in the 1960s more than 40% of the wells drilled in the US were dry, this percentage had dropped to 10% at the beginning of the millennium.

With this increasing information about the field, a production development plan and a reservoir model are completed, enabling a simulation of oil and natural gas production from the field. Following the acquisition of all necessary data, an FDP (Field Development Plan) is done, followed by a FEED study (Front End Engineering Design) and leading to an FID (Final Investment Decision) for field development. This field development typically costs 10–100 times more than exploration and appraisal.

4.2.2.2 Production

Production is the process of lifting natural gas (and other resources) to the surface. It further includes gathering, treating, field processing, and field storage. Production typically starts after exploration, appraisal, and development have been finished.

Each reservoir is a unique system that must be understood before production is planned or started. Petroleum engineers use all the available data to develop a mathematical model of the reservoir and simulate production to find the most effective setup for recovery of reserves.

With start and ongoing production, further information on the behavior of the field is gathered, making possible stepwise refinement of the reservoir model. The commercially recoverable content of a natural gas field will depend on various factors:

- reservoir conditions, *inter alia*:
  - extension,
  - porosity of reservoir rock,
  - permeability of reservoir rock,
  - pressure, and
  - temperature,
- reservoir depth;
- field saturation;
- the type of reserve, i.e., being a non-associated or associated natural gas field;
- natural gas quality, i.e., the physical and chemical parameters[62];
- economic and legal conditions, *inter alia*:
  - (obtainable) price(s) for natural gas,

[62]See Sect. 4.1.

- (obtainable) price(s) for other hydrocarbons,
- taxes and royalties,
- regulation (license regime),
- costs for
  - drilling,
  - processing,
  - transport, etc.

Due to the variability, in particular of the economic and legal factors, reserve estimates are usually adjusted over time.

The amount of remaining reserves can be estimated using the following formulae:

$$V_{GIIP} = V_{GRV} * \frac{N}{G} * \Phi * \frac{\sigma_{Gas\ saturation}}{\sigma_{Formation}}$$

$$V_{UR} = V_{GIIP} * \sigma_{Recovery}$$

$$R_{Remaining} = V_{UR} - V_{Produced}$$

with:

$R_{Remaining}$	remaining reserves (volume) [m^3]
$V_{Produced}$	volumes already produced (volume) [m^3]
V_{GIIP}	gas initially in place (volume) [m^3]
V_{GRV}	gross rock volume (volume) [m^3]
N/G	net to gross ratio [%]
Φ	porosity [%]
$\sigma_{Gas\ saturation}$	gas saturation factor [%]
$\sigma_{Formation}$	formation volume factor [%]
V_{UR}	ultimate recovery (volume) [m^3]
$\sigma_{Recovery}$	recovery factor [%]

A typical production profile of a conventional natural gas field starts with a short-lasting build-up phase, followed by a plateau-phase, prolonged as long as possible, and ends with the decline-phase and finally the tail-end. The build-up phase, when the production of a newly discovered gas field increases by investing in additional wells and surface installations, is followed by the plateau-phase where a stable production is maintained over a period. The capacity of processing and transport facilities, typically constructed through consideration of technical and reservoir restrictions with the objective of economic optimization, restricts the maximal production rate. After having produced approximately 30–50% of the recoverable reserves, typically, the decline-phase starts, as, due to the decreasing reservoir pressure, the initial production level cannot be maintained. Generally, natural gas flow to the surface is driven by pressure from the reservoir only. The production is stopped when (variable) production costs exceed revenues. At this point and under favorable geologic and commercial conditions, up to 80% of the natural gas initially present in the reservoir has been produced. An example of a production profile is depicted in Fig. 4.7.

Excursus

Oil production in principle follows an identical scheme: build-up, followed by a plateau-phase and the later decline.

Depending on the technology applied, the following production methods are distinguished:

- primary production: the oil is lifted only by the intrinsic reservoir pressure (this includes lifting by down-hole pumps). Here the recovery factor usually does not exceed 25% of the oil in place;
- secondary production: additional energy is introduced in the reservoir by water or gas injection. The ultimate recovery is increased up to 40%. This is the most frequently applied production technology; and
- tertiary production: the oil displacement in the reservoir is improved by injection of chemicals or heat.[63] In this case and under favorable conditions, the recovery factor might be increased up to 60%. The high costs of tertiary methods however limit their application.

[63] This could be achieved by injecting steam to reduce the oil's viscosity or, e.g., by injection of surfactants (detergents) or carbon dioxide flooding or microbes.

Fig. 4.7 Typical production profile

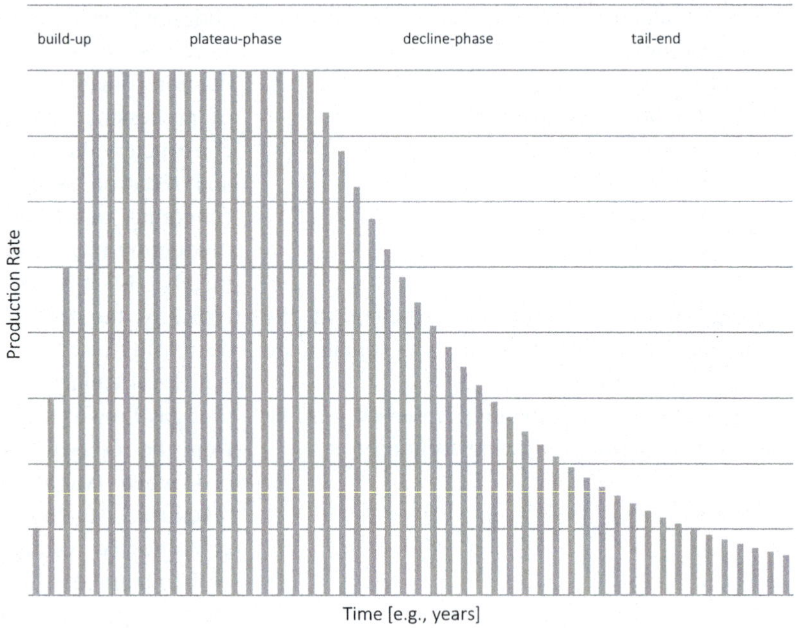

Hydraulic fracturing, also named 'fracing',[64] is a technique to stimulate oil or natural gas wells and applied worldwide since the early 1950s. Today, this method is used for shale gas production. The principle objective is to create cracks in a reservoir to enhance permeability and thus production rates. This is achieved by using high pressure to inject a so-called fracing fluid, which typically contains water, a proppant (mostly sand), and chemical additives. Yet, the latter are less than 2% of the total volume of the fracing fluid.[65] Nonetheless, in particular, chemical additives cause many political discussions on alleged environmental damages occurring in areas of shale gas production.

Unconventional plays became economic in the first decade of the 2000s in North America. Technological progress when producing natural gas, e.g., intensified use of horizontal drilling and the increased use of hydraulic fracturing, brought sources of natural gas to profitability that were regarded as uneconomic only a few years earlier.

Horizontal (or better, directional) drilling enables, e.g., for a longer section of a reservoir to be exploited by drilling, improving the productivity of the respective well. Furthermore, it allows more wellheads on a given piece of land, thus reducing costs of well-completion and production.

Independently whether produced from conventional or unconventional sources, natural gas produced, does not contain only pure methane. Therefore, extensive processing measures are necessary.[66]

A field reaching the end of its economic viability, i.e., the variable costs of the production are permanently above the revenues, the so-called decommissioning and abandonment phase

[64] Or, mostly in non-industry usage 'fracking', sometimes even 'fraccing' or 'frac'ing'.

[65] "*The number of chemical additives used in a typical fracture treatment varies depending on the conditions of the specific well being fractured. A typical fracture treatment will use very low concentrations of between 3 and 12 additive chemicals depending on the characteristics of the water and the shale formation being fractured. Each component serves a specific, engineered purpose.*", Ground Water Protection Council, ALL Consulting (2009), p. 61 and "*Overall the concentration of additives in most slickwater fracturing fluids is a relatively consistent 0.5% to 2% with water making up 98% to 99.5%.*", Ground Water Protection Council, ALL Consulting (2009), p. 61.

[66] See also Sect. 4.1.2.

begins. Depending on the local environmental legislation, the area of production and processing must be returned to its original state. Consequently, at this stage, holes are filled and aboveground facilities are dismantled.

Thönse

Natural gas production does not always take place in giant fields in Russia or large offshore systems like in Norway, the UK Continental Shelf, or the Gulf of Mexico. One of the oldest German natural gas fields Thönse is a small site near Hanover between the two villages Großburgwedel and Thönse. Natural gas was found by chance when drilling for oil in 1952—long before the 'natural gas era' started. The well is still in operation. Until today, in total 4 bcm have been produced; the current annual production is at 60 million m^3/a. When production started in 1952, the reservoir pressure was at 200 bar; today it is not more than 25 bar. Nevertheless, production is expected to continue for another ten years.

4.2.2.3 Upstream Contracts

In most countries, the state is the owner of hydrocarbon reserves. Before private companies are allowed to explore for and produce hydrocarbons, agreements between the government and respective companies are required. In practice, two types of contracts have been developed:

- Production Sharing Agreements (PSA) or Production Sharing Contracts (PSC), and
- License Agreements.

In a PSA in its purest form, the private company and the government or, on its behalf, the state-owned national oil and gas company (NOC) or the respective ministry agree that the private company is allowed to explore, in a defined region and at its own expense and risk, for hydrocarbons. In the event of a successful discovery, the private company will have the right to produce the hydrocarbons. However, the production quantities have to be shared according to a pre-agreed plan with the NOC or the government.

In Europe, license agreements prevail. Here, the government tenders, in so-called license rounds, predefined regions (so-called 'blocks') for exploration, and private companies bid for these blocks. Usually, the award of a license is linked to a direct payment of the company and/or an obligation to perform a certain exploration 'work program' (seismic, exploratory drilling). If the company is successful in finding hydrocarbons, it is allowed to produce these hydrocarbons, which typically pass to the ownership of the company at the wellhead. For the quantities produced, however, a royalty has to be paid to the state. Moreover, taxes (mostly on revenues from sales of the hydrocarbons) will be due. Royalties and taxes are set independently and differently by the respective countries. In addition, in some countries, participation of the NOC in the economically successful blocks is foreseen. In most countries, it is possible that several companies jointly apply for a block, and one company acts as operator.

After the award of a license, companies might decide—for reasons of cost and risk sharing—to transfer parts of the license to other companies, usually against payment of a certain sum of money or the adoption of certain works (assignment). Such contracts are called farm-in or farm-out agreements.

As the state at the time of licensing does not know the exact outlines of a reservoir, it occurs that a reservoir is located in multiple blocks with different license holders. In this case, the parties negotiate a so-called unitization (usually by request from the state, which is interested in an optimal technical production). A precise knowledge of the reservoir—or better knowledge of the exact location of the hydrocarbons in the reservoir—is essential for these negotiations.

Besides that, there are a number of other types of contracts like service contracts in the upstream sector. These are concluded with companies having specialized in providing services, e.g., drilling or carrying out borehole-measurement services for producers.

In addition, field development is handled by an EPC contract (Engineering Procurement Construction). This is a contract, typically stipulating that the EPC Contractor is responsible for, e.g., design, procurement, construction, and commissioning.

4.2.3 Anthropogenic Production of 'Natural' Gas

There are primarily two ways to produce gas that resembles natural gas anthropogenically:

- biogas, and
- coal gas.

4.2.3.1 Biogas

Producing biogas follows a similar process that formed natural gas. Through anaerobic (i.e., oxygen-free) digestion of organic matter, so-called biogas is produced. The prefix 'bio' is used because it is classified as a renewable energy source—other than the 'fossil' fuel natural gas.

Its classification as a renewable energy source is mainly attributed to the fact that CO_2 produced from combustion of biogas would have been anyway released to the atmosphere during the decomposition of the input material for biogas production. Therefore, biogas is regarded as being CO_2 neutral. Compared to other sources of renewable energy like wind and solar power, it is relatively easy to store biogas with today's available technologies. In addition to storing the raw material before production of biogas, the produced biogas can either be stored in aboveground gas tanks on the site of the biogas plant or —after some treatment—it might be injected into the natural gas grid where it can be stored using the same infrastructure as for natural gas. Further, biogas production can be flexible, i.e., different quantities can be produced in different periods—flexibility, which is more or less unavailable to wind and solar electricity production.

Anaerobic microbes are used to produce biogas from organic matter in biogas plants.[67] Thereby, organic waste, waste from food industry, manure, sewage, or specially produced crops (e.g., silo maize) can serve as raw material. In airtight tanks, anaerobic organisms are deployed to start a fermentation process. Biogas is produced after the anaerobic digestion process consisting of hydrolysis, acidogenesis, acetogenesis, and methanogenesis is completed.[68]

Raw biogas contains approximately only 45–70% methane.[69] The rest is mainly carbon dioxide (CO_2), but also nitrogen, water vapor, hydrogen, and hydrogen sulphide (H_2S). Some of these components are detrimental and need to be removed before biogas can be used. In many cases, biogas is used on-site to produce heat and electricity in CHP plants.[70] In case biogas production is not used on-site but intended to be injected into the (natural gas) pipeline grid, usually the content of methane has to be increased, i.e., the biogas is upgraded to so-called biomethane.

As soon as biogas has been purified and upgraded to biomethane and fulfils pipeline specifications of the local or regional grid operator, biomethane can be injected into the pipeline grid. After such injection, it commingles with natural gas and has thus the same usages.

As biogas and biomethane are classified as a renewable energy source, many governments worldwide (intend to) subsidize or make their production tax exempt. Furthermore, biogas is mostly regarded as an indigenous source of energy, which is not to be imported and, therefore, considered as a way to increase the security of supply.

According to the World Bioenergy Association, global biogas production reached 1.28 EJ in

[67]Further, direct recovery from landfills can be regarded as an additional source of biogas.

[68]Another potential source of biogas is the direct transformation of CO_2, water, minerals, and light into methane by using micro-algae for photosynthesis. This technology is still emerging and not available on an industrial scale.

[69]See Rasi (2009), p. 9.

[70]See Sect. 4.3.3.

2013.[71] This is equivalent to approximately 59 bcm/a, whereby 45% of this quantity is produced in Europe, 31% in Asia, and 22% in the Americas.

4.2.3.2 Coal Gas

Other than coal-bed methane, coal gas[72] is gas manufactured from coal, but not directly produced from a coal seam or a coal stratum. Coal gas, mainly a by-product of coke production, is a mixture of various gases, mainly hydrogen (H_2), methane (CH_4), nitrogen (N_2), ethylene (C_2H_4), and carbon monoxide (CO).

Historically, the coal industry was one of the kick-starters of industrialization in Europe and North America. Coal was produced and converted to coke long before natural gas was produced on an industrial scale. As a by-product of coke production, coal gas was mainly used for running gas engines or for cooking, lighting, and heating. It was distributed primarily through local town grids. In some instances, in so-called gas works, the coal gasification process was also done to produce town gas as the main product and not as by-product only.

Due to its relatively low methane content of around 25%, coal gas has a gross calorific value of approximately $GCV_n = 5.5$ kWh/Nm^3 and a Wobbe index of approximately $W_{G,n} = 8.7$ kWh/Nm^3 only.

With the growing success of natural gas, town gas grids were converted into natural gas grids.

4.3 Usages

Natural gas offers a wide range of applications. Typically, users and usage of natural gas are segmented. The segments, also referred to as 'sectors', are:

- residential and commercial: In this segment natural gas is used primarily for cooking and heating;
- industrial: Natural gas is used again for heating but also as feedstock;
- electricity generation: The primary energy natural gas is 'transformed' to the secondary energy 'electricity';
- transportation: To a yet limited extent, natural gas is used as a transportation fuel; and
- others.

As an example, the inland sales of natural gas in the European Union for the year 2014 are shown in Table 4.9.

The most important sectors of natural gas use are the residential and commercial sector, where demand is strongly weather dependent, the industrial sector, and the electricity-production sector. Consumption figures in the US shows a pattern with less use in the residential and commercial sector but a higher share in the electricity generation. Inland sales of natural gas in the United States for the year 2015 are shown in Table 4.10.

4.3.1 Residential and Commercial

Natural gas offers a multitude of residential and commercial uses. Traditionally, natural gas is used in this sector for:

- cooking;
- water heating; and
- room temperature regulation (heating or cooling) by using boilers, furnaces, and natural gas-powered air-conditioning systems.

Usually, natural gas systems are regarded as more efficient, environmentally friendly, and easier and more convenient to use than competing fuels including electricity.

Recent developments also show that CHPs are used to produce electricity and heat in a house, a small group of houses, or small businesses. Other than large-scale CHPs[73] these CHPs have a lower installed capacity and can be used as a distributed energy resource, i.e., heat and electricity are produced near the location of their consumption.

[71]See World Bioenergy Association (2016), p. 57.

[72]Sometimes coal gas is also referred to as town gas.

[73]See Sect. 4.3.3.

Table 4.9 Inland sales of natural gas in the EU-28 in 2014[a]

	Residential and commercial [TWh_{GCV}]	Industry [TWh_{GCV}]	Power plants and heat plants [TWh_{GCV}]	Transport [TWh_{GCV}]	Other uses [TWh_{GCV}]	Total inland sales [TWh_{GCV}]
Austria	19.5	37.0	19.0	0.2	7.9	83.6
Belgium	79.5	41.2	39.7	0.3	–	160.7
Bulgaria	1.0	12.0	9.9	0.9	3.6	27.4
Croatia	7.3	9.9	5.1	–	3.7	26.0
Cyprus	–	–	–	–	–	–
Czech Republic	31.7	43.8	–	0.3	1.6	77.4
Denmark	10.8	9.1	7.4	–	3.8	31.1
Estonia	1.0	0.5	3.8	–	0.2	5.5
Finland	0.7	17.4	14.2	–	–	32.3
France	241.2	135.5	31.3	1.2	12.2	421.4
Germany	357.3	352.6	111.2	2.3	1.2	824.6
Greece	4.6	9.3	17.6	0.2	–	31.7
Hungary	46.1	22.8	18.1	–	5.1	92.1
Ireland	13.3	7.7	26.1	–	0.8	47.9
Italy	273.3	161.6	187.6	11.1	21.6	655.2
Latvia	2.6	1.6	9.1	–	–	13.3
Lithuania	3.8	14.7	7.8	0.1	0.4	26.8
Luxembourg	4.2	2.8	3.9	–	n.a.	10.9
Malta	–	–	–	–	–	–
Netherlands	172.5	134.1	61.2	–	7.8	375.6
Poland	61.9	89.2	16.0	–	5.7	172.8
Portugal	4.3	37.8	3.2	–	–	45.3
Romania	38.3	42.4	30.8	–	16.1	127.6
Slovakia	20.8	17.1	8.7	0.1	0.1	46.8
Slovenia	2.2	4.0	0.6	–	–	6.8
Spain	41.5	122.3	136.4	1.2	–	301.4
Sweden	1.9	3.0	3.1	0.6	1.2	9.8
UK	374.5	141.9	244.5	–	12.3	773.2
EU-28	1815.8	1471.3	1016.3	18.5	105.3	4427.2
	41.0%	33.2%	23.0%	0.4%	2.4%	100.0%

[a]Eurogas (2016), p. 5

Table 4.10 Inland sales of natural gas in the US in 2015[a]

Residential	Commercial	Industry	Electricity	Transport	Other Uses	Total
130.6	90.8	212.6	273.9	1.0	69.0	777.9
16.8	11.7	27.3	35.2	0.1	8.9	100.0

[a]See Energy Information Administration (2016b)
Figures in [bcm] (first row), [%] (second row)

This theoretically reduces the need for long electricity-transmission lines and thus contributes to reduced related losses. Installed capacities of these small CHPs are usually less than 10 kW_{el} (micro-CHP), or even in the range of 1–5 kW_{el} (nano-CHP).

Apart from smaller versions of the technologies used for large-scale CHPs, *inter alia*, stirling engines and fuel cells are used in this sector, whereby fuel cells produce heat and electricity by converting natural gas and air through an electrochemical process, i.e., rather than in other CHPs where natural gas is not combusted.

'Gas-powered' heat pumps achieve very high efficiency rates by using natural gas in an engine to 'produce' mechanical energy, which is then used to compress (and thereby heat) a so-called refrigerant. By using a refrigeration cycle, thermal energy is absorbed from a cold space and transported and released to a warm space, which is opposite to the typical flow of heat.

4.3.2 Industrial

4.3.2.1 Process Heat

Natural gas offers a multitude of uses for industry. Similar to the residential and commercial sector, it might be used for heating or cooling. Industrial combined heat and power (CHP) plants are common, too.[74] In addition, natural gas can be used for co-firing where natural gas is used in addition to other fuels. Such co-firing improves the combustion of other fuels and therefore helps to reduce emissions and increase efficiency.

Apart from utilization in the chemical industry and petroleum refining, natural gas is heavily used in industries like steel, aluminum, glass, and pulp and paper. The glass industry, for example, uses natural gas for the melting of glass. During this process, temperatures above 1500 °C are reached, and it is crucial to avoid variations in natural gas quality, in particular the Wobbe index, as these might lead to a reduced glass quality. In addition to heating, natural gas is also used for drying and dehumidification through desiccant systems, e.g., in the food industry.

[74]See Sect. 4.3.3.

4.3.2.2 Feedstock

The chemical and petrochemical industry uses natural gas also as feedstock to produce, *inter alia*, hydrogen or so-called syngas. Hydrogen is then further used, e.g., for hydrocracking in refineries or for the production of ammonia. Syngas, or synthesis gas, is a mixture consisting of hydrogen (H_2) and carbon monoxide (CO), and producing it is the first step of (petro-)chemical synthesis. The chemical reaction for the production of syngas in the so-called catalytic steam reforming process is[75]:

$$CH_4 + H_2O \rightarrow CO + 3H_2$$

Adding more water steam leads to a further increased hydrogen production:

$$CO + H_2O \rightleftharpoons CO_2 + H_2$$

Syngas is afterwards used for production of, e.g., methanol or synthetic fuels. When producing synthetic fuels, the Fischer-Tropsch process might be used. Here, alkanes are produced based on the following chemical reaction:

$$nCO + (2n + 1)H_2 \rightleftharpoons C_nH_{2n+2} + nH_2O$$

Obviously, the production of methane (i.e., n = 1) is not wanted. Instead, higher alkanes are produced that can be used as liquid fuels.[76]

The process was invented during the 1920s by the German chemists Franz Fischer and Hans Tropsch at the 'Kaiser-Wilhelm-Institut für Kohlenforschung' in Mülheim/Ruhr. The process was heavily used by the Germans during World War II to produce liquid fuels from coal. Since the 1950s, the process was also used much by the South African Sasol to produce liquid fuels from coal in order to overcome the oil embargo imposed by OPEC against the, then, apartheid regime.

Today, e.g., Shell produces liquid fuels from natural gas as a building block in its plants at

[75]Other feedstocks for the production of syngas are coal and biomass.

[76]The process of making liquid fuels from natural gas is also referred to as 'gas-to-liquids' or 'GTL'-process. GTL production is also one explanation for the relatively high natural gas consumption in countries like Qatar.

Bintulu, Malaysia, or its Pearl GTL plant in Qatar. The liquid fuels produced at these plants compete with products made from crude oil. Due to the cleanliness of the feedstock methane, these products contain fewer impurities like sulphur than crude oil products. The processes and sometimes even the end-products are commonly referred to as 'gas-to-liquids' (GTL).

The Haber-Bosch process invented by the German chemists Fritz Haber and Carl Bosch at the beginning of the 20th century uses nitrogen and hydrogen, which again could be produced by using natural gas, to produce ammonia. The chemical reaction is as follows:

$$N_2 + 3H_2 \rightleftharpoons 2NH_3$$

Ammonia itself is a raw material for the production of fertilizers or explosives.

Another important output is methanol, which is synthesized using methanol synthesis catalysts by either:

$$CO_2 + 3H_2 \rightleftharpoons CH_3OH + H_2O$$

or:

$$CO + 2H_2 \rightleftharpoons CH_3OH$$

4.3.3 Electricity Generation

In general, three technologies have been developed to produce electricity in centralized power plants:

- steam turbines,
- combustion turbines, and
- combined cycles.

Power plants with steam-turbine technology use a combustible fuel to heat water. By this process, high-temperature, high-pressure water steam is produced, which is used to turn turbine blades and, with them, a turbine shaft which again spins a generator to create electricity. As combustible fuels, not only natural gas is used but also oil, coal, or waste. In addition, nuclear power plants use this technology, whereby the nuclear fuel itself is not combusted but instead water is heated using nuclear fission. The thermal efficiency of modern steam turbines is in the range of 30–35%.[77]

Combustion turbine plants use a similar technology as jet engines. In such a turbine, a fuel like natural gas[78] ignites compressed air. The resulting high-speed gas flow is used to run a turbine and a connected generator in order to produce electricity. The thermal efficiency of combustion turbines is in the range of 35–40%.

As the name suggests, combined-cycle plants combine combustion turbines with steam turbines. In principle, the exhaust heat of a combustion turbine is not wasted but used to heat water to run a steam turbine. If the combustion turbine is fired with natural gas, such plants are also named combined cycle gas turbines or CCGTs. The thermal efficiency of cutting-edge CCGTs is in the range of 55–60%.[79]

Natural gas-fired electricity production has many advantages:

- Other than coal (especially lignite) fired plants or nuclear power plants, natural gas fired power plants are flexible to operate. They can easily provide (relatively high-cost) peak load services, but can also run in baseload operation;
- The ability of combustion turbines and CCGTs to be switched on and off relatively quickly give them an advantage as dependable backup to electricity produced by renewables. Wind and solar power are intermittent only, and therefore need backup services to stabilize the electricity grid. Given

[77] All thermal efficiencies used throughout this book refer to the net calorific value. References to gross calorific value are approximately 10% lower, i.e., a thermal efficiency of 60% NCV_n is equivalent to 54% GCV_n. See also Sect. 4.1.3.1.1.

[78] Diesel, kerosene, or propane might also be used as an ignition fuel.

[79] The EdF-operated CCGT in Bouchain in northern France showed during its test phase an efficiency of 62.22%; Natural Gas Europe (2016).

their excellent load-following ability, natural gas-fired power plants are best-suited to provide such services.[80] From warm conditions, it takes less than 1.5 h to bring a CCGT online[81] (compared to ~3, ~6, or ~40 h for hard-coal plants, lignite fired plants, or nuclear power plants, respectively)[82]; and

- Natural gas combusts cleanly. When combusted, coal and oil release by far more environmentally harmful pollutants than natural gas. Especially with regard to sulphur dioxide, the advantages of natural gas are large because it effectively does not emit any sulphur dioxide.[83] Coal-fired power plants, however, need to be equipped or retrofitted with desulphurization units to meet modern environmental standards. Obviously, the investment in such units worsens the profitability of coal-fired power plants compared to those fired with natural gas.

  As already described in Sect. 4.1.5.1, also with regard to carbon dioxide, natural gas is environmentally friendly, e.g., the carbon-dioxide-emissions coefficients are smaller than for competing fossil fuels, in particular coal or oil.

  Some countries introduced carbon-emissions trading systems by emitting CO_2-certificates, which give its holder the right to 'produce' a certain amount of carbon dioxide. These certificates can be traded and have a positive price. Generating electricity with natural gas requires fewer certificates than generating electricity with coal or oil. Therefore, if the prices were equal, the smaller variable production costs of electricity production based on natural gas would be lower in comparison to electricity production with other fossil fuels.

- Compared to other fossil power plants and—let alone—nuclear power plants gas-fired units can be built relatively quickly and they are usually cheaper to build, i.e., the investment costs are lower; and
- As already described, CCGTs have relatively high thermal-efficiency rates, i.e., in comparison to other fuels less primary energy is needed to produce the secondary energy, i.e., electricity.

Supported by these advantages, electricity production has given rise to a dash for gas in some regions. Obviously, to be a success story, other factors have to be fulfilled as well, such as:

- availability of natural gas,
- availability of tools to provide volume flexibility of the natural gas deliveries, and
- security of gas supply.

The most important factor—of course—is the price of natural gas, or better, the relationship between the natural gas price to prevailing electricity prices, which themselves depend, *inter alia*, on the prices of competing fuels. Here the difference between electricity price and natural gas price during a period in a region is referred to as the spark spread.

Power-Plant Utilization

The utilization of a power plant may be approximated by simple comparison of natural gas prices[84] and electricity prices considering the power plant's efficiency. If the 'virtual' electricity price, i.e., the electricity price calculated based on the power plant's efficiency and the natural gas price, is lower than the electricity price achievable in the market, power plant utilization is viable and vice versa (Table 4.11).[85]

[80]Only pumped-storage power plants provide a more responsive technology.

[81]An example is the RWE operated gas-fired power plant in Lingen, Germany. In summer 2016, it was able to bring 540 MW_{el} online within 45 min. See RWE (2016).

[82]See Eurelectric (2011), p. 19.

[83]Compare also Sect. 4.1.5.2.

[84]Yet these prices, quoted typically based on the gross calorific value, have to be adjusted for the net calorific value. A NCV_n/GCV_n factor of 0.9 has been used in this example.

[85]It has to be noted that this calculation is simplistic and, hence, incomplete, as, *inter alia*, CO_2 certificate prices are not included.

Table 4.11 Example for approximation of power plant utilization for different natural gas and electricity prices and power plant efficiencies

Natural gas price [€/MWh]	Electricity price [€/MWh]	Power plant efficiency [%]	Electricity price equivalent natural gas price [€/MWh][a]	Power plant utilization
20.00	50.00	50	44.44	Yes
20.00	44.44	50	44.44	Indifferent
25.00	50.00	50	55.56	No
25.00	55.56	50	55.56	Indifferent
30.00	50.00	50	66.67	No
30.00	66.67	50	66.67	Indifferent
20.00	50.00	55	40.40	Yes
20.00	40.40	55	40.40	Indifferent
35.00	50.00	55	70.71	No
35.00	70.71	55	70.71	Indifferent
30.00	50.00	55	60.61	No
30.00	60.61	55	60.61	Indifferent

[a]Calculation considering the power plant's efficiency and the net calorific value approximation, see Footnote 77, i.e., (electricity price equivalent natural gas price) = ((natural gas price)/(power plant efficiency))/0.9 (whereby the latter is the approximation for conversion of the gross calorific value to the net calorific value)

Again, in regions that introduced carbon-trading schemes also, prices of CO_2 certificates have to be taken into account.[86]

Electricity production from fossil fuels (and nuclear power plants) necessarily produces heat. CCGTs use the heat of the first cycle to produce more electricity in a second cycle. Other than that, so-called combined heat and power plants (CHP)[87] use the heat of a steam or combustion turbine for other purposes. Such purposes include heat for industrial processes or for heating of buildings. The heating of buildings in an area not in close vicinity to the CHP-plant is also referred to as district heating.

In both cases, fuels otherwise required to produce heat are saved by using the 'waste heat' of the CHP-plant. By this approach, carbon and other emissions go down, and thermal efficiencies go up.[88] Because this process is environmentally friendly, many governments tend to subsidize CHP plants.

[86]Furthermore, fuel taxes or other government-induced factors as preferential treatment of certain fuels or renewables are relevant for investment decisions or—after investment—dispatch decisions.

[87]The term cogeneration is used interchangeably.

[88]Using condensing heat-recovery systems, the overall efficiency might increase to above 100% (NCV).

4.3.4 Transportation

Natural gas is used as an alternative fuel in the transportation sector. Again, as natural gas produces only low emissions compared to its competitive fuels like gasoline or diesel, it is a viable alternative to meet increasing environmental standards.

In the automotive sector, it is used in passenger cars, buses, and trucks either in the form of CNG or LNG. As the energy density of CNG and LNG is lower than that of diesel or gasoline, it is necessary to equip natural gas vehicles (NGVs) with larger fuel tanks to store the same amount of energy. In addition, in the case of LNG, these fuel tanks need to be insulated to keep the fuel cold and liquid.

Mostly, natural gas is used for fuel-intensive fleets such as public buses or courier-delivery services. One of the reasons for the only limited use in passenger cars is the limited availability of petrol stations offering natural gas as a fuel, thus providing fewer options to refuel.[89]

A large growth potential is seen in the use of LNG as a fuel for ships. Again, environmental reasons favor LNG as it emits no sulphur and only negligible amounts of particulates. Further, it produces up to 90% less NO_x and up to 25% less CO_2 than traditional marine fuels like heavy fuel oil or marine gas oil. In certain maritime areas like the Baltic Sea, the North Sea, and the US and Canadian coasts, stricter regulation regarding emissions are in place. One way to meet these stricter regulations is the use of LNG-propelled ships. However, developing the downstream LNG-infrastructure enabling LNG-propelled ships to refuel in various harbors is still a challenge.

4.3.5 Others

Another important user of natural gas is the natural gas industry itself. As described in Sect. 7.2.1, compressor stations are needed to increase pressure either in pipelines or on storage sites. Some of these compressor stations use natural gas as a fuel.

Some countries with a high crude-oil production also show high natural gas consumptions as natural gas is regularly used for re-injection to enhance oil production.[90]

Flaring and venting is another means of consuming, however, not necessarily making use of, natural gas. Whereas flaring is the controlled burning of natural gas, e.g., during oil production, venting is the controlled release of gas into the atmosphere. Venting is applied mostly in those cases when the natural gas contains a relatively high concentration of inert gases, which makes burning impossible or difficult. Especially associated gas, which can be regarded as a by-product of crude-oil production, is flared or vented in some regions of the world, mainly those regions where no developed natural gas market exists and which do not have access to export markets.[91]

[89]For this reason, most NGVs have bi-fuel engines, i.e., they can run on CNG or gasoline.

[90]See also Sect. 4.2.2.2.

[91]Due to environmental concerns, in most developed gas markets, flaring and venting are legally restricted and only allowed in emergency cases.

References

British Petroleum (2011) Guidebook to gas interchangeability and gas quality

Bulgartransgaz (2016) Natural gas transport and storage contract—appendices. Available at: http://www.bulgartransgaz.bg/files/useruploads/files/appendices_1,3,4,4a,5,11,12_en.pdf

Bundesverband Erdgas Erdöl und Geoenergie e. V. (2016) FAQ - Antworten auf häufig gestellte Fragen. Available at: http://www.bveg.de/Themen/Klima-Umwelt/NORM-Stoffe/FAQ-Antworten-auf-haeufig-gestellte-Fragen

Cerbe G, Dehli M, Kätelhön JE, Kleiber T, Lehmann J, Lendt B, Mischner J, Mundus B, Pietsch H, Spohn D, Thielen W (2004) Grundlagen der Gastechnik – Gasbeschaffung, Gasverteilung, Gasverwendung, 6th edn, Munich

Deutscher Verband der Gas- und Wasserwirtschaft (DVGW) (2013) Technische Regel – Arbeitsblatt DVGW G 260 (A) Gasbeschaffenheit (mit Korrektur vom April 2013 in Absatz 4.2.1)

Drasdo P, Karasz M, Pustišek A (2013) Dis-harmony in European Natural Gas Market(s)—Discussion of Standards and Definitions. Zeitschrift für Energiewirtschaft, 37

Dutch Safety Board (2015) Earthquake risks in Groningen. An investigation into the role of the safety of citizens during the decision-making process on gas extraction (1959–2014), pp 1–12

Energy Information Administration (1999) Natural Gas 1998. Issues and Trends. U.S. Department of Energy, 560 (April)

Energy Information Administration (2016a) Development of US gas production. Available at: http://www.eia.gov/forecasts/aeo/MT_naturalgas.cfm#natgasprod_exp

Energy Information Administration (2016b) Natural gas consumption by end use. Available at: https://www.eia.gov/dnav/ng/ng_cons_sum_dcu_nus_a.htm

Eurelectric (2011) Flexible generation: backing up renewables

Eurogas (2016) Statistical report 2015

European Association for the Streamlining of Energy Exchange—Gas (2003) Common Business Practice, 2003-001/02, Harmonisation of Units. Available at: www.easee-gas.eu

European Association for the Streamlining of Energy Exchange—Gas (2005) Harmonisation of natural gas quality

Eustream (2016) Technical conditions of Eustream as a transmission system operator—establishing technical conditions for access and connection to the transmission system and operation of the transmission system. Available at: http://www.eustream.sk/files/docs/eng/technical_conditions.pdf

Gas Connect Austria (2013) Frame capacity contract auctions gas connect Austria GmbH Annex 2—quality and pressure specifications

Ground Water Protection Council and ALL Consulting (2009) Modern shale gas development in the United States: a primer (U.S. Department of Energy et al., eds), 1st edn. Oklahoma City, Tulsa, OK. Available at: https://fracfocus.org/sites/default/files/publications/shale_gas_primer_2009.pdf

Länderarbeitskreis Energiebilanzen (2016) Zur Methodik der CO2-Bilanzen. Available at: http://www.lak-energiebilanzen.de/seiten/co2bilanzenMethodik.cfm

Natural Gas Europe (2016) EDF Heralds "Record-Breaking" CCGT. Available at: http://www.naturalgaseurope.com/edf-heralds-record-breaking-ccgt-30176

Net4Gas (2010) Code of the transmission system operator for transit over the territory of the Czech Republic. Available at: http://www.net4gas.cz/en/media/shippers-news/2010_TR_AJ_podleNC_od13_8_cista.pdf?jis=20160727154221

Rasi S (2009) Biogas composition and upgrading to biomethane. Jyväskylä Studies in Biological and Environmental Science (202)

Reuters (2016) Dutch government lowers Groningen gas output cap to 24 bcm. Available at: http://www.reuters.com/article/netherlands-energy-groningen-idUSL8N19G55U

RWE (2016) GuD-Anlage Lingen – ein echter Sprinter: Weltrekord unter Marktbedingungen

Solomon S et al (2007) Climate change 2007 the physical science basis contribution of working Group I to the fourth assessment report of the intergovernmental panel on climate change. IPCC Wg1, pp 23–78

World Bioenergy Association (2016) WBA Global Bioenergy Statistics 2016, p 80. Available at: http://www.worldbioenergy.org/sites/default/files/WBA%20Global%20Bioenergy%20Statistics%202015%20(press%20quality).pdf

World Nuclear Association (2016) Naturally-occurring radioactive materials (NORM). Available at: http://www.world-nuclear.org/information-library/safety-and-security/radiation-and-health/naturally-occurring-radioactive-materials-norm.aspx

5 Economic and Commercial Fundamentals

Summary

- The 'commercial product natural gas' consists of
  - commodity,
  - flexibility,
  - transportation of the commodity to the location of consumption, and
  - ancillary services.
- Natural gas competes with other primary energy sources or fuels like oil (products), hard coal, lignite, and renewables. There is also competition from electricity.
- Today, natural gas markets worldwide are exposed to regulation, i.e., (sector-specific) market interventions by laws, decrees, guidelines, norms, or orders issued by institutions authorized by the respective governments. This is mainly due to the fact that parts of the natural gas industry can be regarded as a so-called 'natural monopoly'.
- In order to provide reliability to investors and market participants, a legal and political framework for natural gas has to be established.
- Regulators, who are setting or detailing the overarching framework of rules, exist in most of the countries which have a natural gas market.
- Unbundling describes the (vertical) 'disintegration' or disentanglement, i.e., separation of the stages of the value chain, of vertically integrated companies. Primarily, those stages of the value chain regarded as monopolistic bottlenecks, such as transportation, distribution, and (partly) storage, are separated from the entities selling or trading the commodity.
- Non-discrimination, transparency, and enhancement of the so-called security of supply of natural gas are further goals of regulation.
- Dependent on the maturity of natural gas markets, the following 'market models' can be distinguished:
  - traditional markets,
  - transitional markets, and
  - modern traded markets.
- In modern markets, large volumes are traded under standardized contracts. Here, liquidity is crucial. In order to enhance liquidity, quantities to be traded need to be geographically concentrated. This has led to the development of so-called hubs, i.e., points where natural gas trades agreed upon between parties are (physically or virtually, i.e., notionally) delivered.

A. Pustišek and M. Karasz, *Natural Gas: A Commercial Perspective*,
DOI 10.1007/978-3-319-53249-3_5

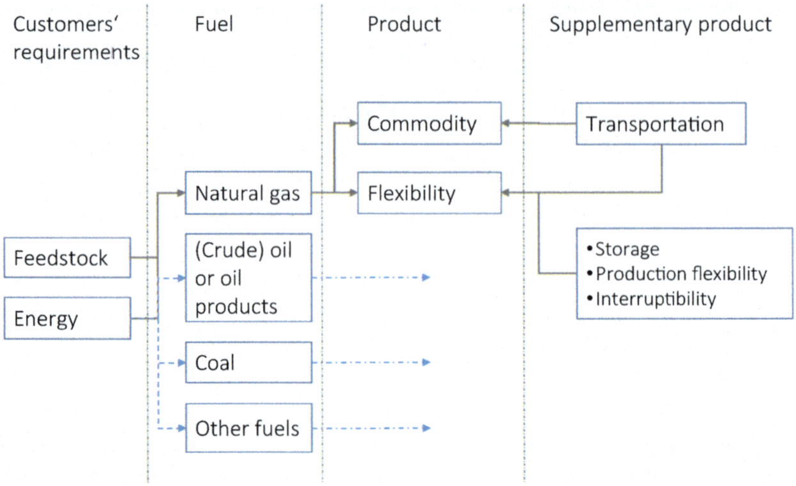

Fig. 5.1 'Product' description

- Further, standardization and an increase of liquidity are necessary to allow the development of successful exchange trading in addition to over-the-counter (OTC) trading.
- In addition, in modern markets, trading instruments like forwards, futures, swaps, and options are used.

5.1 Product

5.1.1 Natural Gas—Commodity and Capacity

Having explained the chemical and physical properties of natural gas, the (commercial) 'product' has not been defined yet. Such a definition is reasonably approached by specifying the end users' demand, a demand, which is characterized by the requirement to have:

- either energy or feedstock available,
- at a given location,
- at defined times, and
- in quantities[1] considered to be sufficient.

This implies that, only if used as feedstock, the ultimately required commodity[2] is natural gas. Otherwise, i.e., if using the commodity for energetic purposes, the end user, at least initially, does not really require natural gas but its energetic content.

Having chosen natural gas as a fuel, the (commercial) 'product' can be divided into, or consists of, the components 'commodity' and 'flexibility'. Then, in order to provide the flexibility, additional services such as 'transportation' and 'storage' (or flexibility from 'production' or from 'interruptible contracts') are required. The 'product' of these supplementary services is capacity (and, additionally, volume in case of storages). Without these supplementary products, neither the components of the product could be delivered nor the requirements of the end user could be met (Fig. 5.1).

In brief, such an end user delivered product consists of:

- commodity,
- flexibility,
- transportation of the commodity to the location of consumption, and
- ancillary services such as measurement, invoicing, consultancy, etc.

[1]For a definition of 'quantity' and other basic term, see Sect. 6.3.1.1.1, Table 6.1.

[2]A commodity is a basic interchangeable good, wherein the quality of the goods interchanged may differ slightly.

This requires a market participant to safeguard access to transportation and flexibility tools compatible in size (i.e., quantities in different periods), time, and location (i.e., at the right delivery points) with the quantities to be purchased and sold.

The physical state (gaseous—if not liquefied) causes a further, economically relevant, particularity of natural gas. Similar to electricity, (gaseous) natural gas is grid-bound. Consequently, natural gas as a source of primary energy and the natural gas industry in general is characterized by high investment, significant economies of scale, and relative inflexibility compared to other energy sources.

In contrast to electricity:

- natural gas in a network is not fully interchangeable, i.e., parts of the network may have to be separated because of different natural gas qualities. Electricity, on the other hand, can be considered exchangeable, as voltage and frequency are defined and adaptable;
- the distribution of sources of natural gas is (geologically) pre-set and cannot be constructed, as for electricity, according to the demand. Consequently, natural gas has to be transported from remote sources to end users; and
- large quantities of natural gas can be stored over long periods.

5.1.2 Competing Fuels and Interdependencies

Effectively, all usages of natural gas described in Sect. 4.3 can also be performed using other fuels, i.e., alternative primary energy sources. For its use as feedstock, natural gas can be replaced by other chemicals. Electricity production can also be performed by burning other fuels like hard coal or lignite, oil products, or by using renewable energy sources or nuclear power. Heating also might be performed by using other fuels or by the use of electricity. The same applies for transportation.

Given this competition with other fuels, demand for natural gas is also (directly and indirectly) influenced by the markets for these fuels.[3]

Substitution of Oil Products with Natural Gas for Heating Purposes

If, for instance, prices for oil products increase, *inter alia*, heating costs for those end users that use oil products for heating will increase. In case these end users had the technical ability to burn natural gas instead, they might switch to natural gas, which in turn might lead to an increase in demand for natural gas, which again might lead to a price increase also for natural gas. This mechanism served (and serves) as an argument to link the prices of natural gas to the prices of the competitive fuels.[4]

Further interdependencies exist since natural gas can be ‘transformed’ into electricity or (yet rarely) into liquid hydrocarbons.[5] *Ceteris paribus*, rising prices for electricity or liquid hydrocarbons lead to an increase in demand for natural gas, and vice versa. Finally, for determining different fuels’ competitive advantages (and disadvantages), the relative prices are decisive. Using coal or natural gas for electricity generation is determined by the so-called ‘dark spread’. Defined analogously to the spark spread,[6] the dark spread is the difference between electricity prices and coal prices during a period in a region. Hence, the difference between spark spread and the dark spread or, simpler, the difference of coal and natural gas prices (taking into account the different efficiencies of power plants) is decisive for the utilization of the fuels. Yet, when considering the scope of fuel switching in electricity generation, not all plants are substitutable. The prices can provide an indication but have to be complemented by additional detailed technical and commercial analysis.

[3]See also Sect. 6.3.2.3.

[4]See also Sect. 6.3.2.2.

[5]See Sects. 4.3.2.2 and 4.3.3.

[6]See Sect. 4.3.3.

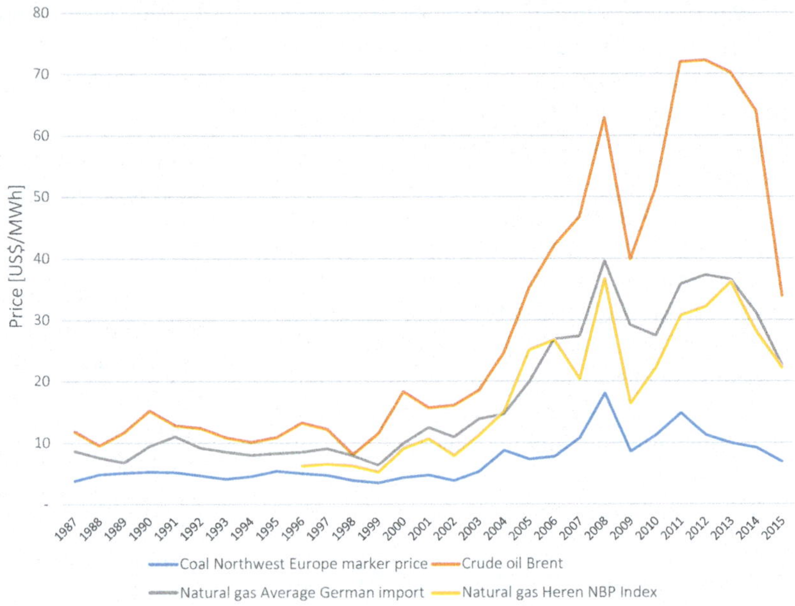

Fig. 5.2 Comparison of primary energy prices

Inter-fuel competition could also be illustrated for the transportation sector. In the long run, whether natural gas is used as a fuel for cars depends primarily on the prices of oil, and consequently gasoline, the prices of electricity and batteries (for electrically powered vehicles), and, importantly, also on taxation or subsidization, frequently diverging between the fuels. In addition, the rate of substitution will depend on factors such as the specific fuel's infrastructure (e.g., filling stations), the age of the existing fleet, and regulation.

Despite such (theoretical) substitutability, prices of competing fuels have developed differently during the last decades. As depicted for Europe in Fig. 5.2, coal, crude oil, and natural gas prices, in particular, as of the beginning of the millennium, started to develop differently, yet not in a totally uncorrelated[7] manner.

Furthermore, natural gas markets are also interlinked to electricity markets. Primarily the products sold to end users are considered complementary. The reasons are: Firstly, traded natural gas markets are assumed to foster trading and the production of electricity. In essence, gas-fired power plants are regarded as options to produce electricity. Secondly, it is assumed that end users perceive natural gas and electricity as comparable or, at least, compatible. Complementary services such as metering and billing are often offered and performed by the same companies. These observations and expectations of profiting from synergies prompted several electricity and natural gas companies at all market levels to merge.[8]

5.2 Value Chain

As the value chain shown in Fig. 2.1 is simple and illustrative, but essentially does not reflect the reality of natural gas markets properly, a more detailed, yet despite of its alleged complexity, still simplifying value chain is depicted

[7]See Fig. 3.6 and also Fig. 3.5 for the US.

[8]Examples that may serve are: the merger of the Belgian and French energy groups Electrabel and GdF; the acquisition of Ruhrgas by the German utility E.ON; and the merger of the Danish utilities DONG, Elsam, Energi E2, Nesa, Københavns Energi, and Frederiksberg Forsyning. All these mergers occurred in the first decade of the new millennium and were fostered by the belief that a 'multi-utility' or 'multi-energy' approach was necessary to survive in the changing European energy environment.

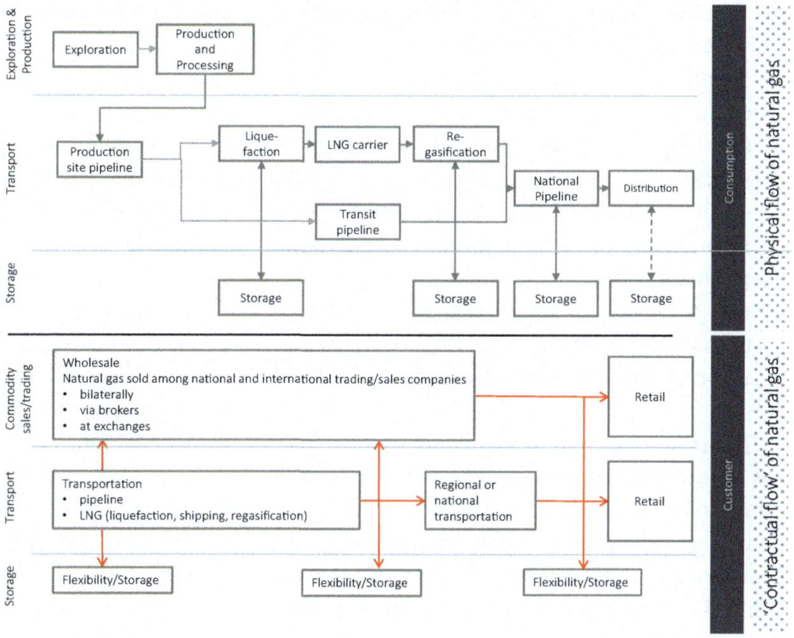

Fig. 5.3 Value chain—contractual and physical

in Fig. 5.3. It is important to note that the physical flow is distinguished from the contractual flow.

Physically, exploration, production, and processing have to be separated from transportation and storage. Even if exploration is strictly speaking not an integral part of the value chain, it is typically included in the case of hydrocarbons. Production and processing are separated from transportation, as a transportation system is not necessarily connected to single production sites only. For LNG, it is necessary to distinguish between liquefaction, (mainly seaborne) transportation, and regasification. Storage can be constructed at each stage of the physical chain, but, primarily for security of supply reasons, is preferred downstream, i.e., in the vicinity of the end users.

Contractually, the value chain has (at least in a modern market environment) to observe the unbundling requirements. Therefore, as depicted in Fig. 5.3, the commodity trading, at all levels of the natural gas industry, is to be regarded separately from the transportation and storage sector. For LNG, it is possible to distinguish between the respective contracts with the liquefaction plant operator, the shipping company, and the regasification plant operator. Again, storage contracts can be concluded at each level of the market, while transportation contracts have to follow the entire chain.

Dependent on the market model applied, companies might act on different stages of the value chain. In the past, integrated companies were, e.g., active as wholesale traders and carriers at the same time. Yet, in modern markets, unbundling provisions might prohibit such 'bundled' activities. However, on the LNG side, it is quite common that producers are active as sellers of LNG via affiliated companies that own and operate liquefaction plants or vessels.

5.3 Regulation

Today, natural gas markets worldwide are exposed to regulation, i.e., (sector-specific) market interventions by laws, decrees, guidelines, norms, or orders issued by institutions authorized by the respective governments.

The quasi-symbiotic relationship between the gas business (i.e., at that time dominated by town gas) and regulation started to evolve in the US already in the 19th century.[9] This was based on:

[9]See example on FERC in this section.

- the observations that the natural gas markets were (or still are) controlled by monopolies, being either vertically integrated or acting on distinct or several stages of the value chain;
- the (general) economic rule according to which a monopolist will produce and offer smaller quantities and charge higher prices than players in a competitive, i.e., polypolistic, market; and
- the governments' main objective to protect consumers' interests, i.e., to foster competition or, at least, to support competitive pricing.

The main objective of regulation in the natural gas industry is to protect consumers' interest by establishing a competitive market environment. However, regulation of gas transportation has been structured differently in the US and Europe. While US regulators aimed at facilitating competition at the wellhead, i.e., upstream, EU legislation, on the background of declining reserves, established rules for a single European natural gas market, thus focusing on the downstream sector. Further objectives are to:

- ensure transparent and non-discriminatory access to essential facilities for all market participants;
- grant the most efficient use of infrastructure (i.e., maximizing utilization);
- eliminate monopolistic inefficiencies, i.e., minimization of deadweight losses;
- ensure security of supply, *inter alia*, by allowing market participants to recover their costs or by incentivizing investments; and
- (particularly in Europe) create or consolidate a single harmonized European market for natural gas (as opposed to several distinct national markets).[10]

The underlying assumption is that the (at least national, if not regional) economic benefits of regulation (are supposed to) supersede costs, i.e., that the costs of the introduction and perpetuation of 'market liberalization' by regulators are less than the benefits of mitigated deadweight losses. Though, because prices, markets in general, and even policy frameworks change, the detailed quantification of the accomplishment of this welfare 'gain' has not been achieved yet.[11]

A competitive market requires opening all stages of the value chain.[12] Regulation, therefore, might be extended to one or more of the following 'objects of regulation', i.e.:

- production,
- pipeline transportation,
- LNG transportation,
- storage,
- distribution,
- sales and trading, and
- mergers in the industry—internationally and/or nationally.

In this context, particularly, pipeline transportation and distribution have been identified as 'natural monopoly'[13] and, in addition, as 'monopolistic bottleneck(s)'. Thus, primarily these sectors have been exposed to regulation. In some countries, however, also storages are identified as natural monopolies and regulated. Other countries choose to regulate also commodity prices, primarily in the end user segment.

[10]The often quoted 'sustainability' objective is not observed being of primary importance for regulators of the natural gas industry.

[11]The process of regulating natural gas markets with the objective to establish competitive markets is often referred to as 'liberalization'. As this term is not considered to describe the concept appropriately, it shall not be used subsequently.

[12]See also the example of the US market regulation and deregulation below.

[13]A natural monopoly exists, if a single supplier can serve a market more efficiently, i.e., at lower costs than various suppliers can. This implies that, in general, it is economically not feasible to duplicate the assets used for its production—in the case of natural gas, the transportation and distribution infrastructure. The market structure will 'naturally' be monopolistic. A natural monopoly can be interpreted as a consequence of the respective sector's cost function. That is, if the sector's cost function is sub-additive, primarily because of inherent economies of scale or economies of scope, then (potential) competitors (rationally) will have no interest in entering such market. Nevertheless, competing transportation (and storage) infrastructure has been built. One example took place in Germany in the early 1990s. See Sect. 2.3. This so-called, 'pipeline-to-pipeline' competition can also be observed in the US. Here, often, interstate pipeline investments are made based on so-called 'open seasons', i.e., basically auctions in which the various carriers compete.

The regulatory approach, i.e., the forms, objects, and methods of regulation, differ regionally and nationally. Whether:

- cost oriented,[14] i.e.,
  - rate-of-return regulation or
  - cost-plus regulation, or
- incentive oriented,[15] i.e.,
  - price-cap, revenue-cap, or
  - yardstick competition

regulations are implemented and which parts of the business (i.e., stages of the value chain) are regulated is a decision taken at the sole discretion of the respective legislative authority.

Despite such differences in regulation, some main principles applied to the regulation of natural gas markets could be identified[16]:

- Framework

To provide reliability to investors and market participants, a legal and political framework for natural gas is established. The laws enacted, *inter alia*: constitute the objectives for the development of the natural gas sector in general: determine the stages of the value chain which shall become subject to regulation; set the tasks and competences of the regulatory body; and, in some countries, also prescribe the level of unbundling, details of pricing for transportation, storage or commodity sales, capacity and commodity allocation mechanisms, security of supply requirements, and rules, as well as the rights and obligations of (selected) market participants.

[14]Following this approach, the total costs of the regulated asset, including capital and operational costs, as well as a reasonable rate of return on invested capital, are calculated and a utilization is assumed. The regulated price (i.e., in case of transportation capacity, the transportation price/fee) is set to enable the owner of the regulated asset to cover these costs.

[15]This form of regulation is deemed more 'modern' and aims at decoupling costs from allowed revenues in order to prevent unwanted incentives to be given to market participants by cost-oriented regulation. Based on (statistical) data, a (virtual) 'best-practice operation' is defined and used as a benchmark. This is set as a target for the operators of a regulated asset. Such a target, typically, has to be reached within a defined period.

[16]See Corbeau et al. (2012), p. 33.

- Regulators

Regulators, setting or detailing the overarching framework of rules, are established in most of the countries where a natural gas market exists. While their competence, organization, and functions differ, as a general observation, regulators:

- are commonly equipped with sufficient authority to fulfil their tasks;
- are established as independent 'bodies' or 'institutions', independent from industrial and (sometimes also) political stakeholders in order to ensure an efficient decision-making process and non-discrimination among all market participants;
- decide by bringing on a consultative, transparent process, involving all relevant market participants and publishing evidence and final decisions; and
- allow decisions to be appealed by independent courts.

Some Regulatory Authorities

In Europe, regulation of the natural gas sector is based on the directives of the European Union.[17] Each European country has implemented these rules (differently) in national legislation. That is, neither the rules for access to the transportation systems, nor the distribution systems nor the storages are fully harmonized, despite the common regulatory framework.

In the US, the Federal Energy Regulatory Commission (FERC) has been established as the regulatory authority, and, in Australia, the Australian Energy Regulator.

In China, the National Development and Reform Commission (NDRC) focusses on macroeconomic planning, *inter alia*, of the energy sector. In addition, other ministries such as the Ministry of Commerce, the Ministry of Land and Resources, the

[17]See the first directive European Parliament and the Council of the European Union (1998) and the third directive European Parliament and the Council of the European Union (2009a).

Ministry of Finance, and the Ministry of Environmental Protection assume various responsibilities, equivalent in toto to the regulation of the Chinese natural gas sector.

FERC's Competences

"*The Federal Energy Regulatory Commission, or FERC, is an independent agency that regulates the interstate transmission of electricity, natural gas, and oil. FERC also reviews proposals to build liquefied natural gas (LNG) terminals and interstate natural gas pipelines as well as licensing hydropower projects. The Energy Policy Act of 2005 gave FERC additional responsibilities as outlined and updated Strategic Plan. As part of that responsibility, FERC:*

- *Regulates the transmission and wholesale sales of electricity in interstate commerce;*
- *Reviews certain mergers and acquisitions and corporate transactions by electricity companies;*
- *Regulates the transmission and sale of natural gas for resale in interstate commerce;*
- *Regulates the transportation of oil by pipeline in interstate commerce;*
- *Approves the siting and abandonment of interstate natural gas pipelines and storage facilities;*
- *Reviews the siting application for electric transmission projects under limited circumstances;*
- *Ensures the safe operation and reliability of proposed and operating LNG terminals;*
- *Licenses and inspects private, municipal, and state hydroelectric projects;*
- *Protects the reliability of the high voltage interstate transmission system through mandatory reliability standards;*
- *Monitors and investigates energy markets;*
- *Enforces FERC regulatory requirements through imposition of civil penalties and other means;*
- *Oversees environmental matters related to natural gas and hydroelectricity projects and other matters; and*
- *Administers accounting and financial reporting regulations and conduct of regulated companies.*

What FERC does not do

Many areas outside of FERC's jurisdictional responsibility are dealt with by State Public Utility Commissions. Areas considered outside of FERC's responsibility include:

- *Regulation of retail electricity and natural gas sales to consumers;*
- *Approval for the physical construction of electric generation facilities;*
- *Regulation of activities of the municipal power systems, federal power marketing agencies like the Tennessee Valley Authority, and most rural electric cooperatives;*
- *Regulation of nuclear power plants by the Nuclear Regulatory Commission;*
- *Issuance of State Water Quality Certificates;*
- *Oversight for the construction of oil pipelines;*
- *Abandonment of service as related to oil facilities;*
- *Mergers and acquisitions as related to natural gas and oil companies;*
- *Responsibility for pipeline safety or for pipeline transportation on or across the Outer Continental Shelf;*
- *Regulation of local distribution pipelines of natural gas;*
- *Development and operation of natural gas vehicles;*
- *Reliability problems related to failures of local distribution facilities; and*
- *Tree trimmings near local distribution power lines in residential neighborhoods."*[18]

[18]Federal Energy Regulatory Commission (2016).

Most of the regulators are responsible for natural gas and electricity, as interdependence of electricity and natural gas markets is often assumed (or considered being proven), primarily because the electricity (or power) sector acts as a driver for natural gas demand[19] and because electricity grids are also regarded as a natural monopoly.

- Non-discrimination

As new entrants to the natural gas (commodity) market depend on getting access to transportation and storage infrastructure, an integrated system operator owning the infrastructure might have an incentive to discriminate the new entrant, e.g., by denying access to its grid or by asking for prohibitively high fees for using the infrastructure. Regulators, promoting and supporting competition, are aware of potential anti-competitive incentives of integrated companies and require non-discriminatory access to monopolistic bottlenecks, independent of the form of unbundling. That is, e.g., carriers and storage system operators are not allowed to offer different prices for identical services to different users, or, to put it more simply, the price must be linked to services only, not to the customer or its characteristics.

- Transparency

In the (prevailing) context of regulation and legislation, transparency denotes the ease of understanding, as well as the openness, public availability, and (instantaneous) accessibility of all relevant information, presumed to be correct. Whether the commodity market or the markets for transportation and storage capacity, all rules are supposed to be 'transparent',[20] as secretive behavior might support or even encourage discrimination. Therefore, transparency is deemed essential for building confidence among market players. This is not restricted to prices only, but also includes availability of terms and conditions for sales and purchases of the commodity, as well as transportation and storage capacity. Transparency also implies simplification of rules and regulations. However, in some cases, simplification of rules might contradict the requirement of non-discrimination.[21]

- Unbundling

Unbundling describes the process of disintegration or disentanglement, i.e., separation of stages of the value chain, of vertically integrated companies. Primarily those stages of the value chain regarded as monopolistic bottlenecks, such as transportation, distribution, and (partly) storage, are separated from the entities selling or trading the commodity. This is performed in order to prevent integrated companies from using their potential market power to obstruct the access of their competitors to their network and thus to compete in commodity sales. Unbundling requires the effective separation of activities performed by integrated companies at the stages of the value chain. Such separation could be achieved by:

- Unbundling of accounts
  The integrated company owns, manages, and operates the network. Separate accounts for unbundled services are introduced and monitored;
- (Internal) separation of organizational entities
  The integrated company owns, manages, and operates the network but maintains a transparent and clear separation of the organizational structure with clear division of responsibilities, including, but not limited to, profit and loss responsibility;
- Transfer of unbundled services to separated companies
  The subsidiary either performs management of the network and the ownership of the network remains with the integrated company, or the management and the ownership

[19]Nevertheless, regulation of power markets in the European Union has preceded regulation of natural gas markets. Also in the US the development was not in parallel, differing even between the states.

[20]See, e.g., European Parliament and the Council of the European Union (2009b).

[21]For example, a postage-stamp system for pricing of transportation capacity in an extended, e.g., national network, can be considered as being a simplification in comparison to a distance-related pricing system, but (some) shippers may be discriminated against, as equal prices for different services (i.e., transportation distances) are paid.

of the network are transferred to the subsidiary. This subsidiary has to be financially and operationally independent from the parent company;

or

- Ownership unbundling,

  The transportation and distribution system have to be sold by integrated companies to independent owners.[22]

An important consequence of unbundling is that carriers and storage system operators may purchase and own natural gas only to the extent necessary to safeguard and maintain the technical operation of the respective system.

Regulation in Europe

Targeting a competitive, internal, harmonized, and single market in the European Union, the European Commission started market integration and 'liberalization' of the energy sector principally already during the 1980s. In the 1990s, the upstream sector was opened[23] by setting the rules of non-discriminatory access to exploration and production of natural gas and oil.

As of 1998, the European Commission started efforts to create the single internal market for natural gas.[24] In this context, owners of transportation systems, storages, and LNG facilities were required to grant third party access to their systems to customers other than their parent or affiliated companies. Yet, member states were still allowed to choose between negotiated and regulated access, and legal unbundling was not mandatory, i.e., transportation and storage assets could remain part of vertically integrated companies. In addition, larger end users, as, e.g., power generators, were granted the right to choose their supplier.

Even before implementation of the first directive into national laws of member states, pushed by the 'Madrid Forum',[25] some member states and the European Council urged an acceleration of market opening. To this end, rules for congestion management, capacity allocation, and balancing, as well as the requirement of legal unbundling and improved third-party access (*inter alia*, by setting the rules for transportation price calculation) and free choice of suppliers for non-residential end users, were set.[26]

Remarkably, the second directive included provisions, which may be used to invoke an exemption for major new infrastructure, such as pipelines connecting member states, LNG facilities, or storage facilities, from third-party access and regulation, aiming to mitigate the risk for new infrastructure and consequently to incentivize investments in such infrastructure. Such an exemption is to be decided by regulators on a case-by-case basis.

Inter alia, further improvement of third-party access, strengthening the independence of regulators, promoting ownership unbundling, and demanding increased regarding the access to storages were to be transposed to national law by the member states by 2011.[27] In parallel, the Agency for the Cooperation of Energy Regulators (ACER) was established.[28]

[22]Even without regulatory interference, e.g., the operating company of the Czech transit system, Net4Gas, and Open Grid Europe, the major long-haul transportation company in Germany were sold to financial investors.

[23]See European Parliament and the Council of the European Union (1994).

[24]See European Parliament and the Council of the European Union (1998) (the so-called 'first directive').

[25]The Madrid Forum is a group of national regulatory authorities, EU national governments, the European Commission, carriers, gas suppliers and traders, consumers, network users, and gas exchanges. It was set up to discuss issues regarding the creation of the internal gas market.

[26]See European Parliament and the Council of the European Union (2003) (the so-called 'second directive').

[27]See European Parliament and the Council of the European Union (2009a) (the so-called 'third directive').

[28]ACER supervises the market and advises the European Commission. ACER is supposed to foster cooperation among national regulators and harmonization of the European natural gas market, preparing guidelines for (national) network codes.

In January 2009, supplies from Russia transited via the Ukraine have been interrupted. As a reaction, the European Commission endeavored to strengthen the security of supply.[29] The new regulation requires clear roles and responsibilities for market players, risk assessment, (potentially) investments in infrastructure, and stronger cooperation among member states (at the regional level), regulators, and market participants.

Regulation and De-regulation in the US[30]

Regulation of the (natural) gas industry in the United States dates back to the mid-1800s, when (natural) gas was predominantly manufactured from coal. Already by then, local governments in the US had recognized the natural monopoly characteristics—primarily for distribution networks. Hence, the local governments' intention was to prevent these natural monopolies from abusing their market power.

Because, by the beginning of the 20th century, municipalities started to 'exchange' natural gas and constructed so-called 'intrastate' pipelines[31] to be able to transport natural gas from one city to another, state governments considered it necessary to intervene by determining the rates (i.e., prices) that could be charged for transporting and selling natural gas. This was done for the first time already in 1907 in New York and Wisconsin. Several years later, the next problem arose: companies started to transport and sell natural gas between the states using so-called 'interstate' pipelines, crossing the borders of the states' jurisdiction.

Following a series of complex legal discussions and cases concerning the regulatory competence and authority of state governments and the market concentration, which took place in pipeline network ownership and operation, the US federal government initiated the Natural Gas Act in 1938. By this act, the Federal Power Commission (FPC, created already back in 1920) was given jurisdiction over regulation of interstate natural gas sales (including transportation). However, in the Natural Gas Act, the prices for natural gas sales at the wellhead were not specified (or regulated). This was changed in 1954 by the Supreme Court's decision in Phillips Petroleum Co. v. Wisconsin (347 US 672 (1954)). According to this decision, natural gas prices at the wellhead, i.e., for sales to interstate pipelines, were regulated analogously to the prices for sales from interstate pipelines to local distribution utilities. The consequence of such a decision was that natural gas prices were cost-based (including a profit considered adequate), but did not reflect the market value of natural gas. This cost-based regulation was regarded to be the principal cause for the shortages of natural gas in the US in the 1970s. In addition, the FPC did not succeed in keeping pace with the developments of a steadily growing market. Both individual and regional rate settings were not accomplished in time. By the mid of the 1970s, the FPC started to introduce price ceilings for natural gas. Yet, the market value of natural gas, in particular in competition with the significantly increasing prices of oil and oil products, was not reflecting the introduction of such ceilings. *A fortiori*, natural gas had to be sold considerably below its value, despite ceilings having been increased by the end of the decade.

The system of price controls was in place until the Natural Gas Policy Act (NGPA) was issued in 1978, which was part of the National Energy Act (NEA) and which was enacted by the US Congress in reaction to supply shortages and demand surges. This NGPA aimed at:

[29]See European Parliament and the Council of the European Union (2010).

[30]See Naturalgas.org (2013b).

[31]Note that, in the US, the term 'pipeline' refers to physical facilities through which natural gas is moved but also to companies transporting or selling natural gas.

- creating a single national market for natural gas,
- supporting market participants to match supply with demand, and
- allowing competitive (free, i.e., non-regulated) wellhead prices of natural gas.

The FPC was abolished and replaced by the Federal Energy Regulatory Commission (FERC). However, FERC did not have jurisdiction over the import and export of natural gas, which was given to the Department of Energy.

While the NGPA kept the principle of ceiling prices, it included increases of these prices, setting economic incentives for producers. These ceilings and the mechanisms for increasing rates were set out in the statute and intended to be phased out. Already, by 1985, the goal was that wellhead prices should be completely deregulated.

In addition, barriers between interstate and intrastate natural gas were broken. The market and market participants reacted by:

- concluding long-term agreements (with take-or-pay clauses);
- expanding exploration and production;
- increasing wellhead prices;
- increasing of end user prices; and
- decreasing demand, which finally resulted in oversupply between 1980 and 1985.

Despite the NGPA's deregulation of wellhead prices, the prices for bundled products sold by interstate pipelines to local utilities and local distribution companies were still regulated. It was essentially FERC Order No. 436 issued 1985, the 'open access' order, that introduced 'unbundling'. That is, pipelines were permitted to act as carriers only, offering transportation services to all customers non-discriminatorily. These customers could realize cost savings by purchasing natural gas directly on spot markets, established in parallel. At such spot markets, often natural gas was sold at prices lower than the ones agreed upon in long-term, take-or-pay contracts. Consequently, pipelines and producers frequently conducted arbitration proceedings to resolve the negative implications of long-term contracts (to mitigate such problems, FERC issued Order No. 500 in 1987 encouraging pipelines to terminate long-term contracts by buying them out and passing part of the costs to their customers). Complete deregulation of wellhead prices was ordered by the Natural Gas Wellhead Decontrol Act (NGWDA) in 1989, so that, by 1993, all price regulations of wellhead prices were eliminated.

In 1992, Order No. 636 changed the voluntary offering of transportation services according to Order No. 436. Unbundling became a requirement; it was not an option anymore. Pipelines had to separate transportation and sales services. Transportation services had to be offered transparently and non-discriminatorily. Thus, sales affiliates of pipeline could not have, or gain advantages over, competitors. According to Order No. 636, pipelines were also required to offer:

- 'no-notice' transportation service;
- access to storage facilities;
- increased flexibility; and
- 'capacity release' programs including the setting-up of electronic bulletin boards.

Despite the fact that the FERC issued several other orders after 1992, the final restructuring of the US natural gas market was perceived to have been completed with Order No. 636.

5.4 Security of Supply

Security of supply is not defined uniformly. The definitions range from broad definitions, indirectly including the demand side:

> *By security of supply we mean that all the technical and organisational requirements for*

uninterrupted operations are met, and all the necessary action has been taken to maintain the continuous physical availability of sufficient quantities of natural gas of defined quality at all times and at affordable prices.

Gas supply security begins with infrastructure operation, maintenance and development in accordance with good technical practice. Long-term supply security depends on developing the gas grid to meet future demand.[32]

to more narrow, even indirect definitions

Given the importance of gas in the energy mix of the Union, this Regulation aims at demonstrating to gas customers that all the necessary measures are being taken to ensure their continuous supply, particularly in case of difficult climatic conditions and in the event of disruption. It is recognised that these objectives should be achieved through the most cost-efficient measures in order not to affect the relative competitiveness of this fuel compared to other fuels.[33]

to short ones as:

the uninterrupted availability of energy sources at an affordable price[34]

or definitions specifically focused on diversification of supplies as:

Security of gas supply, expressed as current and projected national or collective dependence of European countries on supplies from individual suppliers (or groups of suppliers) over the next 15–25 years [35]

Here, security of supply should be understood as safeguarding the availability of natural gas according to anticipated demand at each point in time. As such, it is restricted to the supply side.[36]

Supply interruptions could be caused by any non-availability of infrastructure, e.g., breakdown of transportation and storage, or interruption of production. These interruptions can occur at each point along the physical value chain, i.e., anywhere between production and final use. There is no evidence confirming the often implicit, rarely explicit, presupposition of governments, market participants, and end users that indigenous supplies are more secure than imports. Consequently, reduction of import dependence, despite being a political objective, cannot be regarded as a proven remedy to increase security of supply.[37] However, indigenous supplies are—other than imports—not exposed to potential threats of foreign governments to interrupt supplies for purely political reasons.

Measures to mitigate the risk of interruption can be partial at best; they are eventually manageable but not completely plannable. In case supplies are interrupted, the time, extent, and duration of interruption in the system will depend on:

- the degree of diversification, of
  - sources:
    Receiving deliveries from several sources at one or different delivery points, transported via one or on different routes, from one or from several suppliers, grants the customer diversification of supplies, i.e., diversification of sources;
  - contractual partners:
    Receiving deliveries from one source at one delivery point from several suppliers grants the customer no physical, but a contractual, diversification.

[32] Energie-Control Austria (2016).

[33] European Parliament and the Council of the European Union (2010).

[34] International Energy Agency (2014), p. 13.

[35] Stern and Honoré (2009), p. 407.

[36] If the demand side were included in such a way that security of supply compensated or satisfied each, even extraordinary, demand situation, it would—again—necessitate the exact definition of such increased demand situation to be covered. Otherwise, no measure for its compensation could be prepared. Yet, such (increased) demand, would have to be defined, i.e., limited. This implies that again security of supply would not be guaranteed or include 'each' demand situation. Consequently, the problem is reduced again to safeguarding the availability of natural gas according to anticipated demand at each point in time, i.e., to security of supply, not demand.

[37] This is confirmed and emphasized by Stern and Honoré (2009), p. 412: *Indeed, history suggests that all serious security incidents, i.e. where customers have lost gas supplies for a considerable period of time, have stemmed from failure of indigenous supplies or facilities. No empirical experience would lead to the conclusion that a country with substantial dependence on imported gas supplies would be necessarily less secure, i.e. more prone to disruption, than one which was self-sufficient..*

Troll Sales

After the Norwegian GFU was dissolved, natural gas from the Troll field has been sold by the producers separately, yet at the same delivery point. This implied that the traditional buyers of the natural gas production from the Troll field have been engaging in several contracts, instead of one single contract, i.e., they diversified contractually.

- transportation routes:
  Receiving deliveries from one source at one or different delivery points, transported on different routes from one or from several suppliers, grants the customer diversification of transportation routes.
  and
- storages:
  Storing natural gas in more than one storage facility grants contractual or physical diversification of storages.

- the customers' consumption pattern;
- the cooperation of owners of adjacent infrastructure (to compensate interruptions by, e.g., alternative routing;
- the weather during the interruption; and
- available options for usage of alternative fuels.

In case of interruptions of supply, only limited tools for compensation exist. These are mainly:

- increase of supply from other sources;
- increased utilization of storages;
- modification of transportation routes; or
- reduction of natural gas consumption—either by interruption of interruptible customers or by effective demand-side management.

In any case, the undisturbed performance of the technical system is decisive. This requires regular and appropriate maintenance, investments, and adaption of standards. Hence, security of supply is not available at zero cost, neither regional, nor national nor international.

Whether or not security of supply requires legislative or regulatory intervention is discussed worldwide. However, at least for interconnected markets, as, e.g., the European Union, measures of security of supply have to be coordinated and harmonized internationally, i.e., between all stakeholders of such a market, including but not limited to regulators, traders, governments, and utilities. Otherwise the enhancement of one's security of supply might deteriorate the others' security of supply.

Irrespective of the regulatory interventions, in several European countries, most of the market participants regard security of supply as a 'free good'. Despite its indispensability and acknowledgement of importance, the willingness to pay merely exists.

Interruption of Supplies to Europe in January 2009

In early 2009, Gazprom delivered approximately 26% of the European natural gas demand. Seventy-six percent of these exports were transported via the Ukrainian transportation system. By end of 2008, discussions between 'Russia' and the 'Ukraine'[38] were intensified. By the end of 2008, Ukrainian debt accumulated to more than 2 billion US $; nevertheless, the parties negotiated a prolongation of the existing supply contract. Russia demanded immediate payment of debts and increased the natural gas price to a 'West-European' level —as a pre-requisite for the continuation of natural gas deliveries in 2009. Ukrainian representatives asked for an increase of transit fees for exports to Europe. The parties failed to agree upon a contract by the end of 2008. On January 6, 2009, 'Russia' suspended all natural gas deliveries to the Ukraine. Consequently, on January 7, 2009, the natural gas flow from Russia to Central and Western Europe via the Ukraine ceased.

[38]Actually, it started with discussions on a company level rather than between governments.

Further negotiations between Russia and the Ukraine failed at the beginning of January 2009. The EU got involved; representatives were sent to the Ukraine and Western European importers started to supply natural gas to utilities in east and southeast Europe. In addition, measures were taken to increase supplies from other sources, 'optimize' the utilization of infrastructure, i.e., transportation and storage systems, and prepare customer management. In mid-January 2009, some Western European importers elaborated and presented a proposal for financing the fuel gas required by the 'Ukraine' for re-establishing the transit. On January 20, 2009, 'Russia' and the 'Ukraine'[39] agree upon the terms and conditions for a natural gas delivery contract as well as a transit contract for 2009. On January 21, 2009, natural gas from Russia was flowing again via the Ukraine and 'arriving' in Western Europe.[40]

Consequences for the natural gas sector in Europe were partly severe: e.g., one practical and seemingly minor consequence was that, in some regions for months after this 'crisis', not a single new built house was constructed using natural gas for heating purposes. In a broader sense, the EU started focusing on security of supply, and companies evaluated further diversification of supplies.

5.5 Market Models

Natural gas markets are categorized predominantly based on:

- market structure, referring, e.g., to the organizational characteristics of a market, such as the number and relative strength of participants, i.e., buyers and sellers, the degree of (potential) collusion among these participants, the competition, and ease of entry into and exit from the market;
- maturity of the market, i.e., whether it is an existing, developing, or even not yet established market; and
- regional clusters.
  Often natural gas markets are separated nationally or supra-nationally. Nonetheless, depending on the level of aggregation, regional groups based either on:
  - transportation (shipping) routes (e.g., the Atlantic and Pacific basins, Russian export pipelines); or
  - the main regions for natural gas trades (e.g., North America, Europe, and Asia); or
  - common sources of supply (e.g., Northwest Europe); or
  - (even) pricing principles (crude-oil-indexed LNG supplies in Asia) etc.

  can be identified.

Subsequently, the categorization based on the market structure is chosen, referring to different market structures in accordance with the historic development. An overview is depicted in Table 5.1.

5.5.1 Traditional

In Europe and in other countries, the traditional market model prevailed for decades (and partly still prevails) and is characterized by:

- vertical integration,
  i.e., companies in the natural gas business offer services and activities along different, yet not all, elements of the value chain;
- different levels,
  i.e., notwithstanding (potential and partial) vertical integration, companies act at 'successive' (regional) levels, basically following the flow of natural gas from the source to the sink. These levels are internationally not identical. In some countries, they do not exist

[39] Meanwhile supported, or at least represented, by governments.

[40] See Pustišek (2009).

Table 5.1 Principal characteristics of various market structures (simplified, abridged)

	'Traditional'	'Transitional'	'Competitive'
Market structure	(regionally) Monopolistic (horizontal 'demarcation')	Oligopolistic	Polypolistic
Regulation/government interference	Low	Initiated—high	High
Market level ('layers')	Several clearly distincted (in some countries vertical 'demarcation')	Several clearly distincted	Few
Market participants' business model (organization)	Vertical integration	Vertical integration with partly regulated sectors	Unbundled (along the value chain)
Services	Offered integrated (bundled)	Partly disintegration of services	Offered disintegrated (unbundled)
Participants' relation	Direct contract	Direct contract, OTC and anonymous contract	Direct contract, OTC and anonymous contract
Contract structure	Mainly long-term	Long-term and short-term	Mainly short-term
Pricing concept	Substitutable product indexation (often: oil products)	Hybrid pricing	Gas-to-gas competition
Price volatility	Dampened	Medium	High

at all. At lower levels, the number of players may increase[41];

- vertical demarcation,
  i.e., the companies acting at one level of the market do not enter into competition with their customers, i.e., the company at a superordinate level would (typically) not endeavor to capture the customers of its customer;
  and
- horizontal demarcation,
  i.e., players at the same level do not compete with their peers.

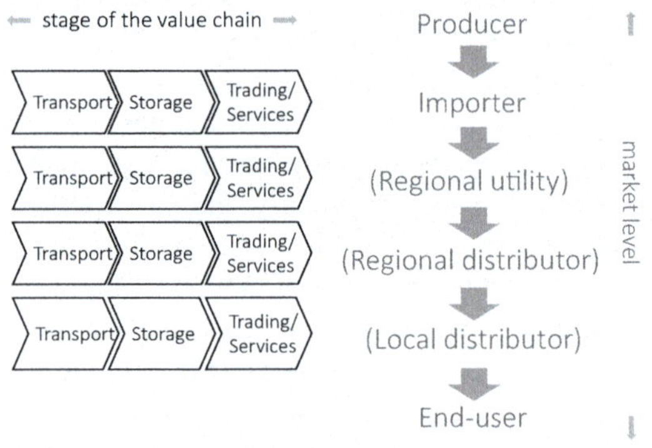

Fig. 5.4 Traditional market model

In the early days of the traditional market, the producers delivered natural gas at the border of the country of destination (or, rarely at another border being upstream) to the importer. Often only one or just a few importing companies existed in a country. The importers (at least usually) did not have interests in the upstream business, i.e., in the exploration and production of natural gas. Yet, these companies owned and operated storage facilities and natural gas pipelines. The importers transported the natural gas in their pipeline system to the delivery point of the companies constituting the next level downstream.[42] These companies themselves might have operated storage facilities and pipelines and provided additional services. They delivered the natural gas to the subsequent downstream level —and so on.

Such market structure is depicted in Fig. 5.4. (Note that the levels in brackets do not exist in every country.)

[41]See also 'horizontal demarcation'.

[42]For example, regional utilities in Germany or local distributors in Italy.

Consequences of this market structure were

- price differentiation,
  i.e., the (commodity of) natural gas is sold in different segments at different prices. The consumers' rent is absorbed by charging (approximately) the consumers' marginal willingness-to-pay. As vertical integration and demarcation together with non-existence of natural gas trading hindered competing suppliers to offer natural gas, the end users' alternative, e.g., gasoil or coal, was used as a benchmark for the determination of the maximum price the end users in each segment would be willing to pay for the next unit of gas[43];
  and
- regionally, monopolistic structures at each level effecting double marginalization,
  companies acting as (regional) monopolists, primarily because of the horizontal demarcation, charge monopolistic prices. If this is the case at different levels, then the prices set are higher than those chosen by a single company (or if the two companies are merged). The consequence (also observed in natural gas markets) is that the final price charged to the end user is higher than it would be in the case where the market was served by a single company, a monopolist. *A fortiori*, the single companies' profit, as well as the joint profit of the companies, is lower than it would be in a monopolistic market, served by one company only, and the price to the end user is higher.[44]

Abandoning the Traditional Market Model

The traditional market model was favored predominantly by suppliers. Yet, regulatory unbundling requirements supported market change. In Germany, the shareholders of BEB[45] were among the first that realized the impact of such change. BEB was an importer and producer of natural gas; it sold gas amounting to 180 TWh in 2002 mainly to German utilities; and it owned and operated a high-pressure network of 2900 km and storage facilities with a working gas volume of almost 2.5 bcm. The company was among the top three German market players. The shareholders were ExxonMobil and Shell. In 2004, even before legally obliged, the shareholders carved out the sales business, split every contract into halves, i.e., between them, and allocated the respective half to the newly founded sales affiliates of Shell and ExxonMobil. BEB remained a network and storage company. In 2007, even the network was sold. The sales process at that time created huge attention across Europe because it was one of the first networks to be sold. It was acquired by the Dutch state-owned infrastructure operator Gasunie. Four years later, ExxonMobil and Shell also sold the storage facilities. The buyer was the (partly) French state-owned GdF.[46]

5.5.2 Transitional

Mainly as a consequence of regulation, *inter alia*, vertically integrated companies were unbundled. In parallel, hubs emerged gradually. The sequential organization of deliveries, prevailing in the past, dominated by distinct market levels, widely disappeared in Europe and North America. The new, yet transitional, market structure was characterized by a multitude of relationships between participants of different levels.

Players at all levels were forced to adjust their products. In general, these new products showed similarities to hub-traded products. Usually, they

[43]See also Sect. 6.3.2.2. For further explanation of price differentiation, see, e.g., Pindyck and Rubinfeld (2012), Sect. 11.

[44]For further explanation of double marginalization, see, e.g., Pindyck and Rubinfeld (2012), Sect. 11. Note that double marginalization does not require a perfect monopoly. It suffices that both companies have some market power (i.e., be able to set the price above marginal cost).

[45]BEB stands for "BEB Gewerkschaften Brigitta, Elwerath Betriebsführungsgesellschaft". It was an important part of German mining history.

[46]Meanwhile called 'Engie'.

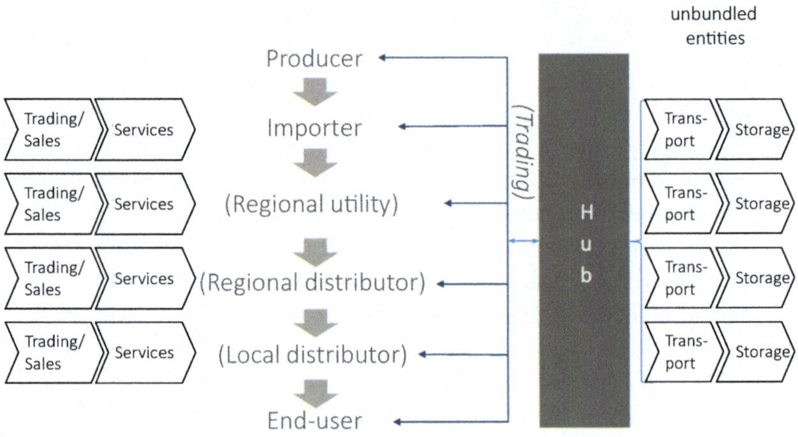

Fig. 5.5 Modern (transitional) market model

had a shorter term than traditional products and, in some cases, they did not foresee the delivery of the entire natural gas quantity needed by the customer. Furthermore, often these products increasingly included less or no flexibility.

As the products showed similarities to hub-traded products, so started the prices and pricing structures. Oil price indexation was partly replaced by fixed prices or prices linked to natural gas indices. Yet, the products and pricing structures offered to small end users, e.g., households, did not change in a significant way. However, also on the wholesale level, oil price indexation remained part of some contracts, even after their renegotiation. In this new market structure, risk- and portfolio management required an extended view and a higher level of sophistication.[47] While in the traditional market environment different oil-indexed contracts were optimized (against each other), in the new market environment, hub-based prices became the reference. As a consequence, price risks started to increase while volume risks were still existing, as markets were not (yet) liquid. Such a market structure is depicted in Fig. 5.5. (Note that the levels in brackets do not exist in every country.)

The Changing Role of the British Incumbent

In the UK in the period until the mid-1990s, British Gas as the state-owned natural gas monopoly buyer, later privatized, agreed to individually negotiate supply contracts mainly with producers of the UK Continental Shelf. The contracts were predominantly based on the producers' cost of natural gas production. Consequently, these contracts typically did not include a mechanism for market value based price review.

The basic concept of British Gas was to contract as much long-term gas as needed for the UK market at the lowest costs. The target was to keep the over-all weighted-average cost of gas (called WACOG) reasonably low. Thus, over time, due to different price indexation factors and structures, individual prices could deviate significantly from contract to contract.

After market 'liberalization' in the UK in the 1990s, British Gas's portfolio prices were 'out of the money', i.e., higher than natural gas prices quoted at the hubs. In parallel, end user prices became hub based.

British Gas approached, in a concerted action, all its suppliers and negotiated over a two-year period, predominantly on a commercial basis, an adjustment of the supply contracts with its suppliers. The target was to reduce take-or-pay quantities and to reduce contract prices. Centrica, the successor of British Gas as natural gas supplier after the demerger of British Gas,

[47] See Sect. 9.

reported that these renegotiations resulted in exceptional charges, representing the net present value of payment that were due to the suppliers as consideration for the renegotiations, of £573 million in 1997 and £601 million in 1996. New natural gas contracts were generally agreed to, with hub indexation, but, in existing contracts, price indexations to oil product prices and inflation remained unchanged.

5.5.3 Competitive—'Modern' Markets

Modern markets for natural gas, often also referred to as traded gas markets, developed in North America in the 1980s and 1990s. This transition to modern markets started in the UK in the 1990s, and in other parts of Continental Europe approximately at the turn of the century. In other regions, e.g., in Africa, the implementation of modern markets did not start yet.

However, what is regarded to be modern for natural gas is and was common in other markets since years. Markets for other commodities like oil, some metals, or some agricultural products used the same tools by far longer than the natural gas industry had. Even further advanced are markets for, e.g., stocks or currencies. Therefore, it was not necessary to invent new tools but only to adapt tools that were already known and available elsewhere. Nevertheless, certain specifics of natural gas have to be taken into account, e.g., while evaluating options on natural gas.

Liquidity is crucial for traded markets. In order to enhance liquidity, quantities to be traded need to be geographically concentrated. This led to the development of so-called hubs,[48] i.e., a point where natural gas trades agreed upon between parties are (physically or virtually, i.e., notionally) delivered.

Further standardization and an increase of liquidity are necessary to enable the development of a successful exchange trading,[49] in addition to over-the-counter (OTC) trading.

[48]See Sect. 5.5.3.1.

5.5.3.1 Hubs

A hub describes a (physical or even virtual, sometimes also referred to as 'notional') 'location' ('space'), at which natural gas is traded, i.e., sold and purchased. Hubs have become more than a simple 'standardized delivery point'. Here also additional services are offered.

A physical hub designates a well-defined, constrained area at which several pipelines come together. In order to be accepted by the market participants, it is important that among these pipelines several 'input' and several 'output' lines exist. Otherwise, the market may suffer illiquidity.[50] A physical hub may be illustrated as shown in Fig. 5.6. Examples of physical hubs are the Henry Hub (US) or the Zeebrugge Hub (BE).

A virtual hub is not located at a specific (geographical) point, but is assumed to be any (virtual) point within a well-defined market area or pipeline network. That is, already by entering a market area or pipeline network, the natural gas is regarded to be available at such a hub. This is illustrated in Fig. 5.7. Examples of virtual hubs are the NBP (UK) or the TTF (NL).

To foster trading, hubs should fulfil the following[51]:

- transparent and simple access rules;
- (often) hub operator established as an independent entity;
- third party access to transportation system, non-discriminatory third party access (TPA) to the network often requires legislative changes to have been implemented in order to avoid hoarding of transportation capacities that may prevent competitors to enter the incumbents' markets. Such

[49]See Sect. 5.5.3.2.

[50]This has been observed, e.g., at the CEGH in Austria, near the Austrian/Slovakian border. Despite several pipelines to the markets in Austria, Germany, Hungary, Slovenia (and Croatia), and Italy, the hub's natural gas supply relied mainly on Russian natural gas supplies.

[51]These topics will be discussed in the subsequent sections on sales, transportation and storage.

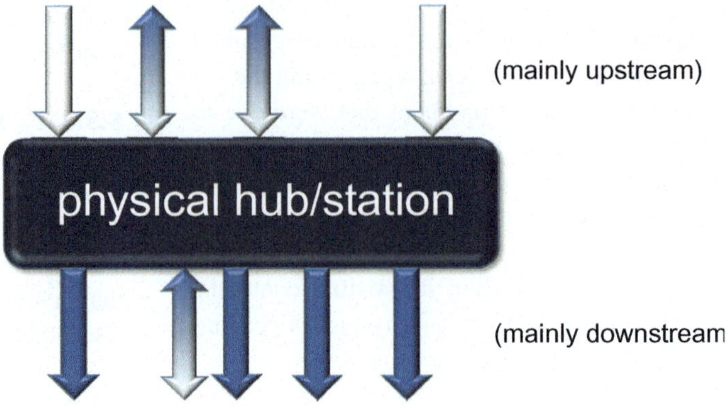

Fig. 5.6 Schematic illustration of a physical hub

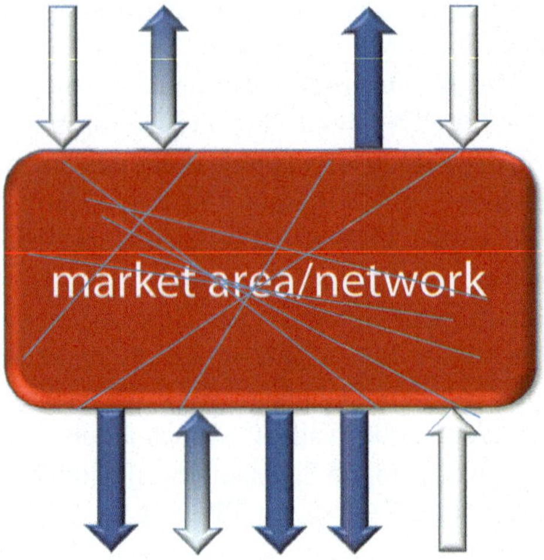

Fig. 5.7 Schematic illustration of a virtual hub

non-discriminatory access may be supplemented, *inter alia*, by 'use-it-or-lose-it' rules;

- non-discriminatory third party access to storage capacities;
- non-discriminatory treatment of market participants;
- standardized purchase and sale (trading) contracts;
- anonymity of trading for the traders;
- transparent and (as far as possible) simple pricing system for the commodity of natural gas, as well as the transportation and storage capacities;
- simplification and standardization of technical rules;
- standardized (and potentially extended) balancing rules;
- customers in the market shall be granted the free choice of supplier;
- transparent process for changing gas suppliers; and
- if not only a liquid spot market is required but also forward market: incentives for financial players to participate in the market.

When a hub is in a well-developed stage, the following phenomena will be observed:

- liquidity will increase,
  whereby market liquidity[52] shall be defined as a measure applied to a specific natural gas market indicating market participants' ability to purchase or sell natural gas without causing significant change in the price;
  Indicators for such liquidity are considered[53] to be:
  - quantity,[54] i.e., liquidity is high if a substantial quantity is traded by a large number of different[55] market participants with high churn rate;
  - costs, i.e., liquidity is high if the transaction costs and the bid-ask spread are low; and

[52]In contrast to a company's liquidity.

[53]See GasTerra (2015).

[54]Sometimes liquidity is distinguished from availability of quantities. While liquidity refers to the price effects of individual transactions in a market, the availability of quantities, despite being interdependent with liquidity, designates the mere existence of supplies regarded as sufficient in quantity by the market participants.

[55]With regard to their individual business objective and activity, e.g., upstream companies, banks, energy traders, or local distributors.

– time, i.e., liquidity is high if transactions can be performed quickly, at any time and over a long-time horizon;
- volatility[56] of prices itself becomes more volatile.

In the US, market centers and hubs evolved as a consequence of restructuring the natural gas market. At market centers and hubs, traders could receive physical services and administrative support. These services include[57]:

- transportation/wheeling: transfer of natural gas from one pipeline to another;
- hub-to-hub transfers: arranging the transfer (= virtual transportation) between two hubs;
- parking: a short-term storage service using either (fast churn) underground storage facilities, line-pack, or aboveground storage facilities;
- loaning: a short-term advance of natural gas to a trader; to be redelivered in kind. (Basically 'loaning' is a form of reversed storage, as the natural gas may be withdrawn before injection. This is also referred to as 'advancing', 'reverse parking', or 'drafting'.);
- peaking: short-term (usually less than a day and perhaps hourly) sales of natural gas;
- balancing: a short-term interruptible sales or purchase (i.e., often in conjunction with parking and loaning);
- electronic trading: providing services for anonymized trading services and data exchange between buyers and sellers; and
- administration: providing traders assistance for administrative services as, e.g., nominations.

Henry Hub

Perhaps the most prominent pipeline interconnection point is the Henry Hub in Erath, Louisiana (US). Nine pipelines interconnect at the small village, where two compressor stations operated by Sabine Pipe Line LLC manage gas flows in different directions. 'Hub' is the term used for such an interconnection point of various pipelines. Operations started in the 1950s, but the name 'Henry Hub' became famous for market participants in the natural gas industry when, in the 1990s, NYMEX started to offer standardized gas trading products with the delivery point 'Henry Hub'. Although there is still no world reference price for natural gas, NYMEX Henry Hub prices are one of the leading global price indicators.

By the way, William Henry Kittenkat is responsible for the name. He was the sponsor for Henry High School that gave the name for the Henry hamlet of Erath.

NetConnect Germany

In Germany

> *NetConnect Germany was formed as of 1 October 2008 ... Today under the umbrella of NetConnect Germany there cooperate the transmission system operators bayernets GmbH, Fluxys TENP GmbH, GRTgaz Deutschland GmbH, Open Grid Europe GmbH, terranets bw GmbH and Thyssengas GmbH and operate a common market area...*
>
> *In its role as market area manager NetConnect Germany ... handles the operational management of the market area cooperation. Its business activities involve the management of balancing groups and system balancing activities, the provision and operation of the market area's Virtual Trading Point, and the provision of data, for example settlement and balancing information.*[58]

As operator of a (virtual) hub, NetConnect Germany is allowed to charge a fee, so:

> *The German federal regulator Bundesnetzagentur (...) has provided that the market area manager may levy a fee for use of its Virtual Trading Point for periods beginning on or after 1 October 2011.*

[56]A definition of volatility is provided in Appendix D.

[57]Particular reference is made here to the US market, as not all of the services are offered in other markets as, e.g., the European ones.

[58]NetConnect Germany (2016a).

According to the regulator's decision, the market area manager is required to set the fee at a level that enables it to recover the direct and indirect costs it incurs in connection with its Virtual Trading Point. The VTP fee is set annually with effect from 1 October of each year and applies for a period of 12 months, respectively.
... From 06:00 h on 1 October 2016, the VTP fee applied by NetConnect Germany will be 0.0012 €/MWh.[59]

5.5.3.2 OTC Versus Exchanges

In general, all sorts of transactions at a hub or any other delivery point, e.g., spot deals, forward deals, or deals with options can be traded over-the-counter ('OTC'), i.e., directly between counterparties, partly supported by brokers. Apart from market risks, e.g., caused by price changes, OTC transactions inherently adhere to credit risks as at any time one of the counterparties could default.

To reduce credit risk, transactions could be made via exchanges. Exchanges use so-called clearinghouses as intermediaries that guarantee the performance of the counterparties to each individual transaction.[60] As there is no such thing as a free-lunch transaction, costs for exchange-traded deals are higher than for OTC deals. Further, only standardized products can be traded via exchanges. Hence, e.g., contract periods, contract quantities, delivery points, lot sizes, etc. are pre-defined for exchange-traded products. This standardization is mainly to allow for a higher liquidity for these products, and such liquidity is pivotal to the success of exchanges and their respective products.

Economies of scale are a very important factor in the exchanges' business, which have led to several mergers of exchanges during the last decade. Frontrunner exchanges like the New York Mercantile Exchange (NYMEX) or the International Petroleum Exchange (IPE) in London are meanwhile part of larger exchange groups. Exchanges also compete with each other, i.e., various exchanges offer the trading of products for the same hubs, e.g., NBP products might be traded not only on the Intercontinental Exchange ICE, but also on CME and PEGAS.

CME Group,[61] which includes the NYMEX, offers trading for futures and options for several North American hubs including Henry Hub. Its affiliate CME Europe provides a future trading platform for the UK, the Netherlands, Belgium, Germany, Austria, and Italy.

The Intercontinental Exchange ICE[62] provides a trading platform for spot trades, futures, and (partly) options and spreads for several hubs in North America, including Henry Hub, but also for e.g., the UK, the Netherlands, Belgium, Germany, and Italy.

PEGAS[63] is a cooperation between EEX and Powernext, and it operates all business activities of EEX and Powernext on the European gas markets. PEGAS provides a market place for the following products:

- spot products delivered in/at:
  - Germany (GASPOOL and NCG market areas),
  - Denmark (ETF market area),
  - France (PEG Nord and TRS market areas),
  - Belgium (ZEE and ZTP market areas),
  - The Netherlands (TTF market area), and
  - UK (NBP market area)
- natural gas futures delivered in/at:
  - Germany (GASPOOL and NCG market areas),
  - France (PEG Nord and TRS market areas),
  - Belgium (ZEE and ZTP market areas),
  - Italy (PSV market area),
  - The Netherlands (TTF market area), and
  - UK (NBP market area)

[59] NetConnect Germany (2016b).

[60] Clearinghouses use the so-called margining process as a means to achieve this guarantee of performance. For further reading, reference is made to, e.g., European Commodity Clearing (ECC) (2016), the central clearing unit of the European Energy Exchange EEX.

[61] See CME Group (2016).

[62] See ICE (2016).

[63] See PEGAS (2016).

In addition, PEGAS offers the opportunity to trade spread products between these market areas.[64]

Furthermore, there are several exchanges that provide trading platforms on a more regional basis. Among these are the Power Exchange Central Europe (PXE) offering natural gas trading at the Czech virtual trading point operated by OTE, a.s., and the Central Eastern European Gas Exchange (CEEGEX) offering a platform for the Hungarian market.

European Energy Exchange

The European Energy Exchange is both a company operating the exchange, as well as the exchange itself, i.e., the market place for trading energy products in general and natural gas products in particular.

As a company, the European Energy Exchange AG has the following business model:

> *EEX develops, operates and networks secure, liquid and transparent markets for energy and commodity products. Clearing and settlement of all transactions are provided by the European Commodity Clearing AG (ECC) clearing house. EEX Group pursues a "one-stop shop" business model under which the customers are offered all services from one source—from exchange trading, trade registration, clearing (i.e., physical and financial settlement) to services, such as market and transparency data as well as data reporting services.* [65]

and

> *Our core business is the operation of electronic marketplaces for energy and commodity products. ... By publishing reference prices in addition to all traded prices and volumes, we create transparency and confidence in the market. We connect various European and international markets and, in doing so, provide our participants with the greatest possible freedom of action.*[66]

5.5.4 Excursus: Trading Instruments

Trading instruments are essential tools of competitive natural gas markets. Such instruments, to be described subsequently, are:

- forwards and futures,
- swaps, and
- options.

5.5.4.1 Forwards and Futures

A forward contract is a contract between a seller and a buyer to deliver and off-take a defined quantity (of a commodity) at a delivery point during a future period (or at a future point in time). The buyer will purchase and the seller will sell the specified commodity at a price determined at the time of entering into the (forward) contract. Thus, a forward is a bilateral obligation and right.

A special case of a forward contract is a future contract. The main difference is that future contracts are exchange traded and therefore standardized. This leads to a relatively high fungibility.[67] Furthermore, credit risks are nearly eliminated because a central clearinghouse covers this risk, and the products are marked to market on a regular (usually daily) basis. On the other hand, standardization precludes tailor-made solutions.

Forwards and futures are, *inter alia*, used for hedging[68] purposes. Market participants that use forwards or futures to lock in prices for their physical production of a commodity or their physical needs of a commodity are called hedgers.

Hedging Using Futures

Suppose that, in January, a natural gas producer expects to produce in April 7200 MWh of natural gas at a constant hourly rate of 10 MW. As the producer loses money if spot prices fall, it might be

[64]With the exception of the Danish market area.

[65]European Energy Exchange AG (2016), p. 48.

[66]European Energy Exchange AG (2016), p. 43.

[67]Fungibility designates a good's interchangeability or substitutability with other similar or identical goods.

[68]A 'hedge' is an investment to reduce the (price-) risk of existing trades or assets. A 'hedge' is effectively a sort of insurance against price movements—risks and chances are reduced.

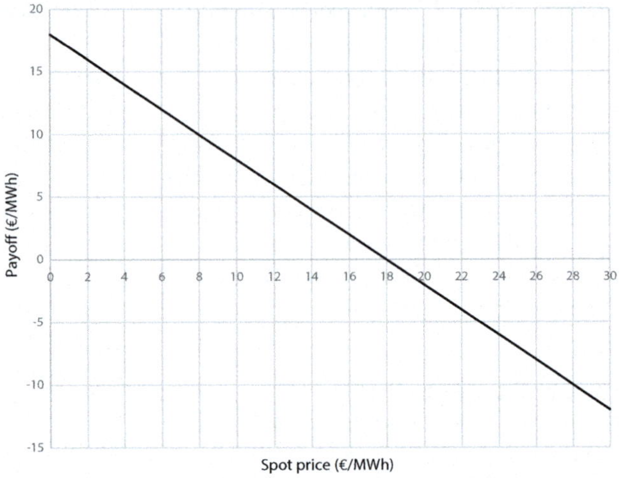

Fig. 5.8 Payoff profile of a short hedge with futures

inclined to mitigate that risk by entering into a futures transaction. In order to achieve this risk mitigation, the producer might sell futures contracts for 18 €/MWh for the expected production quantity (short hedge). Assuming that the production costs were 15 €/MWh the producer wants to achieve with this transaction a secure profit of 3 €/MWh. In April, the producer would then sell natural gas for the then applicable spot price and at the same time close out its financial futures position by buying futures. It can be shown that futures' prices converge to the spot price as the delivery period is approached, and effectively become equal once the delivery period is reached.

Assuming that the April spot price was just 12 €/MWh the producer would:

- 'pay' production costs of 15 €/MWh;
- receive 12 €/MWh for its natural gas;
- receive 18 €/MWh for its original futures position; and
- pay 12 €/MWh for closing out its futures position.

In total, this would lead to a profit of 3 €/MWh[69] for the producer.

Assuming that the April spot price was 25 €/MWh, the producer would:

- 'pay' production costs of 15 €/MWh;
- receive 25 €/MWh for its natural gas;
- receive 18 €/MWh for its original futures position; and
- pay 25 €/MWh for closing out its futures position.

Again, in total, this would lead to a profit of 3 €/MWh[70] for the producer. The payoff profile of the (pure) futures transaction is depicted in Fig. 5.8.

The example shows that a hedger, in this case the producer, achieves the same profit independent from spot price development. On the one hand, it is insured against falling prices; on the other hand, it does not participate in rising prices, and, therefore, incurs opportunity costs. This balance of insurance and opportunity costs is inherent in any hedging transaction using (plain vanilla) futures or forwards.

5.5.4.2 Swaps

In general, swaps are financial or physical transactions in which two parties agree to

[69](12 + 18 − 12 €/MWh minus 15 €/MWh production costs.

[70](25 + 18 − 25) €/MWh minus 15 €/MWh production costs.

exchange various cash-flow streams or (physical) natural gas or other commodities. Thus, a swap is —similar to forwards and futures—a bilateral obligation and right.

The transaction can consist of various payments made by the buyer or received by the seller, such as periodic payments, upfront payments, or termination payments. A periodic payment is generally based upon an interest rate or index factor multiplied by a (notional) quantity. An upfront payment can be an agreed-upon lump sum payment, paid to enter into a swap contract. A termination payment is a payment made or received to terminate or assign the rights of a swap contract.

A swap may:

- be purely financial;
- involve an exchange of natural gas for other commodities;
- be physical, i.e., involve exchange of natural gas either
  - in time, or
  - in place;

or be made up of any combination of the above.

5.5.4.2.1 Financial Swaps

Typically, buyer and seller agreeing on a so-called 'financial swap' do not exchange natural gas, but rather defined amounts of money calculated based on natural gas prices and quantities. The swap transaction is made based on a notional quantity.

> **Financial Swaps**
> The seller of a swap receives a fixed price for a perceived quantity of natural gas, e.g., 20 €/MWh for a quantity of 10 million kWh = 10,000 MWh, i.e., the seller receives 200,000 €. In return, the seller pays to the buyer an index price for the same quantity during a defined period. This price is, of course, not known at the time of conclusion of the swap contract, but forecasted differently by the parties. The seller expects the indexed price to be lower than the fixed price (including interest rates), and the buyer expects the indexed price to be higher than the fixed price (including interest rates).

The swap explained in the preceding example is commonly referred to as a 'fixed for floating price swap'. Alternatively, the fixed price leg of the swap can be replaced by another floating price. In such a case, the swap is referred to as 'floating for floating price swap'.

In case the floating prices reflect prices at two different hubs, the swap is named 'basis swap'.[71] Basis swaps are one of the most common types of swaps traded on the market. They are used to hedge the price difference between natural gas prices at different locations (yet, not to eliminate such differences).

5.5.4.2.2 Inter-commodity Swaps

Purchasing for a period at a location one commodity and selling another commodity during the same period (typically) at the same location (or vice versa), equals an exchange of natural gas for another commodity. The other commodity is typically associated to the first. Crude oil as the first and gasoil as the second commodity serve as likely examples.

Inter-commodity swaps may be performed between a buyer and a seller, but equally, in liquid traded markets, by one party with anonymous market participants. Essentially, parties performing such swaps intend to monetize or hedge changes of the prices between different commodities, the so-called inter-commodity spread.

5.5.4.2.3 Physical Swaps

5.5.4.2.3.1 In Time

For a physical swap in time (time swap),[72] natural gas is delivered by the buyer of the swap to the seller of the swap at one point in time and

[71]In many cases, a fixed amount is agreed to be added to (or deducted from) one of the two floating prices.

[72]See Sect. 8.6.1 for further explanation of storage capacity reservations.

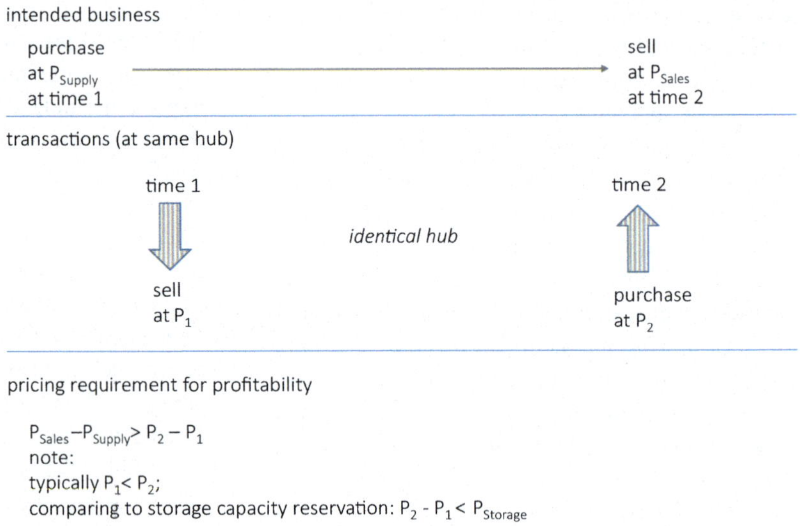

Fig. 5.9 Time swap—business model, transactions, and pricing requirements

redelivered by the seller of the swap to the buyer of the swap at another point in time. To simplify, it is assumed that such deliveries take place at the same location. If, however, a liquid traded market exists, a party intending to perform a time swap can arrange it directly by selling and purchasing in the market, i.e., the liquid market assumes the role of the swap partner. This is depicted in Fig. 5.9.

The intended transaction, for a given quantity, consists of purchasing (or having to purchase) a quantity of natural gas at 'time 1', which is needed at 'time 2' only.[73] To accomplish this, at 'time 1', the party sells in the liquid market at a price P_1 this quantity of natural gas already purchased from a seller at a price P_{Supply}. At 'time 2' the same party purchases, in the same liquid market at a price P_2, the same quantity of natural gas to be sold to a customer at a price P_{Sales}.

Only if $P_{Supply} - P_1 < P_{Sales} - P_2$, or $P_{Supply} + P_2 < P_{Sales} + P_1$,

i.e., if the sales margin at 'time 2' exceeds supply costs incurred at 'time 1' or, to refer to the second equation, the total sales prices, (P_{Sales} + P_1), exceed supply prices, (P_{Supply} + P_2), the party has an incentive to conduct the transaction.

In practice, such a deal would be concluded *ex ante*. That is, at or until a given point in time before 'time 1', the party endeavoring to conclude the time swap would conclude:

- the purchase at P_{Supply} at 'time 1',
- the sales at P_{Sales} at 'time 2',
- the sales at P_1 at 'time 1', and
- the purchase at P_2 at 'time 2'.

The prices P_1 and P_2 are the forward (or future) quotations for deliveries at 'time 1' and 'time 2', quoted at a hub at any applicable point in time before the beginning of 'time 1'. If P_1 refers to natural gas prices for deliveries during summer and P_2 refers to natural gas prices for deliveries during the following winter, the difference of prices, i.e., ($P_2 - P_1$), is referred to as 'summer-winter spread'. Hence, in case the summer-winter spread decreases, it will become more favorable for market participants to perform such a time swap.

A time swap is often interpreted as a tool to replicate reservation of storage capacity[74]: Any buyer of storage capacity uses such capacity to 'dispose' of, i.e., store or inject, natural gas purchased in addition to the prevailing immediate requirements in summer, i.e., at 'time 1', and 'retrieve', i.e., withdraw, such natural gas in winter, i.e., at 'time 2'. As, from this buyer's perspective, 'disposing of' natural gas, i.e., injecting it

[73]Note that both 'time 1' and 'time 2' typically designate periods, not points in time.

[74]See, e.g., Burger et al. (2007), p. 83 et seq.

into storage, is physically equivalent to selling and 'retrieving' is physically equivalent to purchasing, at least these physical operations of a time swap can be considered equivalent to the physical operations performed by buyers of storage capacity.

The buyer of storage capacity will compare, at each point in time, the price to be paid for the reservation of such storage capacity with the difference of forward (or future) prices. Therefore, this difference ($P_2 - P_1$), the summer-winter spread, is considered a first, yet incomplete, approximation of the value of storage. It is often referred to as the 'intrinsic value' of storage capacity. In a liquid traded market, this value can be locked-in at any point in time. A time swap can be concluded at each point in time at the prevailing summer-winter spread—replicating (or replacing) a storage capacity reservation, yet a storage capacity reservation with restrictions.

Despite such partial equivalence, there is an important difference between a storage capacity reservation and a time swap. A storage capacity reservation (contract) grants the buyer the right, but not the obligation, to utilize the capacity reserved. It is a (real) option. A time swap, in contrast, implies rights and obligations. Both, the sales at 'time 1' as well as the purchase at 'time 2' are unconditional, i.e., these contracts have to be fulfilled at any point in time, the natural gas has to be delivered, at 'time 1', and off-taken, at 'time 2', during each period in 'time 1' and 'time 2', without any flexibility. The time swap does not include any optionality, for either of the parties involved. Hence, the buyer of storage capacity may utilize the optionality to extract a further value. This additional value is determined, *inter alia*, by the volatility of the forward (or future) quotations, P_1 and P_2, before the beginning of 'time 1'. It is commonly referred to as the 'time' value of storage capacity.[75] This time value cannot be replicated by a time swap.

[75]The exact determination of the time value requires application of option theory, see, e.g., Hull (2014). Also, the time value is sometimes, wrongly, referred to as extrinsic value.

5.5.4.2.3.2 In Place (location)

In analogy to the time swap,[76] (partially) replicating storage services, a location swap can be used as (partial) replication (or replacement) of transportation services.

For a physical swap in place (location swap[77]), natural gas is delivered by the buyer of the swap to the seller of the swap at one point (location) and redelivered by the seller of the swap to the buyer of the swap at another point (location). To simplify, it is assumed that such deliveries take place during the same period and that the transactions take place at two liquid hubs. If, however, a liquid traded market exists, a party intending to conduct a location swap can arrange it directly by selling and purchasing it at these hubs, i.e., the liquid market hubs assume the role of the swap partner. This is depicted in Fig. 5.10.

The intended business, for a given quantity, consists of purchasing (or having to purchase) a quantity of natural gas at 'hub 1', which is needed at 'hub 2' only, during one period. To this, at 'hub 1', the party sells at a price P_1 this quantity of natural gas purchased from a seller at a price P_{Supply}. At 'hub 2', the same party purchases during the same period at a price P_2 the same quantity of natural gas to be sold to a customer at a price P_{Sales}.

Only if $P_{Supply} - P_1 < P_{Sales} - P_2$, or $P_{Supply} + P_2 < P_{Sales} + P_1$,

i.e., if the sales margin at 'hub 2' exceeds supply costs incurred at 'hub 1', or, to refer to the second equation, the total sales prices, ($P_{Sales} + P_1$), exceed supply prices, ($P_{Supply} + P_2$), the party has an incentive to conduct the transaction.

In practice, such a deal would be concluded *ex ante*. That is, at or until the period of delivery and off-take, the party endeavoring to conclude the location swap would conclude:

- the purchase at P_{Supply} at 'hub 1',
- the sales at P_{Sales} at 'hub 2',
- the sales at P_1 at 'hub 1', and
- the purchase at P_2 at 'hub 2'.

[76]See Sect. 7.5 for further explanation of transportation capacity reservations.

[77]Also referred to as 'buy/buy back deal'.

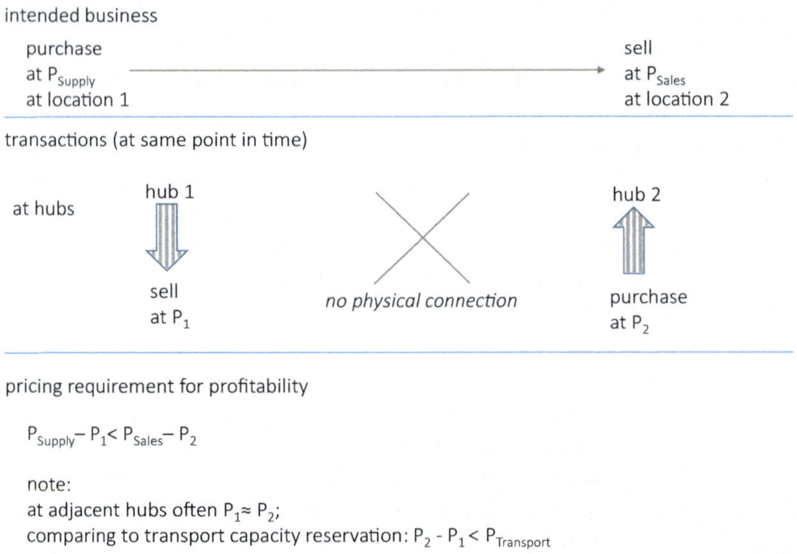

Fig. 5.10 Location swap—business model, transactions, and pricing requirements

The prices P_1 and P_2 are the forward (or future) quotations for deliveries at 'hub 1' or 'hub 2', respectively, at any applicable point in time before the beginning of the delivery period. The difference of prices, i.e., $(P_2 - P_1)$, is referred to as 'location spread' between 'hub 1' and 'hub 2'. Hence, in case the location spread decreases, it will become more favorable for market participants to perform such a location swap.

A location swap is often interpreted as tool to replicate reservation of transportation capacity[78]: Any buyer of transportation capacity uses such capacity to 'dispose' of, i.e., deliver, natural gas purchased in addition to the prevailing immediate requirements at a location, i.e., at 'hub 1', and 'retrieve', i.e., receive, such natural gas at another location, i.e., at 'hub 2'. As, from this buyer's perspective, 'disposing of' natural gas, i.e., delivering it at the delivery point for transportation, is physically equivalent to selling and 'retrieving' is physically equivalent to purchasing, at least these physical operations of a location swap can be considered equivalent to the physical operations performed by buyers of transportation capacity.

The buyer of transportation capacity will compare, at each point in time, the price to be paid for the reservation of such transportation capacity with the difference of forward (or future) prices at the two locations. Therefore, this difference $(P_2 - P_1)$, the location spread, is considered a first, yet incomplete, approximation of the value of transportation. It is often referred to as the 'intrinsic value' of transportation capacity. In a liquid traded market, this value can be locked-in at any point in time. A location swap can be concluded at each point in time at the prevailing location spread—replicating (or replacing) a transportation capacity reservation, yet a transportation capacity reservation with restrictions.

Despite such partial equivalence, there is an important difference between a transportation capacity reservation and a location swap. A transportation capacity reservation (contract) grants the buyer the right, but not the obligation, to utilize the capacity reserved.[79] It is a (real) option. A location swap, in contrast, implies rights and obligations. Both, the sales at 'hub 1' as well as the purchase at 'hub 2' are unconditional, i.e., these contracts have to be fulfilled at any point in time, the natural gas has to be

[78]Note that, contrary to other commodities, transportation of natural gas is perceived to take place immediately, i.e., by entering into a swap transaction, no advantages in timing compared to a transportation contract can be accomplished by the buyer.

[79]This statement is correct only under the assumption that the so-called 'use-it-or-lose-it' rule does not apply.

delivered, at 'hub 1', and off-taken, at 'hub 2', during each (sub-) period of the period agreed, without any flexibility. The location swap does not include any optionality, for neither of the parties involved. Hence, the buyer of transportation capacity may utilize the optionality to extract a further value. This additional value is determined, *inter alia*, by the volatility of the forward (or future) quotations, P_1 and P_2, before the beginning of period of delivery. It is commonly referred to as the 'time' value of transportation capacity.[80] This time value cannot be replicated by a location swap.

5.5.4.3 Options[81]

5.5.4.3.1 Plain-Vanilla Options

An option contract is a contract between a seller and a buyer providing the right (but not the obligation) to purchase or sell a commodity (also referred to as 'underlying') at a delivery point during a future period (or at a future point in time) at a price determined at the time of entering into the (forward) contract. The price in the contract is referred to as strike price or exercise price. Options are typically either European style and can be exercised only at the exercise date or they are American-style options, which can be exercised at any time until the exercise date. The fact that options give the right to do something, but not the obligation to do so, distinguishes them from forward or futures contracts. There are two basic types of options:

- call options, and
- put options.

A call option on natural gas gives its holder (i.e., the buyer of the option) the right, but not the obligation, to purchase a specified quantity of natural gas at or by a certain date ('the exercise date') for a specified price from the writer of the option (i.e., the seller of the option).

[80]The exact determination of the time value requires application of option theory, see, e.g., Hull (2014).

[81]Option-like instruments such as caps, floors, and collars are described in Sect. 6.3.2.4.

A put option on natural gas gives its holder (i.e., the buyer of the option) the right, but not the obligation, to sell a specified quantity of natural gas at or by a certain date ('the exercise date') for a specified price to the writer of the option (i.e., the seller of the option).

Option markets, whether OTC or exchange traded, know four types of participants:

- buyers of call options ('Long Call'),
- sellers of call options ('Short Call'),
- buyers of put options ('Long Put'), and
- sellers of put options ('Short Put');

As options provide a right, but not an obligation, to do something, they always have a positive (commercial) value. For this reason, the buyer of an option pays a so-called option premium to the seller when entering into the option contract.

Hedging Using Options

Suppose that in January a natural gas producer expects to produce in April 7200 MWh of natural gas at a constant hourly rate of 10 MW. As the producer loses money if spot prices fall, it might be inclined to mitigate this risk by entering into an options transaction. In order to achieve this risk mitigation, the producer might buy physical put options with an exercise price of 18 €/MWh for the expected production quantity (long put). The option premium shall be 1 €/MWh. Assuming that the production costs were 15 €/MWh, the producer wants to achieve with this transaction a secure profit of 2 €/MWh.

In April, the producer would then sell natural gas for the then applicable spot price and, at the same time, close out its financial futures position by buying futures. It can be shown that futures' prices converge to the spot price as the delivery period is approached, and effectively become equal once the delivery period is reached.

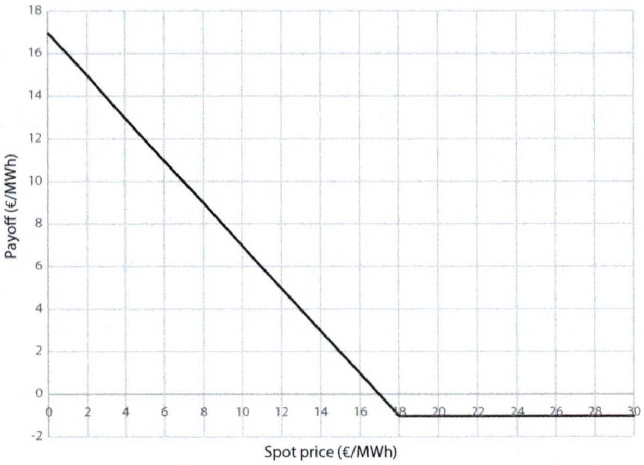

Fig. 5.11 Payoff profile of a long put

Assuming that the April spot price was just 12 €/MWh, the producer would:

- exercise its put option and receive 18 €/MWh; and
- pay 1 €/MWh as option premium.

In total, this would lead to a profit of 2 €/MWh[82] for the producer.

Assuming that the April spot price was 25 €/MWh, the producer would not exercise his option and

- receive 25 €/MWh for its natural gas; and
- pay 1 €/MWh as option premium.

This is depicted in Fig. 5.11.

As with a futures-hedge the producer is hedged against losses, i.e., is insured against falling prices. However, different to a hedge using futures, it does participate in rising prices. The producer does not incur opportunity costs. Instead, it bears the costs for the option premium, independent of whether spot prices rise or fall.

Figure 5.12 shows payoff profiles for calls and puts, as well in their respective short position as long position.

For a call option, the option is regarded to be 'in the money' if the price of the underlying is higher than the exercise price. It is regarded 'at the money', if the price of the underlying equals the exercise price; it is regarded to be 'out of the money', if the price of the underlying is lower than the exercise price.

For a put option, the option is regarded to be 'in the money', if the price of the underlying is lower than the exercise price. It is regarded 'at the money', if the price of the underlying equals the exercise price; it is regarded to be 'out of the money', if the price of the underlying is higher than the exercise price.

There are always two sides to every option contract. The payoff of the holder of the option (long position) is always the opposite of the payoff of the writer of the respective option (short position). This is true independently of whether call or put. The profit the holder receives when exercising the option is the loss of the writer. The premium the holder has to pay is the premium the writer receives.

The size of the premium or better the value of the option depends on several factors:

- spot price of the underlying:
  Ceteris paribus, call options become more valuable as spot prices increase. This is obvious as the payoff (when the option is exercised) is defined as the difference

[82](18 − 1) €/MWh minus 15 €/MWh production costs.

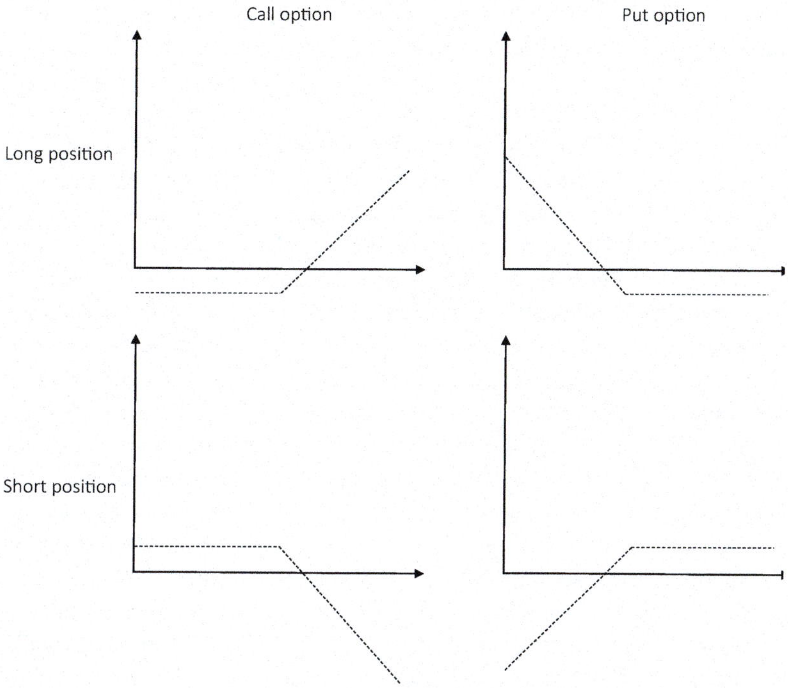

Fig. 5.12 Payoff profiles of calls and puts

between spot price and exercise price.[83] Put options behave the opposite way;

- exercise price of the option:
  Ceteris paribus, call options are less valuable when exercise prices are high. Again, this is obvious as the payoff (when the option is exercised) is defined as the difference between spot price and exercise price. Put options behave the opposite way;
- time to expiration:
  At least American-style options, whether call or put, increase in value with a long time to expiration, as the likelihood that an option comes into the money increases with long maturity, i.e., time to expiration. If the time to expiration is nearly zero, the overall value of the option is very near to the intrinsic value as the time remaining during which the intrinsic value could increase diminishes;
- volatility of the underlying:
  All options whether call or put increase in value with increasing volatility. When volatility, considered as a measure how uncertain future spot price movements are, is high, the likelihood that an option comes into the money is also high, which, again, leads to higher overall value of the option (in contrast, the option on a commodity with the certain expectation of a permanently constant price in the future, i.e., expected volatility being equal to zero, would be worthless); and
- risk-free interest rate:
  Interest rates influence not only the value of an option but also the value of the underlying itself. *Ceteris paribus*, the value of a call increases with increasing interest rates; a put option behaves the opposite way.

In 1973, Fischer Black and Myron Scholes developed a closed-form model of how to price stock options based on several restrictive assumptions.[84] Similar restrictions apply for the Black model, developed in 1976, which enables pricing of European-style options on futures.

In the basic form, the options described so far in this section are also referred to as

[83] The non-negative value of these two prices is also referred to as intrinsic value.

[84] For further reading, reference is made to Hull (2014).

'plain-vanilla' options. Other than exotic options, they do not have complex features.

5.5.4.3.2 Exotic Options

All options which depart from standard (plain vanilla) features regarding payoff definition, exercise date, or underlying can be regarded as exotic options. They are quite common in the natural gas business as the options are traditionally embedded in contracts for, e.g., sale and purchase or storage of natural gas. Within this book, it is impossible to cover all exotic options developed so far, let alone valuation models for them.[85] Therefore, only the most important ones are described in more detail. Among those not described in detail belong:

- Bermuda options: an option that can be exercised only on more than one predetermined dates[86] during a period;
- Perpetual American options: options without a maturity date;
- Ladder options: an option with a payoff depending on whether certain price levels, so-called 'rungs', have been reached during the lifetime of the option;
- Basket options: options with a payoff depending on the prices of a 'basket of commodities';
- Compound options: options on options, i.e., the holder has the right but not the obligation to buy another option;
- Digital options[87]: options offering a predetermined fixed-payoff amount; and
- Swaptions: options to enter into a swap agreement.[88]

5.5.4.3.2.1 Path-Dependent Options

These options offer a payoff structure that is not (only) dependent on the difference between exercise price and the price of the underlying at a certain date.

'Asian options', for example, offer a payoff that depends on the average of the underlying price during a specified period. Asian options tend to be cheaper than respective European- or American-style options, since the volatility of an averaged price is usually lower than the volatility of the price itself.

Other than Asian options where a fixed exercise price is compared with an average of prices of the underlying, 'average strike options' compare the price of the underlying with an exercise price that is calculated as an average of prices during a certain period within the maturity of the option contract.

'Lookback options' allow for the purchase or sale of the underlying commodity at the best price that occurs during the life of the option.

'Barrier options' (also known as 'knock-in' or 'knock-out' options) are de-activated or activated if certain price levels are reached during the life of the option. For calls as well as for puts, four different combinations of price and event can be classified:

- up and out: price rises and option is de-activated,
- up and in: price rises and option is activated,
- down and out: price falls and option is de-activated,
- down and in: price falls and option is activated.

5.5.4.3.2.2 Spread Options

As the name suggests, multiple commodity options offer a payoff that depends on the spread between underlying prices. Most common are spread options based on prices of:

- natural gas at two different locations but at the same point in time, i.e., a location spread option;
- natural gas at two different points in time but at the same location, i.e., a calendar spread option;
- natural gas and electricity, i.e., a spark spread option.

---

[85] For most of them, no closed-form solution is available. For further reading, reference is made to Kaminski et al. (1999) and Mack (2014).

[86] Bermuda options are effectively a combination of European- and American-style options.

[87] Sometimes also referred to as binary options.

[88] See also Sect. 5.5.4.2.

The spread options described previously show similarities to physical assets within the natural gas industry. A pipeline is similar to a location-spread option, a natural gas storage is similar to a calendar-spread option, and a power plant is similar to a spark-spread option. These similarities explain why these physical assets are from time to time also referred to as real options.

Pricing spread options is challenging, e.g., because a price spread can have negative values, a phenomenon that normally[89] does not occur for the prices of the underlying itself. Even more challenging is the pricing of real options as many physical restrictions have to be taken into account.[90]

A special case of a spread option is an option to exchange one asset for another. This option gives the holder the right to choose the better of two assets. One example is the right of a customer to pay for the natural gas purchased the cheaper of two prices (e.g., a natural gas indexed price or an oil product indexed price). Such options may also be restricted with regard to timing and other contractual conditions.

5.5.4.3.2.3 Embedded Options

Traditional agreements for natural gas sales,[91] transportation,[92] or storage[93] usually do contain a combination of exotic options. Such options are embedded in the agreements.

Take-or-Pay as an Option

Consider a traditional long-term take-or-pay contract with a price indexation to the price developments of oil products. To name just a few option components:

- natural gas prices depend on the development of oil product prices. Hence, the exercise price is not constant but dependent on oil product prices. The contract contains a multi-commodity component;
- the buyer has the right but not the obligation to off-take, e.g., a daily quantity within certain restrictions, which is a call option[94];
- the buyer holds a distinct call option for each individual day in the future, which gives the contract a so-called 'forward start' component, which complicates pricing; and
- the call options for future days are restricted by how often the daily options have been exercised in the past, giving the contract a component of path dependency.

This example shows that traditional gas sales agreements do contain a multitude of embedded exotic options. Contracts like these do have a long history, and most participants did not see the option characteristics in the past. On the other hand, not valuing options in a liquid traded market implies potential future losses. In addition, without liquid natural gas markets available, it is not only difficult to fully use the respective options, let alone to evaluate them properly.[95]

References

Burger M, Graeber B, Schindlmayr G (2007) Managing energy risk—an integrated view on power and other energy markets. Chichester

Corbeau A-S, Volk D, Sinton J, Jiang J, Ping J, Teng T, Boshu L, Fen Y (2012): Gas pricing and regulation: China's challenges and IEA experience, pp 1–119. Available at: http://www.iea.org/

[89]Consciously neglecting the fact that recently negative electricity prices existed in the German electricity market. During a few windy and sunny hours, subsidized electricity production from renewables (wind and photovoltaics) pushed electricity spot prices down below zero.

[90]For further reading, see, e.g., He (2007), Secomandi and Seppi (2014), Zhou (2010).

[91]See also Sect. 6.3.

[92]See also Sect. 7.6.

[93]See also Sect. 8.6.

[94]In this case; also named a swing option.

[95]Even relatively simple option pricing models like Black's require as a precondition for their application, *inter alia*, the existence of liquid traded markets.

publications/freepublications/publication/ChinaGas Report_Final_WEB.pdf
CME Group (2016) Homepage. Available at: http://www.cmegroup.com/trading/energy/
Energie-Control Austria (2016): Security of supply. Available at: https://www.e-control.at/businesses/natural-gas/security-of-supply
European Commodity Clearing (ECC) (2016) Margining. Available at: https://www.ecc.de/ecc-en/risk-management/margining
European Energy Exchange AG (2016) Growth in new dimensions—annual report 2015, Leipzig. Available at: http://www.eex.com/download/en/16500
European Parliament and the Council of the European Union (1994) Directive 94/22/EC of the European Parliament and of the Council of 30 May 1994 on the conditions for granting and using authorizations for the prospection, exploration and production of hydrocarbons. Available at: http://eur-lex.europa.eu/LexUriServ/LexUriServ.do?uri=CELEX:31994L0022:EN:HTML
European Parliament and the Council of the European Union (1998) Directive 98/30/ECof the European Parliament and of the Council of 22 June 1998 concerning common rules for the internal market in natural gas. Available at: http://eur-lex.europa.eu/legal-content/EN/TXT/PDF/?uri=CELEX:31998L0030&from=EN
European Parliament and the Council of the European Union (2003): Directive 2003/55/EC OF the European Parliament and of the Council of 26 June 2003 concerning common rules for the internal market in natural gas and repealing Directive 98/30/EC, Official Journal of the European Union. Available at: http://eur-lex.europa.eu/legal-content/EN/TXT/PDF/?uri=CELEX:32003L0055&from=GA
European Parliament and the Council of the European Union (2009a) Directive 2009/73/EC of the European Parliament and of the Council of 13 July 2009 concerning common rules for the internal market in natural gas and repealing Directive 2003/55/EC. Official Journal of the European Union, L211, pp 94–136. Available at: http://eur-lex.europa.eu/LexUriServ/LexUriServ.do?uri=OJ:L:2009:211:0094:0136:en:PDF
European Parliament and the Council of the European Union (2009b) Regulation (EC) No 715/2009 of the European Parliament and of the Council of 13 July 2009 on conditions for access to the natural gas transmission networks and repealing Regulation (EC) No 1775/2005, European Union: Official Journal of the European Union. Available at: http://eur-lex.europa.eu/legal-content/EN/TXT/PDF/?uri=CELEX:32009R0715&from=en
European Parliament and the Council of the European Union (2010): Regulation (EU) No 994/2010 of the European Parliament and of the council of 20 October 2010 concerning measures to safeguard security of gas supply and repealing Council Directive 2004/67/EC, Official Journal of the European Union. Available at: http://eur-lex.europa.eu/LexUriServ/LexUriServ.do?uri=OJ:L:2010:295:0001:0022:EN:PDF
Federal Energy Regulatory Commission (2016) What FERC does. Available at: http://www.ferc.gov/about/ferc-does.asp
GasTerra (2015) What is liquidity and how do you measure that? Available at: http://www.gasterra.nl/en/news/what-is-liquidity-and-how-do-you-measure-that
He Y (2007) Real options in the energy markets. Available at: http://doc.utwente.nl/58482/1/thesis_He.pdf
Hull JC (2014) Options, futures, and other derivatives, 9th edn. Prentice Hall, USA
ICE (2016) Homepage. Available at: https://www.theice.com/index
International Energy Agency (IEA) (2014) Energy Supply Security 2014—Part 1. Available at: https://www.iea.org/media/freepublications/security/EnergySupplySecurity2014_PART1.pdf
Kaminski V, Gibner S, Pinnamaneni K (1999) Energy exotic options. In: Managing energy price risk, pp 57–94. Risk Books
Mack I (2014) Energy trading and risk management. Available at: http://www.scopus.com/inward/record.url?eid=2-s2.0-84927028907&partnerID=tZOtx3y1
Naturalgas.org (2013b) The History of Regulation. Available at: http://naturalgas.org/regulation/history/
NetConnect Germany (2016a) Profile of NetConnect Germany. Available at: https://www.net-connect-germany.de/en-gb/About-us/Company-Profile
NetConnect Germany (2016b): Publication of the VTP fee. Available at: https://www.net-connect-germany.de/en-gb/Transparency-information/Prices/VTP-Fee
PEGAS (2016) Homepage. Available at: http://www.pegas-trading.com/en/
Pindyck R, Rubinfeld D (2012) Microeconomics, 8th edn. The Pearson Series in Economics
Pustišek A (2009) Experiences of the gas crisis—a view on Security of Supply. In: Siófok: E.ON Hungary. Available at: http://2009.gazkonferencia.hu/dl/share/mp_gk_pustisek_andrej_ENG.pdf
Secomandi N, Seppi DJ (2014) Real options and merchant operations of energy and other commodities, pp 1–169
Stern J, Honoré A (2009) The future of gas supply in Europe. In: Bausch A, Schwenker B (eds) Handbook utility management. Berlin, Heidelberg
Zhou J (2010) Real options valuation in energy markets

Sales Along the Value Chain

6

Summary

- Players in the natural gas (trading) market are either producers, natural gas traders, banks, utilities acting along the whole value chain, or end users.
- Supply and demand of natural gas depend on several 'natural' factors like geological and geographical conditions or weather conditions, but also on economic activity or political and regulatory decisions taken.
- A commodity sales contract is a binding agreement to purchase and sell defined quantities of natural gas during defined periods at defined locations at defined prices.
- As a buyer of natural gas often lacks certainty about his requirements, it will require in each period an option to take natural gas within limits either from a delivery contract or, alternatively, e.g., from a storage contract. The options for all periods, including their interdependence, are referred to as 'flexibility'. A natural gas delivery often includes commodity and flexibility.
- Agreements for sales and purchases of natural gas transported in pipelines ('GSA') have many stipulations in common with agreements for sales and purchases of natural gas transported in the form of LNG ('LNG SPA').
- Basic terms to describe quantity and flexibility provisions in GSAs are, e.g., Total Contract Quantity (TCQ), Annual Contract Quantity (ACQ), Daily Contract Quantity (DCQ), Hourly Contract Quantity (HCQ), and their respective norm, minimum, and maximum values.
- In LNG SPAs, parties agree on a Contract Quantity as well as a delivery schedule in which details, e.g., regarding the vessel used and its capacity and the arrival time at the port of destination, are agreed upon.
- In a natural gas sales and purchase contract, the price is one of the main elements. Commonly, applied pricing models are: indexed pricing, gas-on-gas pricing, and regulated pricing.
- Not only the commodity delivered but also the capacity, partly corresponding to the flexibility, provided might be priced.
- Other main issues to be agreed upon in a GSA or LNG SPA are: term, hardship, delivery point, quality, and nominations.
- To facilitate the negotiation process for GSAs or LNG SPAs, many standard contracts have been developed.

A. Pustišek and M. Karasz, *Natural Gas: A Commercial Perspective*,
DOI 10.1007/978-3-319-53249-3_6

6.1 Players

The most 'natural' participants in the market for selling and purchasing natural gas are the producers[1] and end users of natural gas. Different than in the past when natural gas was (dis-) regarded by many producers as an (undesirable) by-product of crude oil production, in today's developed natural gas markets, natural gas has become an important revenue source on its own, and producers, obviously, intend to achieve natural gas prices as high as possible.[2] Not surprisingly, end users that use natural gas, e.g., for heating purposes or within an industrial process, want the opposite.

Instead of producing natural gas, producers might either leave the gas in the ground, which, of course, is financially challenging once investments have been made, or try to re-inject natural gas into oil reservoirs in order to enhance liquid hydrocarbon production. However, this alternative is restricted technically, geologically, and commercially.

Some end users have the possibility to use other fuels like gasoil[3] instead of natural gas. In order to do so, the end user does not need only physical access to alternative fuels but also technical facilities (e.g., dual fuel burners). As many end users do not have this possibility, at least in the short run, they are so-called 'captive customers'.

In most cases, producers and end users of natural gas do not have a direct contractual relationship. Instead, many intermediaries, i.e., companies that purchase natural gas from one party to sell it to another party, participate in the market. All these intermediaries are—in the widest sense—traders. Dependent on the prevailing regulatory environment, these traders might also be active in other business segments of the natural gas value chain, e.g., as operator of natural gas transportation or storage systems. In a market environment requiring unbundling of transportation, storage, and trading, these intermediaries, supplying (directly or indirectly) end users, participate in the transportation and storage markets as customers of (unbundled) transportation to bring natural gas from one location to another, and might use storage capacities to structure it.[4] Intermediaries as profit maximizing companies, intend to maximize their margins between sales and purchase prices. Hence, other than producers or end users, they are—in general—not necessarily interested in the absolute price level, but only in price differences between their sales and purchase prices.[5]

In addition to physical players, also pure 'financial' players participate in highly developed natural gas markets, as, e.g., the US. Other than physical players, which are part of the natural gas value chain by either producing, or consuming or transporting (or having transported) or storing (or having stored) natural gas, pure financial players try to avoid 'physical' deliveries of natural gas by trading financial products[6] or forwards, which they net, i.e., purchase or sell, by opposing transactions before the natural gas is physically delivered.

To the group of physical intermediaries belong importers of natural gas and local or regional distribution companies that were part of the natural gas business also before liberalization.[7] Trading companies, depending on their strategies, may hold physical capacities in storage or transportation and may supply natural gas to larger end users. Some can, therefore, be seen as physical intermediaries and some like most of the banks or hedge funds belong to the group of financial players.

[1]Effectively, everything that is said in this section for producers of natural gas also applies to producers of anthropogenic gas like biogas.

[2]Of course, the price achievable depends on many factors, e.g., whether the producer is able to offer flexibility. See also Sect. 6.3.

[3]Or, to a limited extent, other feedstocks.

[4]However, in some markets, it is the customer not the trader who is responsible to book transportation capacity.

[5]As a matter of fact, the margin achievable depends on many factors, e.g., the products traded, e.g., baseload or flexible deliveries, firm or interruptible deliveries etc., and the level of risk that the market participant is prepared to take.

[6]See Sect. 5.5.4.

[7]Sometimes referred to as 'incumbents'.

6.2 Supply and Demand

Supply and demand of natural gas are not fully independent 'variables', influenced by market participants only. Both depend on a number of exogenous variables.

Supply of natural gas depends primarily on:

- the availability of reserves (ideally in the vicinity of consumption or at locations at which economically viable transportation to region(s) of demand can be set up);
- (advances in) exploration, development, and production technology;
- availability of exploration, development, and production services (influencing the prices to be paid for such services);
- the possibility of transporting and structuring the produced natural gas quantities according to the demand of the end customers and availability of companies (network, storage, and sales) performing these services;
- (stable) political, regulatory, and legal framework;
- prices of natural gas and the expectation thereof; depending, *inter alia*, on prices of alternative fuels, primarily crude oil and oil products, and the expectation thereof;
- financing possibilities, including the interest rate and guarantee requirements; and
- competitors' behavior and cooperation (e.g., in consortia).

Demand for natural gas depends primarily on:

- availability (in the respective region, primarily determined by the existence of supplies, a network, and a sales organization);
- availability of a natural gas infrastructure;
- the time or period, e.g., demand will differ depending on the weekday, the month, vacation period, holidays etc.;
- weather, in particular, temperature;
- (macro-)economic situation (influencing, e.g., the demand for goods produced by industrial end users, which, in turn, demand natural gas for process heat or feedstock);
- alternative fuels' prices (irrespective of the actual pricing (formulae) of natural gas), as long as these fuels compete with natural gas;
- (stable) political, regulatory, and legal framework and interventions; and
- (perceived) security of supply.

In most of today's liquid traded markets the balance between supply and demand is—when looking at the prompt market only—driven by the activities, capacities, and needs of producers and end users. However, the futures markets are additionally influenced by natural gas traders, banks, and utilities and their respective expectations.

Nonetheless, the pure commodity natural gas can be 'bundled' (or 'assembled') into different 'products', distinguished by delivery features, such as firmness, flexibility, or quantity as a function of time. The governing concepts for 'assembling' such products are either buyers' requirements or sellers' technical or commercial restrictions.

6.2.1 Demand

As described in Sect. 4.3, natural gas is used for different purposes. If used for heating, the demand for natural gas will be primarily (outside) temperature dependent. As average outside temperatures are (at least usually), colder in winter than in summer and as temperatures may vary in short periods as an hour or a day, residential natural gas demand will show seasonal and super-imposed, short-term variations. A schematic consumption pattern with seasonal variation and short-term peaks of a virtual (end-) user is depicted in Fig. 6.1.

However, temperature is predictable in the short-term only.[8] Therefore, the fluctuations in natural gas demand are, in the medium- and long-term, and, to a limited extent, in the short-term, not predictable. Such demand can be

[8]Moreover, the functional relationship between natural gas demand and temperature is based on observations. The real natural gas demand at any point in time may deviate.

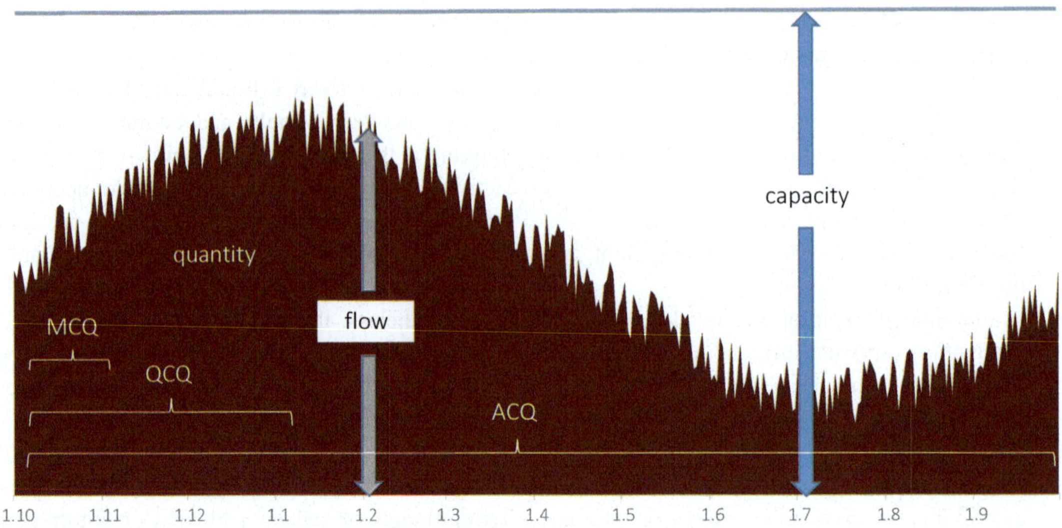

Fig. 6.1 Schematic seasonal and short-term fluctuating natural gas consumption pattern and illustration of selected terms

satisfied only by granting the buyer the right (but not the obligation) to off-take more or less natural gas than forecasted. The buyer gets an option to off-take natural gas. Such option is called 'flexibility'.[9] Typically, flexibility will be not only included in natural gas delivery contracts to residential customers or their suppliers, but also in other contracts, e.g., for deliveries to industrial customers, where it does not serve to mitigate temperature-dependent fluctuations of demand but production-dependent ones.

An industrial customer may have different requirements. Natural gas consumption for industrial purposes is typically not temperature dependent, but, primarily in the long-term, dependent on the economic situation. Hence, the consumption pattern is primarily influenced by fluctuations of demand of the industrial customers' products, as well as (potential) standstill periods at times of plant maintenance or any other demand interruption (e.g., plant holidays, strikes). Such a profile is schematically illustrated in Fig. 6.2.

In its simplest form, the off-take profile shows the same quantities of natural gas to be consumed at each point in time. Such a flat profile, also referred to as baseload profile, is depicted in Fig. 6.3. Such a baseload product is a standard product, which is traded at exchanges or via brokers for different maturities, e.g., days, months, or years.

Combining two different, yet flat, off-take profiles for winter and summer, the customer will be able to cover increased demand in winter and lower demand in summer. This is depicted in Fig. 6.4.

Excursus

Natural gas demand of a segment, a group, or individual customers as a function of outside temperatures, Q(T), is approximated by a sigmoid-function of the form:

[9] See Sect. 6.3.1.1.4.

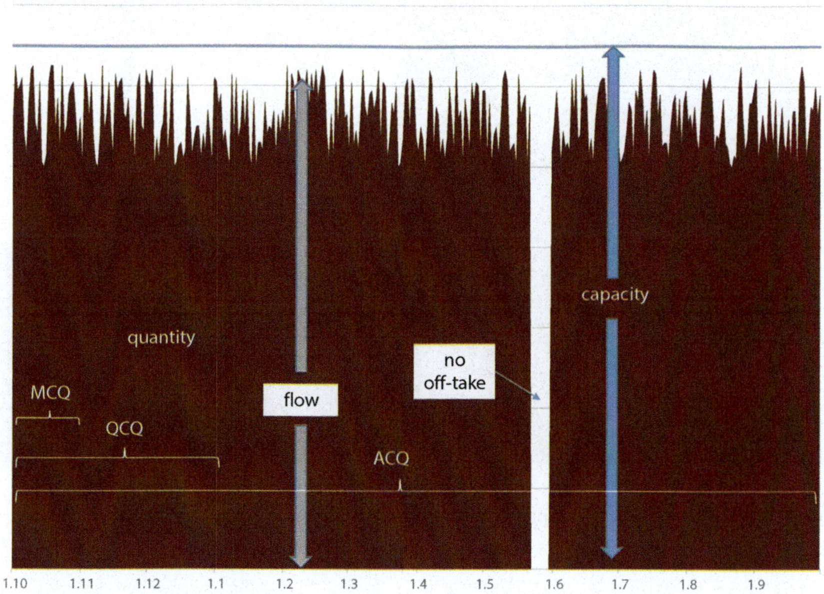

Fig. 6.2 Schematic short-term fluctuating natural gas consumption pattern and explanation of selected terms

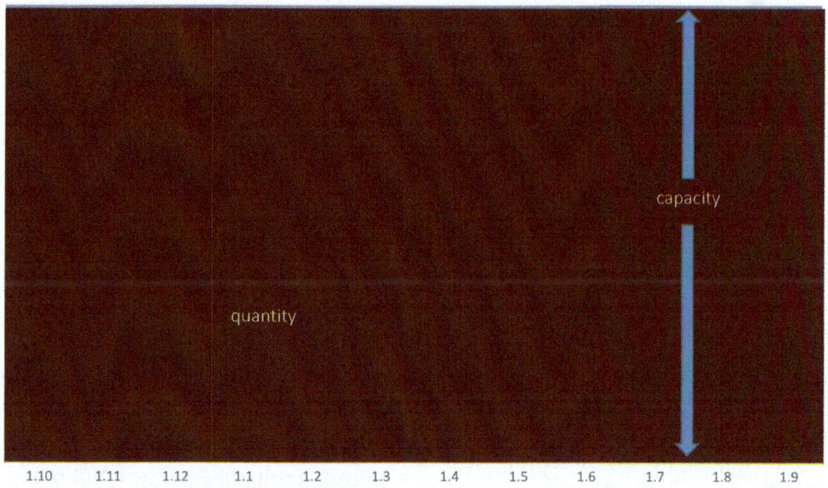

Fig. 6.3 Flat (baseload) natural gas consumption pattern

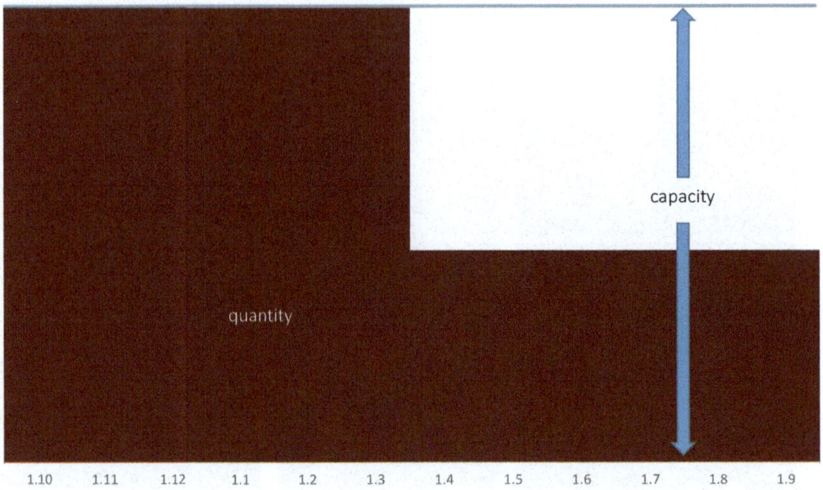

Fig. 6.4 Schematic seasonal band consumption pattern

$$Q(T) = Q_0 * \frac{\beta_1}{1 + \left(\frac{\beta_2}{T-\beta_3}\right)^{\beta_4}} + \beta_5$$

with

$\beta_1, \beta_2, \beta_3, \beta_4, \beta_5,$	parameters,
T	temperature, [K]
Q_0	quantity calibration, e.g., [kWh]
Q(T)	quantity as a function of temperature, e.g., [kWh].

The parameters $\beta_1, \beta_2, \beta_3, \beta_4,$ and β_5 result from regression analysis of (consumption quantity and temperature) data collected in the region or segment analyzed.

According to this (type of) function, natural gas demand will decrease with increasing temperature until a baseload level is reached, required, e.g., for cooking or warm water. For higher temperatures, the quantity consumed will be, more or less, independent of temperature, i.e., in Fig. 6.5, the asymptote at the lower right. Also for decreasing temperatures, the gradient of the function decreases, i.e., the curve is 'flattened'. Once very cold temperatures are reached, heat requirements do not increase as much with temperature as they do at higher temperatures.[10] This is illustrated in Fig. 6.5. Such sigmoid functions are used, e.g., in Germany, to simulate natural gas consumption of small customers without permanent measurement of their consumption.[11]

6.2.2 Supply

The supply of natural gas depends on several geological, geographical, technical, and commercial factors.

Neglecting anthropogenic production of gas, certain geological conditions have to be fulfilled to produce significant quantities of natural gas.[12] If natural gas is not produced in the vicinity of the demand region, natural gas has to be transported in pipelines or as LNG over long distances to supply a region. Reservoir characteristics and

[10]This is since heating installations can only be turned on up to their maximum, which is usually already reached sooner.

[11]See Bundesverband der Energie- und Wasserwirtschaft (BDEW) et al. (2011).

[12]See Sect. 4.2.

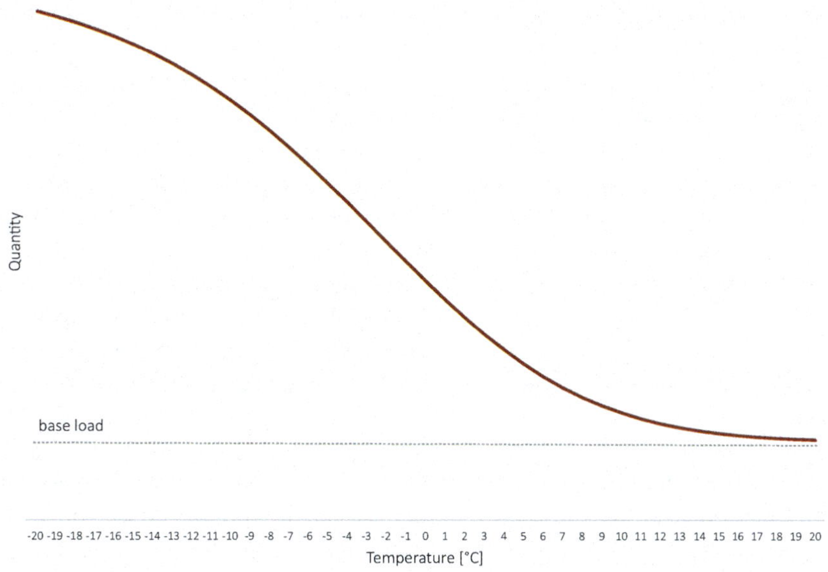

Fig. 6.5 Schematic sketch of the sigmoid function approximating a temperature dependent natural gas consumption

the capacities of production and transportation facilities thereby impose restrictions on using a field (or several fields) to supply a region. Not only is the size of a field (or fields) an obvious limitation factor, reservoir and production characteristics also influence the product and its characteristics. For example, a producer who produces from only one small gas field might not be able to sell its production under a supply-type contract.[13] Instead, it is very likely that this producer will sell a product under a depletion type contract, which, on the one hand, might serve his interest to sell the whole commercially recoverable reserves of his field; on the other hand, however, such depletion-type product is less favored by potential buyers as, e.g., diminishing production over the lifetime of the contract does not support the long-term development of a (downstream) market. The larger the reserve basis and the production capabilities, the easier it becomes to serve products fitting the requirements of a demand region.

Further, transportation infrastructure, i.e., pipelines or LNG facilities, have to be built to transport the natural gas produced from its source to the demand region. Economies of scale thereby favor the building of larger infrastructure, which, again, makes commercial sense only when quantities to be transported are large and presumably available for a long period. Obviously, a demand region benefits from nearby reserves as investments in transportation infrastructure is lowered. Within the demand region, a distribution system needs to be (put) in place to serve a large number of end users. Existing consumers of large quantities of natural gas, i.e., reliable (and preferably base-load) off-take, will support the economics of the infrastructure.

In addition, storage facilities need to be built to provide, e.g., demand balancing services.[14]

[13]As described in Sect. 6.3.10.1.

[14]See Sect. 8.2.

Again, a demand region benefits from nearby production as it might be possible to use production flexibility instead to serve such needs.

Ideally, demand regions are supplied by more than one production region. Diversification of supply is a tool to mitigate supply-side interruptions caused, e.g., by technical problems or political tensions. For producing regions, a strategy of diversifying demand is advisable for the same reasons.

6.3 Contracts

The commercial products are defined in contracts between the parties described in Sect. 6.1. Simply put, a commodity sales contract is a binding mutual agreement to purchase and sell defined quantities of natural gas during defined periods at defined locations at defined prices. During negotiations of such contracts, buyers take into account the needs of the market they want to serve, and sellers (in case they are producers) take into account the (technical) limitations they face when producing natural gas.[15] Usually, contracts are individually negotiated in detail between the parties. Nevertheless, so-called standard or framework contracts[16] have been developed to facilitate these negotiations.

When the traditional market model was still in place, negotiations of international long-term and large-scale contracts took years, primarily because the economic value of such contracts was immense.[17] Both parties, seller(s) and buyers(s) developed a negotiation strategy and carefully prepared for each round of negotiations.

In some cases, so-called memoranda of understanding (MOU), letters of intent (LOI) or heads of agreement (HOA) had been concluded before the final contract was signed. Effectively all had the same intention: to agree upon and record interim results of the negotiations.

Further to this, the negotiations and their results were usually kept strictly confidential, and the parties agreed upon separate confidentiality agreements stipulating rules for confidentiality and penalties in case of their breach.

In general, agreements for sales and purchases of natural gas transported in pipelines (hereinafter 'GSA') have many stipulations in common with agreements for sales and purchases of natural gas transported in form of LNG (hereinafter, 'LNG SPA'). Issues like billing and payment, force majeure, etc., are effectively the same in both types of agreements. They are further described in Sect. 6.3.8.

6.3.1 Quantities and Flexibilities

One of the main issues to be agreed upon in any natural gas sales agreements are the quantities of natural gas to be sold and purchased, or, from a physical point of view, quantities to be delivered by the seller and off-taken by the buyer.

Following the idea

- that 'products' required to cover the customers' demand will be described and defined in contracts;
- that demand varies over time; and
- that, at least the exact, anticipation of physical demand can never be accomplished (as, *inter alia*, temperature is not predictable for the mid- and long-term future),

annual, seasonal, quarterly, monthly, daily and hourly quantities, their minima and maxima, i.e., the components of flexibility are used to 'assemble' and define natural gas commodity products. However, as described in the following section, such elements are also interdependent.

6.3.1.1 Elements and Definitions

To describe the various quantities or quantity components to be delivered in different periods, a

[15]See Sect. 6.2.2.

[16]See Sect. 6.3.9.

[17]Assume a typical long-term import contract with a duration of 20 years and an annual delivery volume of 100 TWh. At a price of 20 €/MWh, this leads to an annual payment of 2 billion €/a, over the whole contract term of 40 billion €.

Table 6.1 Terms used for the description of commodity sales and purchases

Term	Definition	Example
Quantity	Total amount of energy or volume (often with reference to a defined period)[a]	10 TWh(/a) or 5 bcm(/a)
Annual contract quantity (ACQ)	Quantity to be delivered by the seller and taken by the buyer (with contractually defined exceptions) during a (contract) year	8.76 TWh/a
Houry contract quantity (HCQ)	Quantity to be delivered by the seller and taken by the buyer (with contractually defined exceptions) during an hour	2,000,000 kWh/h = 2,000,000 kW
Daily (DCQ), monthly (MCQ), quarterly (QCQ) and seasonal (SCQ) contract quantity	Analogous to HCQ and ACQ	
(Hourly) flow	Quantity actually delivered by the seller and taken by the buyer (during an hour)	993,000 kWh/h = 993,000 kW
Load hours	ACQ/HCQ, i.e., the number of hours required to off-take the ACQ by continuously taking the (maximal) HCQ (or analogously for any other period)	8.76 TWh/a/2,000,000 kWh/h = 8,760,000,000 kWh/a/2,000,000 kWh/h = 4380 h/a
Load factor	Load hours/8760 (or general: load hours/period), i.e., the percentage of the year (or any other period) required to off-take the ACQ by continuously taking the (maximal) HCQ	4380 h/a/8760 h/a = 50%
Structure	hourly or daily off-take profiles during a year (or any other period), which are *ex ante* known and agreed upon between buyer and seller	October 1 to March 31: 500,000 kWh/h; April 1 to September 30: 200,000 kWh/h
Capacity	The (maximum) hourly quantity technically deliverable or to be off-taken during a period.	50,000 kWh/h/a
Flexibility	See Sect. 6.3.1.1.4	

[a]Already the quantity is heterogeneously defined. While in most countries quantity is equated to energy, i.e., expressed in units as, e.g., [MWh], [GJ] or [Btu], elsewhere the volume of natural gas at standardized conditions represents the quantity used as the denominator for pricing and is expressed in units of, e.g., [m^3] or [scf]

nomenclature has been developed and applied—initially to traditional (integrated) natural gas deliveries by pipeline. Yet, some of these definitions are also applied to modern hub-based deliveries or to LNG contracts.

6.3.1.1.1 Basic Terms of Pipeline Deliveries

Specifically, the contractually defined quantity components are[18] (Table 6.1):

- Annual Contract Quantity (ACQ), i.e., minimum ACQ, ACQ_{Min}, standard ACQ, ACQ_{Norm}, and maximum ACQ, ACQ_{Max};
- Seasonal Contract Quantity (SCQ), i.e., minimum SCQ, SCQ_{Min}, standard SCQ, SCQ_{Norm}, and maximum SCQ, SCQ_{Max};
- Quarterly Contract Quantity (QCQ), i.e., minimum QCQ, QCQ_{Min}, standard QCQ, QCQ_{Norm}, and maximum QCQ, QCQ_{Max};
- Monthly Contract Quantity (MCQ), i.e., minimum MCQ, MCQ_{Min}, standard MCQ, MCQ_{Norm}, and maximum MCQ, MCQ_{Max};

[18]Usually only two to three of these components are specified. Note that they designate contractual restrictions, not the actual off-takes.

- Daily Contract Quantity (DCQ), i.e., minimum DCQ, DCQ_{Min}, standard DCQ, DCQ_{Norm}, and maximum DCQ, DCQ_{Max}; and
- Hourly Contract Quantity (HCQ), i.e., minimum HCQ, HCQ_{Min}, standard HCQ, HCQ_{Norm}, and maximum HCQ, HCQ_{Max}.[19]

For each of the above-mentioned periods the following can be defined in contracts:

- Minimum quantity (in any period):
  - minimum take obligation (for hourly and daily quantities: 'minimum flow');
  - if combined with pay obligation: take-or-pay (also referred to as 'minimum bill').

  The minimal quantity to be off-taken during a period is defined as the (buyer's) minimum take obligation. Typically, this minimum take obligation is linked to a pay obligation, i.e., the minimum quantity has to be taken or paid for if not taken by the buyer during a defined period. It represents an expected off-take and a revenue guarantee for the seller.[20] Such minimum take provisions are frequently stipulated for annual quantities, but in some contracts for other periods as, e.g., months or quarters as well. Some contracts foresee that only a certain percentage of the price has to be paid in case the quantities have not been off-taken.[21] Consequently, the minimum restriction for each period is defined by the take-or-pay quantity. Trivially, a minimum quantity of zero [units] during a period is equivalent to the absence of a take-or-pay provision for this period.

- Maximum quantity (in any period): maximum (supply) obligation.

  The maximal quantity to be delivered during a period is defined as (seller's) maximum supply obligation. That is, seller's obligation to make available all contractually defined quantities during each period will be limited by these maximum supply obligations for each period. Only by separate mutual agreement will the buyer be allowed to off-take more than the maximum supply obligation.

For annual quantities, the restrictions are illustrated in Fig. 6.6.

6.3.1.1.2 Interpretation and Consequences

The definitions impose restrictions on both sellers and buyers.[22] As the actual off-take is usually determined by the buyer (with the rare exception of seller-nominated contracts), and leaving everything else aside, sellers will prefer the utmost restriction and definition of the above mentioned quantity components and vice versa, buyers will favor the least possible constraints. Extreme examples may illustrate this: If, e.g., only ACQ_{Max} is defined and all the other components remain undefined, the buyer would, at least theoretically, not only be free to take any quantity between zero and ACQ_{Max}, but also to take the maximum annual quantity within one hour, as no restriction for the hourly quantity is in place. If, on the other hand, all of the above components are defined and each maximum equals the corresponding minimum, the buyer has to take at each point in time the respective quantity—being restricted to the maximum.

The quantity restrictions for intersecting periods are interdependent and have to be defined interdependently.[23] Hence, the following rules

[19]As other periods as, e.g., the second, the minute, the week, or the trimester are typically not used in natural gas contracts, their inclusion can be neglected.Anyway, only the astronomical day, the astronomical month, and the astronomical year are 'natural' periods, i.e., periods 'defined' by the rotation of the earth, the revolution of the moon, and the revolution of the earth. The others, i.e., the hour, the week, and the quarter are (arbitrarily) defined, i.e., 'man-made'.

[20]In GSAs, only rarely take-and-pay obligations are agreed upon instead of take-or-pay obligations. If agreed upon, it does not only guarantee a certain revenue for the seller but also guarantees a firm and guaranteed off-take.

[21]For example, 80% of the average price applicable in the period, during which the natural gas had to be taken (or paid for if not taken).

[22]Some restrictions, however, are subject to third-party requirements, such as transportation capacity availability.

[23]The subsequent observation applies to the periods: hour (H), day (D), month (M), quarter (Q), season (S), and year (Y), but could be extended to multi-annual periods.A period t_2 shall be called 'superordinate' to a period t_1 if its length exceeds the one of t_1 and if t_1 is an integral part of t_2.A period t_1 shall be called 'subordinate' to a period t_2 if its length falls short of the one of t_2 and if t_1 is an integral part of t_2.For example, the month is superordinate to the day but subordinate to the quarter.

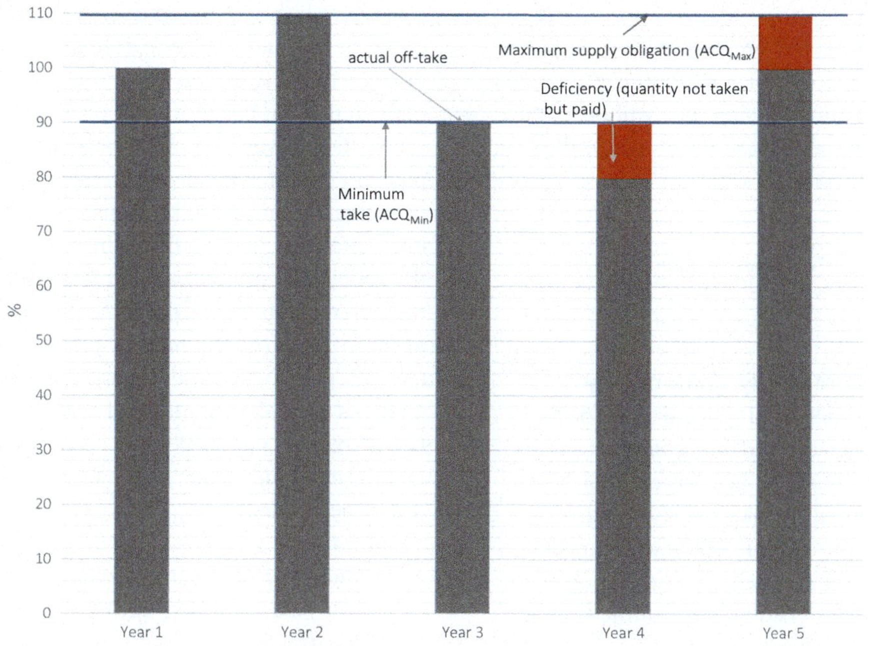

Fig. 6.6 Illustration of (annual) maximum and minimum quantities

have to be observed defining restrictions in contracts:

- The maximum quantity defined in a superordinate period can only be taken if the cumulated maximum quantities of any subordinate period are equal or exceed the maximum of the respective superordinate period. This can be described as follows:

  Let k_i^{i+n} be the number of periods i in period (i + n) (with i being subordinate to (i + n), with n = 0, …, 6 being the sequential numbering of the periods defined), and let Q_{i+n}^{max} and Q_i^{max} be the respective absolute quantity in period (i + n) and i, then:

$$\sum_{j=1}^{k_i^{i+n}} Q_{ij}^{Max} \geq Q_{i+n}^{Max}$$

Interdependence of Hourly and Daily Restrictions

For (i + n) = D and I = H and, consequently $k_i^{i+n} = 24$, $V_i^{Max} = HCQ_{Max}$ and $Q_{i+n}^{Max} = DCQ_{Max}$, this means that the sum of contractually permitted maximum hourly quantities has to be equal or exceed the maximum daily quantity, i.e.,

$$\sum_{j=1}^{24} HCQ_{Max\,j} \geq DCQ_{Max}$$

- Vice versa, the minimum quantity defined in a superordinate period can only be observed if the cumulated minimum quantities of any subordinate period are equal or lower than the minimum of the respective superordinate period. In other words, if the minimum quantity of a super-ordinate period would be lower than the one of a subordinate it would be rendered otiose. Using the symbols defined above ('min' and 'max' exchanged), this can be described as follows:

$$\sum_{j=1}^{k_i^{i+n}} Q_{i_j}^{min} \leq Q_{i+n}^{min}$$

Annual and Monthly Take-or-Pay

If both monthly and annual take-or-pay provisions are defined in a contract, then the annual has to be larger than the cumulated monthly take-or-pay provisions. Otherwise, the annual take-or-pay provision would be fulfilled once the monthly take-or-pay provisions are fulfilled and, thus, rendered otiose.

6.3.1.1.3 Make-Up and Carry-Forward

In addition to the intra-annual quantity restrictions, long-term contracts with a take-or-pay obligation often also contain multi-annual provisions (or rights), named 'make-up' and 'carry-forward' (see Fig. 6.7).

- Make-up rights are buyer's rights to recover quantities paid for, but not taken, in one period in subsequent periods during the term of the contract;
- 'Carry-forward' rights, denote buyer's rights to take quantities above a certain pre-defined level (e.g., the minimum take quantity) for a period in advance and offset them against quantities to be paid for but not off-taken in later periods during the term of the contract.[24]

Consequently, both rights increase volume flexibility but keep the total quantity over the defined contract term of several years in defined limits.

Make-upand carry-forward provisions might be accompanied by some of the following rules:

- The buyer has to reimburse seller for make-up quantities actually taken at times of higher prices. That is, if the price of natural gas delivered is higher at a time when the buyer seeks to recover (i.e., make-up) natural gas paid for but not taken at an earlier point in time than when such take-or-pay payment has been effected, then the buyer has to reimburse the seller. The reimbursement is calculated as the difference between the price of natural gas at the time when the natural gas is made-up and the price when the natural gas has been paid but not taken multiplied with the quantity of natural gas made-up;
- Both, make-up and carry-forward rights might be limited:
  - in time:
    - make-up and carry-forward rights can only be exercised during the term of a contract;
    - during the term of the contract, the period when such rights may be exercised is frequently limited. For example, in a long-term contract with a term of 20 years, the make-up rights could be limited to five years after the natural gas has been paid for but not taken;
    - some contracts may stipulate that make-up gas will be delivered during defined periods of a year only or, vice versa, exclude periods of delivery of make-up gas. For example, no deliveries of make-up gas during January and February or deliveries of make-up gas will take place after the end of June only;
  - in quantity:
    - make-up and carry-forward rights can only be exercised once a (defined) minimum quantity is taken. Whether natural gas is recovered (make-up) or whether carry-forward is built up, the minimum quantity does not necessarily have to correspond either to the take-or-pay quantity or to the ACQ—it can be any quantity below the maximum supply obligation (only in very particular events, exceeding this limit);
    - the total amount of make-up and carry-forward quantities is typically limited.

[24]Even if usually applied only to annual quantities both, make-up rights and carry-forward rights could be defined for any period.

For example, even if a buyer in one year took only ten [units] while the minimum pay quantity was 80 [units], it may be allowed to recover 40 [units] only;

- the (hourly or daily) capacity for deliveries to the buyer may be limited. Hence, make-up and carry forward deliveries will be allowed in a (narrow) range of the capacity only.

Make-up and carry-forward quantities, together with the minimum take quantities and the minimum quantity to initiate make-up or carry-forward, are interdependent. Therefore, in long-term contracts priorities will have to be defined.

Make-Up and Carry-Forward
Assume that: the ACQ is 100 [units]; the minimum take obligation 70 [units]; any quantity taken in excess of 75 [units] may be declared as make-up; and any quantity in excess of 80 [units] may be declared as carry-forward. Assume further that: the buyer's off-take in year two of a contract was at 65 [units] and buyer's off-take in year 3 is at 90 [units]. Then it remains to be defined in the contract if, in such a situation, the buyer in year three may declare five [units] as make-up and an additional ten [units] as carry-forward (logic: [units] 76–80 are make-up and [units] 81–90 are carry-forward) or if the buyer may declare five [units] as make-up and only additional five [units] as carry-forward (logic: [units] 81–85 are carry-forward and [units] 86–90 are make-up (which can only be effected after carry-forward is defined).

Even if both, make-up and carry-forward provisions have originally been included in long-term contracts to foster the development of the natural gas market by eliminating potential (financial) obstacles and disadvantages of the buyer, these contractual provisions, as all flexibility components, are also used by buyers for purposes of pure financial optimization.

Natural gas paid for but not taken and not recovered (made-up) during the term of the contract may be treated at the end of the contract in various ways:

- no reimbursement and no recovery: this is the 'simple' way, based on the argument that the buyer has lost any right to recover natural gas after the contract is finalized;
- extension of the contract: If a long-term contract does not foresee any of the measures explained below and if, by the end of the contract, the buyer paid for natural gas but did not recover all of it, then both buyer and seller often prolong such a contract, including provisions for make-up from the (then) previous contract. The buyer will then have the advantage of being able to recover payments lost otherwise, and the seller might profit from a prolonged contract granting revenues and profits in the future;
- reimbursement: in some contracts, natural gas paid for but not taken until the end of the contract is reimbursed by the seller to the buyer at a (discounted) price; or
- delivery: (parts of) the natural gas paid for but not taken until the end of the contract (are) is delivered within the shortest possible period (potentially also at a different delivery point) after the end of the contract.

6.3.1.1.4 Flexibility

'Flexibility'[25] is a term used frequently—also in other industries—yet, the term is only presumably clearly defined. This also applies to the natural gas industry. More specific, volume flexibility[26] is often defined with reference to single flexibility components only, neglecting interdependencies, and other components. Hence, ambiguity may surface, incompleteness may entail different interpretations, and operability may be constrained.

A definition of volume flexibility should fulfil the following requirements:

[25]From Latin, *flexibilis*, from *flexus*, past participle of *flectere*, to bend.

[26]Typically, 'volume flexibility', described subsequently, is distinguished from 'geographical flexibility' and 'price flexibility'.

- take into consideration all flexibility components;
- be operational, applicable, and practical;
- provide a quantification of flexibility;
- aggregate the numbers of the single flexibility components;
- reduce ambiguity;
- facilitate and ensure comparability of contracts; and
- comply with intuition.

In short, volume flexibility of a physical or contractual natural gas delivery or storage shall be defined as the cumulated difference of maximal and the minimal utilization patterns divided by total quantities. In doing so, all components' restrictions defined for different periods are considered. As such, flexibility is the 'area of permitted off-takes' in all periods.

Given restrictions of hourly, daily, monthly, quarterly, seasonal, and annual components,[27] the customer has during each period the option (=flexibility) to take any quantity which is not lower than the minimum quantity and does not exceed the maximum quantity defined for that period. However, these minimum and maximum quantities are not independent.

The flexibility tools, i.e., the sources, are necessarily physical. They are devices (in the broadest sense) of and for the production of flexibility. These are:

- natural gas production;
- storage (short-, mid- and long-term); and
- switching to, or utilization of, other energy sources.

Contractual tools can only be offered and fulfilled if such physical backup exists. Therefore, a long-term contract, often characterized to be a source of flexibility, is only the contractual manifestation of the flexibility provided by either natural gas production or utilization of storage capacity. The same applies for interruptible delivery contracts. In this case, the physical backups to provide flexibility are other sources of energy, other than natural gas.

This reasoning implies that natural gas hubs cannot be regarded as potential sources of flexibility, even if prices quoted at hubs might be used for the valuation of flexibility.

The flexibility tools fulfil different purposes and are not utilized by the same players. For example, producers may sell production flexibility to traders. Storage system operators will sell flexibility provided by utilization of storages to either traders or end users, such as large industrial customers, based on storage contracts. The physical flexibility tool, however, is the respective storage or production itself (Fig. 6.8).

Seasonal and Annual Flexibility

Let, e.g., the seasonal minimum quantity, $SCQ_{Min} = 80$ (arbitrary units), the seasonal maximum quantity, $SCQ_{Max} = 120$ (arbitrary units), and the annual minimum quantity, $ACQ_{Min} = 100$ (arbitrary units), then, in order to reach the annual minimum, the customer will, after having taken the minimum in the first season, have to take the maximum, i.e., $SCQ_{Max} = 120$, in the second season in order to reach the annual minimum. Hence, during each hour of the first season, the customer will take $SCQ_{Min}/4380$,[28] i.e., the minimum hourly quantity to reach the minimum seasonal quantity,[29] and, during each hour of the second season, the customer will have to take $SCQ_{Max}/4380$, i.e., the maximum hourly quantity to reach the minimum annual quantity.

Interdependence of Flexibility Components

Figure 6.9 shows a more complex interdependence of flexibility components. The restrictions for the quantities in various periods are defined (in percent) as shown in Table 6.2.

Observing all restrictions and their interdependencies, at each hour during the year

[27] See Sect. 6.3.1.1.

[28] With 4380 being the number of hours in a season.

[29] Which again does not necessarily correspond to the minimum hourly quantity, H_{Min} defined.

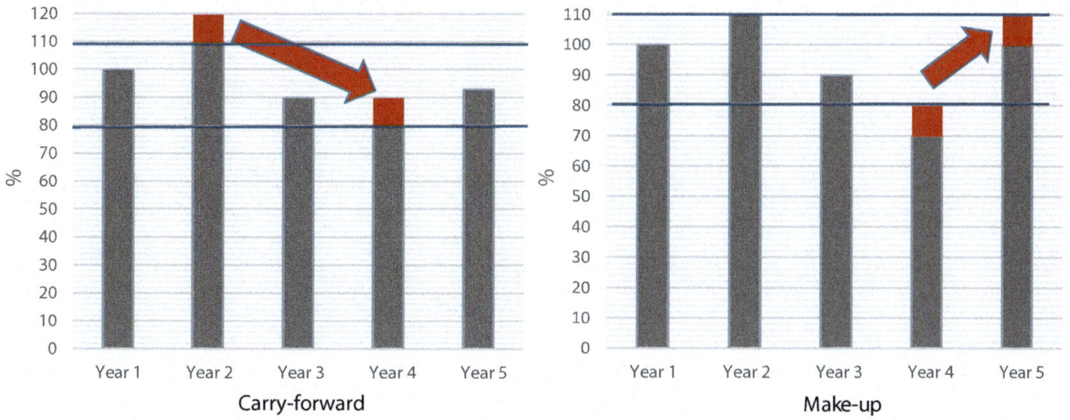

Fig. 6.7 Illustration of annual carry-forward (*left* part) and make-up (*right* part) provision in long-term contracts

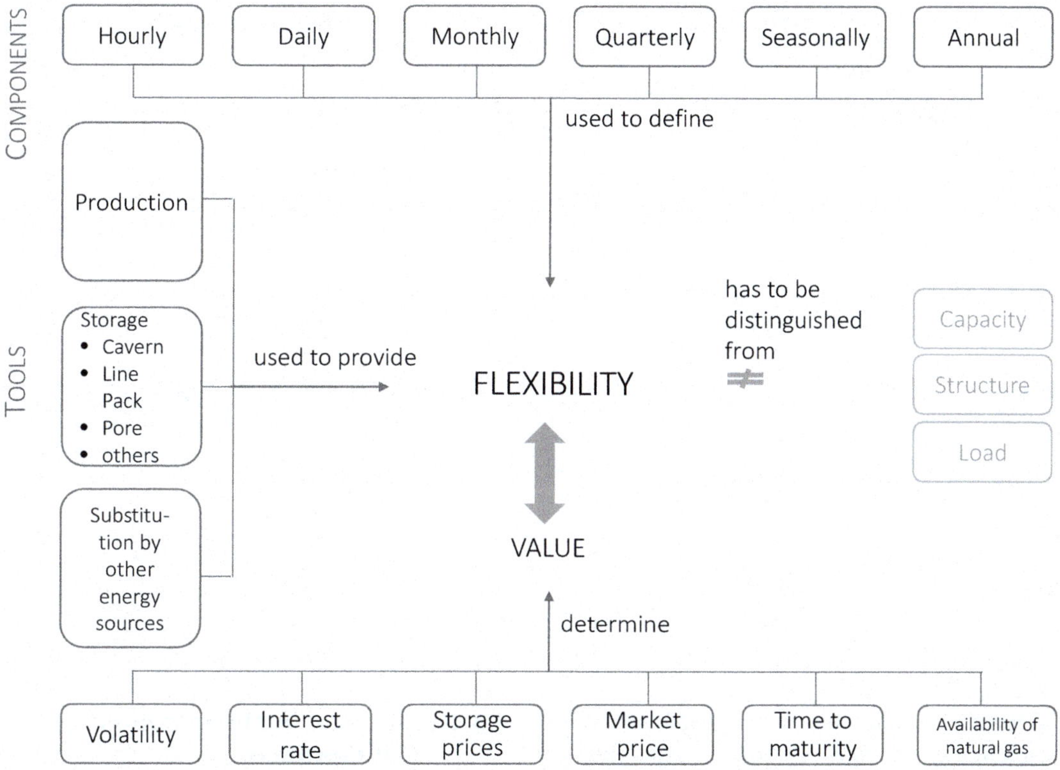

Fig. 6.8 Structure of the flexibility concept

(the horizontal axis), any quantity within the outer boundaries shown, e.g., in Fig. 6.9 could be off-taken. In this example, the off-take in each of the six periods (hour, day, month, quarter, season, and year) is influenced by the superordinate period's restrictions.

If, as shown in Fig. 6.10, hourly, daily, and annual quantity restrictions only are defined, then the picture is simplified. Yet, the customers' volume flexibility is still represented by the area delimited by the outer, i.e., upper and lower, lines (Table 6.3).

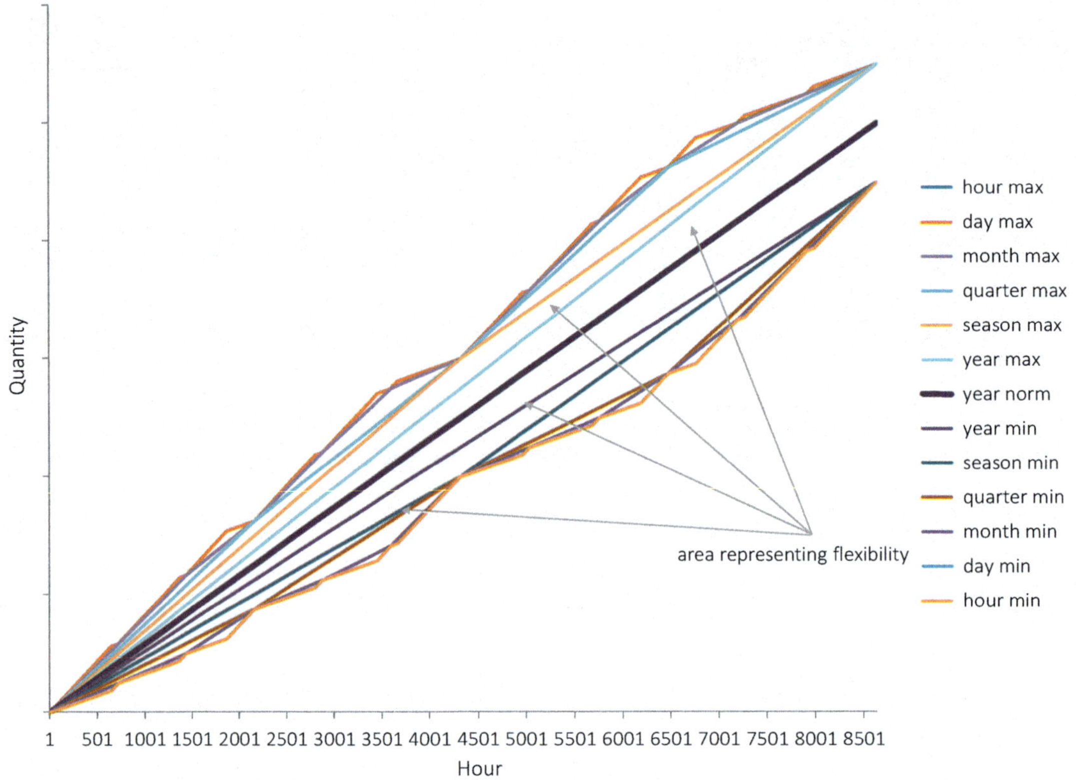

Fig. 6.9 Example of flexibility components defining volume flexibility

Table 6.2 Example for quantity restrictions in various periods

	Minimum [%]	Norm [%]	Maximum [%]
Hour	40	100	160
Day	50	100	150
Month	60	100	140
Quarter	70	100	130
Season	80	100	120
Year	90	100	110

Hence, the minimum annual quantity imposes restrictions on the seasonal quantities—in addition to the restrictions defined. Analogously, this applies to all other components, i.e., the hourly, daily, monthly, and quarterly quantity restrictions.

6.3.1.2 Typical Implementation in Contracts

6.3.1.2.1 Standard Contracts for Deliveries at Hubs

Usually, parties to a standard contract agree on a contractual quantity to be delivered and taken

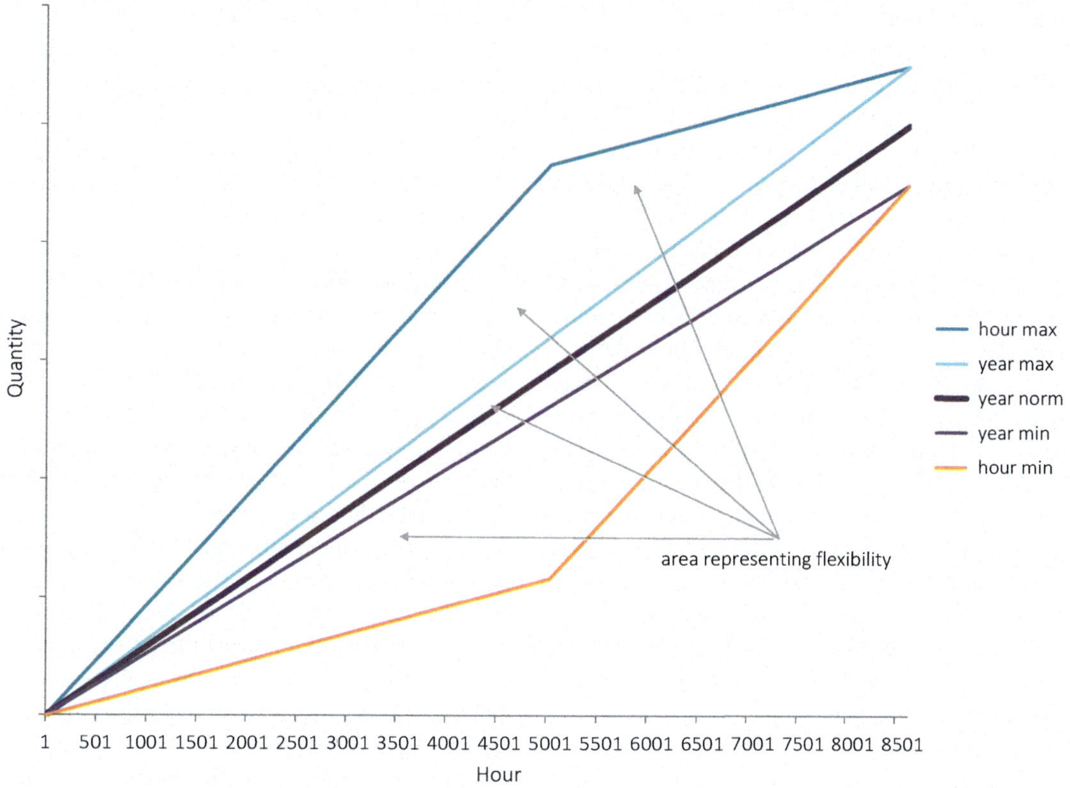

Fig. 6.10 Simplified example of flexibility components defining volume flexibility

Table 6.3 Example for simplified quantity restrictions in various periods

	Minimum [%]	Norm [%]	Maximum [%]
Hour	40	100	160
Day	40	100	160
Month	40	100	160
Quarter	40	100	160
Season	40	100	160
Year	90	100	110

and paid for during a specific period. For example, the EFET standard contract defines the term contract quantity as follows[30]:

> *'Contract Quantity' means, in respect of an Individual Contract, the quantity of Natural Gas, expressed in MWh, to be delivered by the Seller and accepted by the Buyer in a Time Unit pursuant to such Individual Contract as agreed between the Parties.*

Typically, flexibility components are not agreed upon in standard contracts for deliveries at hubs.[31]

Furthermore, exchange traded contracts such as futures do not allow for an individual agreement of a quantity. Instead, standard lot sizes are pre-defined.[32]

[30]European Federation of Energy Traders (2016a), Annex 1–2.

[31]Exceptions to this are obviously standard option contracts.

[32]For instance, 30 MWh/h = 30 MW.

6.3.1.2.2 GSAs

(Natural) gas sales agreements (GSAs) stipulate quantities to be delivered and off-taken during the term of the agreement.[33] Typically, it is not feasible to define just the overall quantity to be delivered and off-taken during the whole lifetime of a contract, also referred to as 'total contract quantity' or 'TCQ'. Instead and in addition, contractual quantities are split into quantities to be delivered and off-taken within a pre-defined period, e.g., a year. Hence, not all GSAs define a TCQ; many GSAs define 'annual contract quantities' or 'ACQs' instead, i.e., quantities to be delivered and off-taken during a contract year.

Some GSAs, in addition, stipulate a contractual right of the buyer or seller to 'nominate' with a certain lead time a future ACQ (within pre-defined limits). In this case, e.g., the buyer might have the option, i.e., the right, but not the obligation, to increase an already agreed ACQ to be delivered and off-taken in a future year by a certain amount or percentage.

Quantity Adjustment in Traditional Norwegian Sales Contracts

One example for such buyers' options to increase delivery quantities were the respective stipulations in the Troll agreement between the Norwegian producers Statoil, Norsk Hydro, Saga, Shell, Mobil, and Conoco and the Continental European buyers Ruhrgas, Thyssengas, BEB, Distrigaz, Gasunie, and Gaz de France.[34]

A special case for such 'ACQ nominations'—in this case to be exercised by the seller—are stipulations to determine the ACQ in so-called depletion-type contracts.[35] In its pure form, under these contracts, seller (being a producer of natural gas) and buyer agree to sell and purchase all economically recoverable reserves of a specified natural gas field. As these reserves are usually not known at the start of production or the start of deliveries, respectively, seller and buyer agree on a procedure how to determine (and re-determine) the ACQ during the lifetime, i.e., the term, of the GSA.

In any case, at the start of a contract year, the ACQ is determined as a fixed contractual quantity of natural gas to be delivered and off-taken during the respective year. As described in Sect. 6.3.1.1, seller might agree not only on a ACQ_{Norm} but also on a ACQ_{Min} and/or ACQ_{Max}. By doing so, the buyer receives volume flexibility, as it gives her/him, on the one hand, the right to take during the respective year up to the ACQ_{Max} and, on the other hand, the obligation to take at least the ACQ_{Min}. The seller has to make available a quantity up to the ACQ_{Max} but receives security that at least the ACQ_{Min} will be off-taken (or paid for if not physically taken) by the buyer. As explained, the ACQ_{Min} is often referred to as 'minimum bill quantity' or 'take-or-pay quantity', as the buyer has to take these quantities during the contract year or, in case of non-take, to pay for them. This mechanism provides some financial reliability to the seller because, independent from buyer's off-take, it can be sure to receive at least a certain amount of money during a contract year. Only rarely, GSAs stipulate for a 'take-and-pay' clause instead of the described 'take-or-pay' clause.

Together with annual take-or-pay commitments, some GSAs stipulate for make-up and/or carry-forward rights as described in Sect. 6.3.1.1.3.

In addition to the definition of ACQs, also quantities for other periods might be agreed upon within GSAs. Typically, 'daily contract quantities' or 'DCQs' or, alternatively, 'hourly contract quantities' or 'HCQ s' might be agreed upon by seller and buyer. In some cases, contractual quantities also are defined for seasons, quarters, or months. For all these periods, minimum and/or maximum values might be defined by the parties. As described in Sect. 6.3.1.1, these quantity definitions and restrictions are interdependent.

Typically, the flexibility that is provided by a seller to a buyer increases the further downstream

[33] This section details the quantity-related contractual rights and obligations sellers and buyers of natural gas might agree upon in a GSA. Deviations from these contractual obligations like over-deliveries or under-deliveries are described in Sect. 6.3.8.

[34] See Mellbye (1996).

[35] See also Sect. 6.3.10.1.

the buyer is 'located' in the value chain. Other than importing companies, which face a clear set of quantity restrictions in their import contracts, local distribution companies might purchase natural gas on the basis of a full supply contract. The extreme case, i.e., household customers have effectively no contractual quantity restriction, as they might take whatever they need. The only restriction households face is a technical one, i.e., the capacity of their grid connection.

6.3.1.2.3 LNG SPAs

As with pipeline gas in modern LNG sale and purchase agreements (SPAs), LNG is sold in terms of its energy content, i.e., the parties agree upon a certain quantity to be delivered.[36] As the energy delivered cannot be measured directly, the following formula has to be applied to determine the net quantity of energy transferred from seller to buyer:

$$Q = (V_{LNG} * \rho_{LNG} * GCV_{LNG}) - Q_{gas}$$

with:

Q	quantity/energy delivered [kWh]
V_{LNG}	volume of LNG loaded or discharged[37] [m^3]
ρ_{LNG}	density of LNG loaded or discharged[38] [kg/m^3]
GCV_{LNG}	gross calorific value of LNG loaded or discharged [kWh/kg]
Q_{gas}	quantity/energy of natural gas from ship to terminal during loading or quantity/energy of natural gas from terminal to ship during discharging [kWh].

In addition to stipulations of quantities to be delivered (the 'contract quantity') under a LNG SPA, the parties also agree on a so-called delivery schedule in which details, e.g., regarding:

- the vessel used and its capacity,
- the arrival time at the port of destination,
- the allowed lay time, and
- the volume of LNG heel[39]

are agreed upon. Once the schedule has been finalized, it might only be changed by agreement between the parties. One-sided options to change the schedule are not common in traditional contracts.

As described in Sect. 6.3.1.2.2, volume flexibility in GSAs depends, on the one hand, on quantity restrictions (i.e., minimum and maximum values of quantities to be delivered) and, on the other hand, on the right for one or the other party to a contract (usually the buyer) to decide which quantity shall be delivered at which point in time. This volume flexibility is not commercially feasible under LNG SPAs, as buyer and seller are not permanently physically connected to each other, as is the case for transportation of natural gas via a pipeline. The physical connection only exists at the points in time when the vessel is either loaded (under a Free on Board (FOB-) based contract) or discharged (under a delivery-at-place (DAP-) (or delivery-ex-ship (DES-)) based contract). Therefore, under an LNG SPA, the timing of the start of discharging (or loading) and their respective duration is crucial for the parties to agree. Usually, agreements on time allow for a certain (not too long) grace period. Physically, delays can be managed by either seller (under a FOB-based contract) or buyer (under a DAP- (or DES-) based contract) because both have access to LNG storage capacities at either the liquefaction plant or the regasification plant, respectively.

Another typical feature of pipeline-based natural gas sales agreements, the so-called take-or-pay obligation[40] is usually not found in LNG SPAs. Instead, most LNG SPAs stipulate for a take-and-pay obligation, i.e., the buyer has

[36] Alternatively, the parties might agree upon selling and purchasing one or more full cargo lot(s), i.e., the LNG transport capacity of any particular LNG vessel.

[37] LNG loaded is used for a FOB-based LNG SPA; LNG discharged is used for a DAT- (or DES-) based LNG SPA.

[38] LNG loaded is used for a FOB-based LNG SPA; LNG discharged is used for a DAT- (or DES-) based LNG SPA.

[39] See Sect. 7.2.2.2.

[40] See Sect. 6.3.1.2.

not only to pay for the natural gas but it also has to take it. In this sense, LNG SPAs are for buyers less flexible than GSAs, as they do not allow for non-take and (possible) subsequent make-up. Deliveries of LNG make-up quantities would require having flexible access to not only to LNG but also to shipping and regasification capacity.

6.3.2 Price

6.3.2.1 General Remarks on Pricing

The price of (the commodity of) natural gas is the monetary value assigned to a 'unit' or 'quantity' of this commodity. Typically, energy units, only rarely volume units are chosen to define the quantity, i.e., natural gas prices are quoted in monetary units per energy.[41]

In GSAs and in LNG SPAs, the price is one of the main elements. Yet, the price and other terms and conditions of the contract are interdependent. That is, no natural gas price can be defined or analyzed in isolation; all other contractual conditions have to be included as well. This results from the product being not the commodity itself but the commodity including the terms and conditions of delivery, i.e., including, *inter alia*, flexibility, delivery point (s), term, payment conditions etc.

Further to this, the price is also basis for determining other price-based components. Penalties or discounts (e.g., for certain quantities to be delivered) might be defined as a percentage of the contract price or as a percentage of an averaged contract price. For instance, it is quite common to value make-up volumes[42] with an average contract price.

Apart of such complexity of the product and lack of homogeneity of definition, three major pricing models, used around the world, can be identified[43]:

- Indexed pricing
  Based on the notion of substitutable goods and the observation that end users are interested in purchasing energy but not a fuel, natural gas was (and partly still is) regarded as a substitute for other energies, such as oil products or coal. Based on this idea natural gas could be priced competitively to alternative fuels by expressing natural gas prices as a function of the alternative fuels' prices in time. This was the traditional pricing method in Continental Europe (and partly still is), in Southeast Asia and, with modifications, in Japan.
- Gas-on-gas pricing
  In (nearly) polypolistic, i.e., competitive, natural gas markets, as, e.g., North America, the UK, or in some parts of Continental Europe, natural gas prices are set in relation to supply and demand of natural gas itself. Large numbers of buyer and sellers compete (often without governmental intervention), but within a clearly defined framework of regulatory rules. Well-established benchmark prices—as, e.g., the Henry Hub price in the US—are set by transparent markets such as New York Mercantile Exchange (NYMEX).[44] This pricing (or market) model is also referred to as 'gas-on-gas' competition.
- Regulated pricing
  In countries where the natural gas market is primarily dominated by the state (and in which the natural gas market may be immature), end user prices for natural gas may be set by a state-controlled institution or by the governmental authorities by decree. The methodology of pricing may or may not be transparent. Regulated (commodity) pricing is based either on costs or on social and political objectives. In cases where the latter is applied, prices are determined ad hoc by a regulatory authority.

Notwithstanding the above, commodity prices have been, and partly still are, supplemented by

[41]The pricing of natural gas by units of energy is another distinguishing element between the commodities oil and natural gas. The former is typically sold by volume or weight.

[42]See Sect. 6.3.1.1.3.

[43]Not only theoretically, but it is also a viable option to put no price on natural gas. This has been applied, e.g., in Turkmenistan. See Corbeau et al. (2012), p. 41.

[44]Meanwhile CME Group; see also Sect. 5.5.3.2.

so-called capacity prices. These were used primarily by vertically integrated companies supplying end users to charge for the reservation of transportation and storage capacity. However, after the unbundling of companies has been implemented, capacity prices often became otiose.

Obviously, both, buyer and seller take price risks when agreeing upon a price, whether fixed or indexed to the development of, e.g., natural gas prices or oil product's prices. Exceptions would be pure synchronized netback-price-formulae, being risk free for the buyer, or pure cost-plus-formulae, being risk free for the seller.[45]

The seller takes the risk that the contract price at some points in time or during some periods will not cover the costs (mainly for production, structuring, and transport). These costs are the sole responsibility of the seller and beyond the control of the buyer. On the other hand, the buyer runs the risk that the sales prices are not high enough to cover the purchase price and additional costs (e.g., for marketing) when selling the natural gas to its customers. These costs are the sole responsibility of the buyer and beyond the control of the seller. Hence, agreed pricing provisions reflect the agreed risk sharing by the seller and the buyer.

6.3.2.2 Indexed Pricing

The traditional form to set a price formula was to define a (linear) equation expressing the price of natural gas at different points in time as a function of prices of other commodities. Usually, these other commodity prices were prices of competing/alternative fuels, e.g., gas oil, heavy fuel oil, or coal. Accordingly, the price of natural gas changes in a defined way over time. As the variables are not predictable, also such a formula does not imply that the natural gas price is predictable.

In order to determine the structure (i.e., the parameters) of indexed-pricing formulae, the parties used and use various approaches. They are either based on[46]:

- analysis of prices in an end user market, to be defined ('indifference' or 'netback' principle)[47]; or
- comparison with other contracts considered by the parties to be equivalent ('landscape' or 'benchmark' principle); or
- (taking the seller's perspective and) all cost elements ('cost-plus' principle).[48]

The first step of such price determination is the definition of price level itself (at a certain point in time), and the second step is to determine the variables/indices and their respective weightings. In analogy to the price level, the variables are determined by analysis of the end user market (indifference principle), or by comparison of comparable contracts (landscape principle), or by adding all seller's costs (cost-plus principle). The indices are then used in the time-dependent, i.e., indexed, part of the price formula.

---

[45]Applying a pure synchronized netback formula means that the price agreed upon between the buyer and the seller will be calculated by deducting costs and a margin from the price(s) prevailing in the buyer's market. In doing so, the buyer will be put in a position to pay at each point in time a price for the natural gas competitive in his market and to receive a margin. As such, price formulae being synchronized will safeguard that no distortions, even at single points in time, during the term of the contract will occur.

On the other hand, application of the cost-plus principle will guarantee that the seller will receive at each point in time a price covering his costs of production and delivery plus a margin.

Consequently, such pricing concept is risk free for the seller, while the first concept is risk free for the buyer.

[46]In case of indexed regulated prices, the indexation could also refer to, e.g., inflation. See Sect. 6.3.2.5.

[47](In German) '*Anlegbarkeitsprinzip*' is often equated with the 'net-back principle'. See, e.g., Melling (2010), p. 17.

[48]These costs can be categorized as:

a) capital costs for constructing the (production and partly transportation) infrastructure
b) operating costs incurred by the seller and
c) expected (or desired) return

However, as the cost-plus principle is no longer applied (see, e.g., Melling (2010), p. 16), the following discussion will be restricted to the indifference principle and the landscape principle.

The application of the indifference principle and the landscape principle, will result in a price formula describing the price as a function of time and other indices or prices (or their respective averages). As a consequence, the interpretation of price formulae does not allow the attribution of the principle of price determination, i.e., the principle of price determination (indifference, landscape or cost plus), to be identified from the design or structure of the price formula.

6.3.2.2.1 Typical Variables and Indices Chosen

The variables/indices used predominantly in natural gas price formulae are:

- oil product prices: Oil products, such as gas oil or fuel oil, are often considered to be competitive fuels to natural gas in the end user market, and their prices are, therefore, the most commonly utilized variables for natural gas price calculation;
- crude oil prices: Such prices are often preferred by sellers being producers of natural gas because of the interdependence of natural gas and oil production (even in cases it is not associated gas). Buyers, however, will usually not be inclined to accept crude oil prices as a variable for the calculation of natural gas prices, primarily as crude oil is not considered a competitive fuel to natural gas;
- natural gas prices for deliveries at hubs have been used more and more as variables of natural gas price formulae in (long-term) contracts in recent years;
- coal prices are mainly used in contracts for the delivery of natural gas for electricity production. Furthermore, as coal prices are typically less volatile than the other prices mentioned, they are and were used to reduce the volatility of natural gas prices;
- electricity prices are also chosen as an index for natural gas deliveries for electricity production. However, using electricity prices as variables for natural gas price calculation is not everywhere a valid option, as linking the price of the input to the output price creates a circularity or a 'commercial paradox', as variable costs are then dependent on the sales price of the end product, i.e., the secondary energy;
- various inflation indices are and were predominantly used for capacity prices but only rarely for commodity prices.[49] However, used as variable for the calculation of natural gas prices, such indices will also help to reduce volatility;
- prices received by the buyer of natural gas for the final product the buyer (or a buyer's customer) produces can be used as a variable. This may be the case for natural gas used as a feedstock in chemical plants, e.g., ammonia or methanol[50]; or
- other variables as, e.g., prices of wood pellets, were discussed but actually not frequently used.

Typically, in order to grant transparency for seller and buyer, the variables/indices will be easily accessible, reliable published data. In addition, natural gas contracts will include provisions applicable in case the data are not published (any more).

6.3.2.2.2 Principles for the Determination of the Price Formula

6.3.2.2.2.1 Indifference Principle

Establishing and selling an alternative fuel in an existing and developed energy market requires such fuel to be priced competitively. Hence, (potential) natural gas sellers endeavoring to capture shares in new (or even established) markets in competition to other fuels have to incentivize customers by competitive pricing. To this, the price of natural gas is defined as a function of the (averaged) prices of alternative fuels,[51] having due regard to all aspects relevant for the value of natural gas like investment costs for installation of new devices, environmental features, maintenance costs, convenience etc.

---

[49] They were used, e.g., in the UK before 'liberalization'.

[50] The use of such variables is frequently categorized as a separate pricing methodology, see, e.g., International Gas Union (IGU) (2015), p. 13.

[51] That is, other sources of primary energy, which might be used as mutual substitutes.

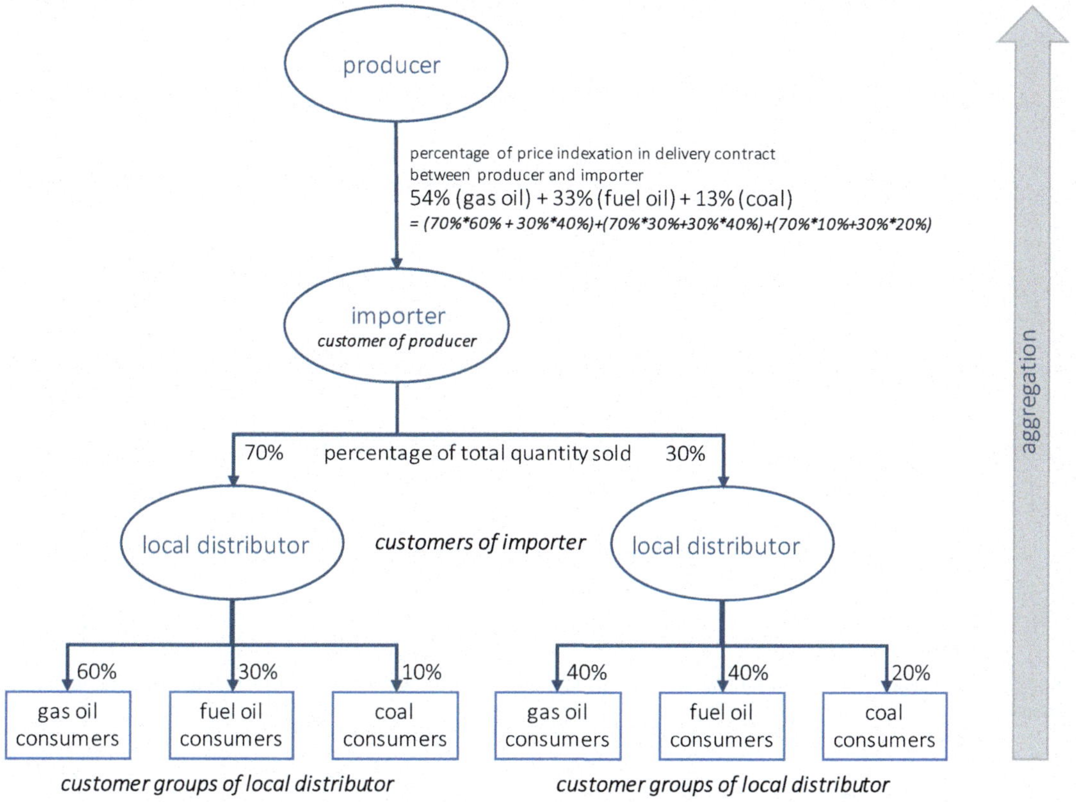

Fig. 6.11 Upstream aggregation of indexed pricing

> Price formulae were designed to ensure that the customer base continued to burn gas rather than returning to oil products since the majority of customers had switched from oil products to gas and, given a price incentive, retained the ability to switch back.[52]

Of course, the application of this principle of determination of the natural gas pricing requires a detailed quantitative analysis of the relevant, i.e., the buyer's, market. To this, the shares of competitive fuels to be replaced by natural gas in the relevant market have to be identified.[53]

Such analysis comprises, *inter alia*, the aggregation of customers in the buyer's market to segments and detailed evaluation of quantities to be sold in such segments. This is required as the competitive fuel is (by definition) different in different segments. That is, the segmentation is performed by determination of the percentage of the total quantity to be sold in competition to the alternative, i.e., the competitive, fuel.

In doing so, the price level in each segment identified is determined based on the analysis of alternative fuels' prices. This includes that for each segment a single 'sub-price-formula' is established.

The price formula agreed upon between buyer and seller could then be regarded as the weighted aggregation of the price formulae of the segments identified. That is, once the prices and quantities for the individual segments are determined, and such determination is based on the analysis of buyer's market as described above, the prices are aggregated and weighted using the quantities to be sold in each segment. The result is a (theoretical) price formula reflecting prices in buyer's market—in structure and level. This procedure is depicted in Fig. 6.11.

[52]Stern and Rogers (2014), p. 9.

[53]Thereafter, such shares are quantified and transformed into equivalent quantities of natural gas.

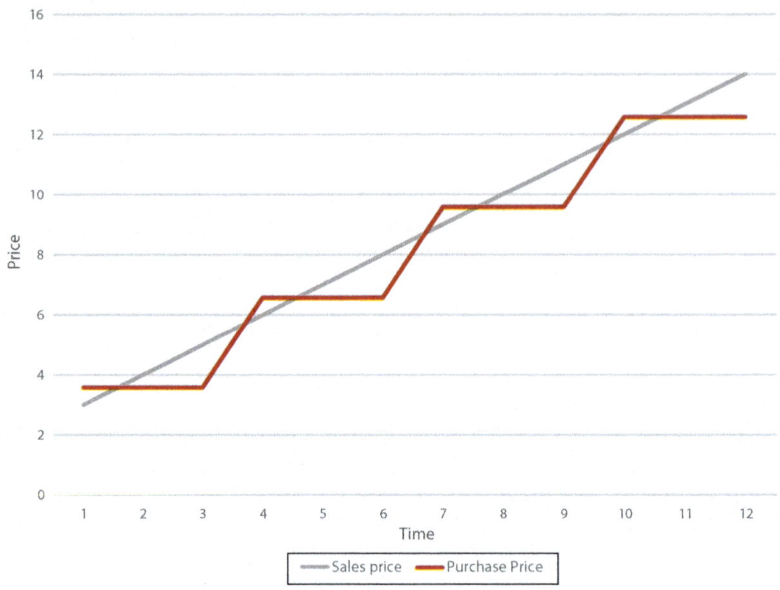

Fig. 6.12 Example of sales and purchase prices, which may 'overlap' for single periods due to differing adjustment periods

However, the price formula determined is used as the basis for the negotiation between buyer and seller and, only after such negotiation, agreed upon and included in the contract.

Therefore, typically a perfect congruence of formula structure with buyer's market was not, and could not be, achieved or guaranteed.

To summarize, the main features to be determined are:

- price level set by the definition of the base price, P_0, whereby such P_0 is often negotiated based on a calculation and not only a result of such calculation;
- price indexation set by the prices of the alternative/competitive fuels chosen; and
- a reference period, a time lag and an adjustment period (incl. price recalculation dates) are chosen, *inter alia*, in order to filter the prices of competitive fuels, i.e., smooth the time dependent price of natural gas in comparison to alternative fuels.

Implications of the above are:

- Based on the pricing system described, natural gas may be sold in some segments below and in some segments above the (average) import price[54];
- There are times of distortion, primarily for buyers reselling the natural gas purchased. If the reference period, the time lag, and/or adjustment period in purchasing and sales do not coincide, then the purchase price may, despite of being in general lower than the sales price, exceed for single points in time the sales price. This is illustrated in Fig. 6.12;
- Once a price formula is established, without additional information, the level of prices intended for each of the different segments cannot be determined by analyzing the price

[54]Such situation is sustainable only, if trading between the customers of the low-price segment and the customers of the high-price segment is prohibited or prevented.

formula alone. In general: knowing the sum does not suffice to identify the addends; and

- Any revision of the price will be based on prices prevailing in buyer's market during the revision period.

The New Dutch Concept of Natural Gas Pricing

The main principles of the Dutch gas policy were established in 1962 in a famous note of the then Minister of Economic Affairs, Jan Willem de Pous, in what became known as the Nota de Pous. In order to generate maximum revenue for the state, the 'market-value' or replacement value principle was introduced as the basis for gas marketing, as opposed to the so-far prevailing principle of cost plus for natural gas. The price of gas was linked to the price of alternative fuels likely to be substituted by the different types of end users—for instance, gas oil for small-scale users and fuel oil for large-scale users. On the one hand, the introduction of the market value principle meant that end users would not have to pay more for gas than for alternative fuels. On the other hand, they would not pay much less.

The market value approach enabled Shell, Exxon, and the Dutch government to obtain much higher revenues than by pricing based on the low production costs of gas from the Groningen field. It also made sure that the growing use of gas did not abruptly jeopardize the past marketing success for oil products.[55]

6.3.2.2.2.2 Landscape/Benchmarking Principle

The benchmarking (or so-called landscape) principle foresees the natural gas price to be determined is based on prices agreed upon in comparable contracts. The benchmark provided by other long-term contracts is thus used to define the price level as well as the price structure.

The logic for buyers is that if the contract price is not higher than the prices of its competitors or peers, e.g., other importers, economical marketability is or—at least: should be—given. The logic for sellers is that, if the price is not lower than which the rest of the market pays, the seller gets fair revenue and, therefore, a fair deal.

Prerequisite for defining the price in accordance with this principle is the (perceived) knowledge of the relevant benchmark contract(s) by both parties. If both parties did not have, or did not assume to have, the knowledge of other comparable contracts (especially the price formulae used in these contracts) at the time of agreement of the price, no benchmarking will be or can be agreed. Indeed, parties agreeing on the application of the landscape principle implicitly acknowledge having knowledge of comparable contracts and their relevant terms and conditions. However, as the knowledge is rather a perceived than a real one, it may differ between buyer and seller.

Consequences and implications of the above are:

- The price formula, i.e., its structure and level, does not necessarily reflect the market price's level and structure in the buyer's market. As it has to be assumed that a price determined on the basis of the landscape principle reflects the 'blended' market prices relevant in all prices and price formulae used for benchmarking, the buyer's single market price structure cannot be matched;
- In the absence of any 'destination clause',[56] prices being the result of price formulae determined by a benchmarking principle will allow the buyer to resell the natural gas purchased in competition to its peers; and
- Application of the benchmark for the determination of a specific price does not necessarily imply equality between the prices. That is, even at the time of conclusion of the contract, the benchmark prices (or their (weighted) average) may deviate from the contract price.

[55] See Energy Charter Secretariat (2007), p. 147.

[56] See Sect. 6.3.8.

6.3.2.2.3 Price Formulae

Application of the principles described above results in a formula describing the natural gas price as a function of indices, which, in turn, may be time dependent. Natural gas prices are, therefore, time dependent, too. Yet, typically the price formulae describe the linkage of the natural gas price to changes of prices of competitive fuels (or any other index), not to the absolute value of the prices of competitive fuels.

6.3.2.2.3.1 Additive Formula

Expressed in general form, the natural gas price at each point in time, $P(t'(t), \overline{\Psi}_i(t'(t)))$ is the sum (or product[57]) of a base price, P_0, and the index-dependent, weighted part(s) of the formula:

$$P(t'(t), \overline{\Psi}_i(t'(t))) = P_0 + \sum_{i=1}^{n} \left[\prod_{j=1}^{m} b_{ij} \{ \overline{\Psi}_i(t'(t)) - \Psi_{i0} \} \right]$$

with

$$\overline{\Psi}_i(t'(t)) = \frac{1}{t_{ref\,i}} \sum_{k=t'-t_{ref\,i}-t_{lag\,i}}^{t'-t_{lag\,i}-1} \Psi_i(k)$$

being the average of the *i*th variable calculated at the adjustment dates over the reference period considering the time-lag.

These averages as well as the prices are calculated for all periods from the beginning of the contract, i.e., $t'(t) = t_{ref\,i} + t_{lag\,i} + 1 + t_{adj\,i} * \text{Integer}\left(\frac{(t-1-t_{ref\,i}-t_{lag\,i})}{t_{adj\,i}}\right)$ periods (typically [months]), for price (re-) calculation, consecutively numbered, i.e.,

$$t' = (\max(t_{ref\,i}) + \max(t_{lag\,i}) + 1), \ldots, t^* \in \mathbb{N};$$

Whereby:

- t periods (typically [months]), consecutively numbered, starting at 1,i.e., $(t_{ref\,i} + t_{lag\,i} + 1)$ periods (months) before the first months of the contract;i.e., $t = 1, \ldots, t^* \in \mathbb{N}(l_i + t_{lag\,i} + 1)$;
- t^* term of the contract, i.e., number of periods (typically [months]), of contract duration;

Further:

- P_0 base price;
- $\Psi_i(t)$ variables, often also referred to as 'indices', defined as a function of time as they typically refer to time-dependent quotations;
- Ψ_{i0} base price for variable i;
- $t_{ref\,i}$ reference period applicable for the calculation of $\Psi_i(t)$; and
  The variables are averaged over the so-called 'reference period'. Typically, the reference period is between 3 and 12 months. Such averaging implies that the volatility of natural gas prices is reduced in comparison to oil product prices, as the moving averages filter the peaks and troughs of the variable.
- $t_{lag\,i}$ time lag applicable for the calculation of $\Psi_i(t)$;
  The 'time lag' defines the gap between the end of the reference period and the beginning of the adjustment period. The reason for the introduction of a time lag in natural gas price calculations was that (often) not all oil product prices were available at the beginning of the immediately following month.[58] If the natural gas price were adjusted with no time lag, this would have implied that, at the beginning of the first month of the adjustment period, the price for the natural gas delivered would have been unknown to (1/reference period) and, therefore, only a preliminary price could have been applied;
- $t_{adj\,i}$ adjustment period applicable for the calculation of $\Psi_i(t)$;
  At the beginning of the adjustment period, typically monthly or quarterly, the averages of the variables are recalculated. Therefore, it is sometimes also referred to as a 'recalculation date'.

(The reference period, the time lag, and the adjustment period are illustrated in Fig. 6.13.)

[57]See Sect. 6.3.2.2.3.2.

[58]For example, at the beginning of December, the average of the gas oil quotations was not yet available. This was published only in mid-December.

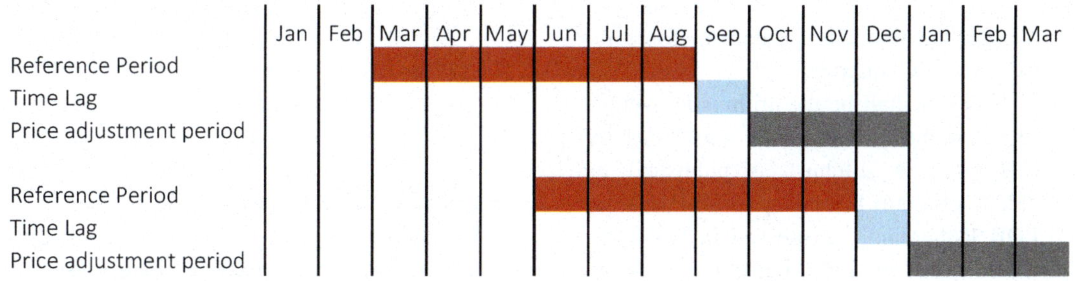

Fig. 6.13 Schematic illustration of reference period, time-lag, and adjustment period

n number of variables;
m number of factors b_{ij} (typically m = 3);
b_{ij} conversion and weighting factors, in particular:

b_{i1} relative weight of variable $\Psi_i(t)$;

i.e., the share of the respective alternative fuel's price changes to the natural gas price, i.e., the relative weight of variable $\Psi_i(t)$;

b_{i2} dampening factor for variable $\Psi_i(t)$;

the pass-through factor of the respective alternative fuel's price changes to the natural gas price. It is typically smaller than 1, i.e., designates a dampening factor,[59] as the changes of the variables $\Psi_i(t)$ will not be entirely passed through; and

b_{i3} unit conversion for variable $\Psi_i(t)$;

the energy conversion factor for variable $\Psi_i(t)$. Converting any unit chosen for $\Psi_i(t)$ to the energetic unit used for price calculation (e.g., [t] to [kWh]).

In practice, the base price, P_0, will be set to reflect the (perceived) value of the natural gas sold at a certain date, e.g., the start of deliveries (if the variables are known or assumed) or the signature of the contract. That is, by setting:

$$\bar{\Psi}_i(t'(t)) = \frac{1}{t_{\mathrm{ref}\,i}} \sum_{k=t'-t_{\mathrm{ref}\,i}-g_i}^{t'-g_i-1} \Psi_i(k) = \Psi_{i0}$$

it follows that $P\left(t'(t), \bar{\Psi}_i(t'(t))\right) = P_0$ at the date of signature.

Simple Fuel Oil and Gasoil Indexation

The following example might be included in a traditional natural gas sales and purchase agreement: The natural gas price to be paid by buyer to seller will be quoted in [€Cent/kWh] and shall be calculated according to the following formula:

$$\begin{aligned} P = P_0 &+ 0.4 * 0.9 * 0.0076 \\ &* (\text{Gas Oil} - \text{Gas Oil}_0) + 0.6 * 0.95 \\ &* 0.0078 * (\text{Fuel Oil} - \text{Fuel Oil}_0) \end{aligned}$$

Let:
P_0 = 2.00 €Cent/kWh
Gas Oil_0 = 400 €/t
Fuel Oil_0 = 200 €/t

and gas oil shall be the arithmetic average of the monthly average of quotations for gas oil (0.2% sulphur) as published by "Platts Oilgram Price Report" for "Barges FOB Rotterdam" (expressed in US $ per ton and converted to € per ton considering the monthly average of the daily US $/€ exchange-rate values indicated by the European Central Bank) of the six months preceding the month of delivery and shall be calculated at the beginning of each

[59]Sometimes, erroneously, also referred to as 'discount' factor. If it is smaller than 1, positive changes of prices for alternative fuels will be decreased (discounted), but negative changes will not be passed through entirely.

quarter and thereafter remain in force for the respective quarter.

Fuel Oil shall be the arithmetic average of the monthly average of quotations for fuel oil (1% sulphur) as published by "Platts Oilgram Price Report" for "Barges FOB Rotterdam" (expressed in US $ per ton and converted to € per ton considering the monthly average of the daily US $/€ exchange-rate values indicated by the European Central Bank) of the nine months period ending one month prior to the month of delivery and shall be calculated monthly and thereafter remain in force for the respective month.

The preceding price formula is an additive price formula with two variables, i.e., gas oil and fuel oil, a six-months' reference period for gas oil, a nine-months' reference period for fuel oil, quarterly price adjustment for gas oil, monthly price adjustment for fuel oil, no time lag for gas oil, and one-month time lag for fuel oil (quotations).

The variables and indices of the general price formula assume the following values:

starting at $t' = 11 = 9 + 1 + 1$ (= max $(t_{ref\,1}, t_{ref\,2})$ + max $(t_{lag\,1}, t_{lag\,2})$ + 1; t' is not defined for $t < 11$, see above)

$P_0 = 2.00$

$\Psi_1(k)$ = gas oil quotation for month k;

$\Psi_2(k)$ = fuel oil quotation for month k;

$Q_{10} = 400$ (Gas Oil_0); $Q_{20} = 200$ (Fuel Oil_0);

$n = 2$; $m = 3$;

$b_{11} = 0.4; b_{12} = 0.9; b_{13} = 0.0076\, t/kWh_{Gasoil};$

$b_{21} = 0.6; b_{22} = 0.95; b_{23} = 0.0078\, t/kWh_{Fueloil};$

$t'(t)$ = month of price (re-)calculation;

$t_{lag\,1} = 0$ month (no time lag);

$t_{lag\,2} = 1$ month (time lag);

$t_{ref\,1} = 6$ (months reference period);

$t_{ref\,2} = 9$ (months reference period);

$t_{adj\,1} = 3$; $t_{adj\,2} = 1$;

as $t_{adj\,1} = 3$, $\Psi_1(t'(t)) = \Psi_1(t'(t+1)) = \Psi_1(t'(t+2))$ and; $t_{adj\,2} = 1$, $\Psi_2(t'(t)) \neq \Psi_2(t'(t+1))$ (Table 6.4).

Japanese LNG Imports

An (additive) price formula is also applicable to the initial form of LNG pricing in Asia. Let JCC be the 'Japanese Crude Cocktail',[60] 'A' the conversion coefficient, converting the JCC quotation of [US $/bbl] to the natural gas price unit of [US $/mmBtu], and 'B' a constant [US $/mmBtu], then the natural gas (LNG) price has been defined by[61]:

$$P(t', JCC) = A * \frac{1}{t_{ref\,1}} \sum_{k=t'-t_{ref\,1}-t_{lag\,1}}^{t'-t_{lag\,1}} JCC_k + B$$

being the reduced form of the additive price formula with:

$P_0 = B$;

$i = 1$;

Q_{1k} = JCC quotation for month k;

$Q_{10} = 0$;

$n = 1$; $m = 1$;

$b_{11} = A\ (\approx 0.172)$

$t_{lag\,1} = 0$ month (no time lag);

$t_{ref\,1} = x$ (defined in individual contract, months reference period);

$t_{adj\,1} = y$ (defined in individual contract).

[60]Japanese Crude Cocktail is an industry-wide used name for JCC. Officially, JCC is an acronym for 'Japan customs-cleared crude oil price'. It is regularly published by the Petroleum Association of Japan, Petroleum Association of Japan (2016).

[61]Such a price formula was frequently supplemented by a collar. See Energy Charter Secretariat (2007), p. 190, for the price formula and Sect. 6.3.2.4.2 for an explanation of the collar.

Table 6.4 Example of a calculation for an oil-product price indexed natural gas price

Date	Exchange rate (monthly average) (random values)	Gas oil monthly average (random values)	Applicable gas oil value	Fuel oil monthly average (random values)	Applicable fuel oil value	Natural gas price
Unit	[€/US $]	[US $/t]	[€/t]	[US $/t]	[€/t]	[€Cent/kWh]
January	0.79	488.00		317.00		
February	0.90	327.00		385.00		
March	0.90	362.00		294.00		
April	0.82	303.00		341.00		
May	0.83	496.00		213.00		
June	0.86	473.00		361.00		
July	0.84	409.00	345.42[a]	391.00		
August	0.90	438.00	345.42[b]	368.00		
September	0.86	495.00	345.42[c]	234.00		
October	0.87	357.00	371.73[d]	399.00		
November	0.81	404.00	371.73	327.00	276.59[e]	2.26
December	0.88	390.00	371.73	311.00	287.33[f]	2.31
January	0.90	469.00	357.42[g]	379.00	278.26[h]	2.23
February	0.81	388.00	357.42	217.00	279.27	2.24
March	0.90	500.00	357.42	232.00	286.10	2.27
April	0.90	351.00	361.24[i]	338.00	285.99	2.28
May	0.84	478.00	361.24	334.00	274.69	2.23
June	0.85	477.00	361.24	308.00	272.00	2.21
July	0.80	456.00	384.88[j]	372.00	266.37	2.25
August	0.83	472.00	384.88	399.00	273.10	2.28
September	0.81	416.00	384.88	301.00	267.60	2.26
October	0.78	321.00	369.40[k]	200.00	274.96	2.25
November	0.91	371.00	369.40	399.00	271.65	2.23
December	0.83	325.00	369.40	319.00	251.08	2.14

Note The oil product-price quotations shown in this example are random values. This time series has not been quoted

[a] Average of quotations January to June (converted monthly to €/t)
[b] Average of quotations January to June (converted monthly to €/t)
[c] Average of quotations January to June (converted monthly to €/t)
[d] Average of quotations April to September (converted monthly to €/t)
[e] Average of quotations January to September (converted monthly to €/t)
[f] Average of quotations February to October (converted monthly to €/t)
[g] Average of quotations July to December (converted monthly to €/t)
[h] … continued analogously
[i] Average of quotations October to March (converted monthly to €/t)
[j] Average of quotations January to June (converted monthly to €/t)
[k] Average of quotations April to September (converted monthly to €/t)

6.3.2.2.3.2 Multiplicative Formula

Alternatively, multiplicative price formulae of the form:

$$P_M(t'(t), \overline{\Psi}_i(t'(t))) = P_{M0} * \left(\sum_{i=1}^{n} (d_i * \overline{\Psi}_i(t'(t))/\Psi_{i0})\right)$$

are or have been used;

with:

$\overline{\Psi}_i(t'(t))/\Psi_{i0}, n, t_{lag\,i}, t_{ref\,i}, t_{adj\,2}, t'(t), t$ defined identically as for the additive formula[62] and

P_{M0} base price;

d_i relative weight of variable $\Psi_i(t)$; i.e., the share ofthe respective alternative fuel's price changes to the natural gas price.

Typically, neither pass-through factors, as sometimes included in the additive formula, nor unit conversion factors, which are not required, as the relation, not the difference of the variable $\Psi_i(t)$ and its initial value, Ψ_{i0}, is applied, are defined.

Further explanations given in Sect. 6.3.2.2.3.1 apply analogously.

Again, in practice, the base price, $P_{0,}$ will be set to reflect the (perceived) value of the natural gas sold at a certain date, e.g., the start of deliveries (if the variables are known or assumed) or the signature of the contract. That is, by setting:

$$\overline{\Psi}_i(t'(t)) = \frac{1}{t_{ref\,i}} \sum_{k=t'-t_{ref\,i}-t_{lag\,i}}^{t'-t_{lag\,i}-1} \Psi_i(k) = \Psi_{i0},$$

it follows that $P_M(t'(t), \overline{\Psi}_i(t'(t))) = P_{M0}$.

It can be shown that—under certain assumptions—additive and multiplicative formulae are equivalent, i.e., that a multiplicative formula could be transformed into an additive formula and vice versa.[63]

6.3.2.3 Gas-on-Gas Pricing

Indexed pricing of natural gas as described in Sect. 6.3.2.2 is mainly used in traditional or immature gas markets, or markets in a transition phase. Once liquid, transparent, and reliable natural gas markets, where natural gas prices are determined by supply and demand are existent, parties to a GSA or LNG SPA, might agree on linking contract prices of (new) contracts[64] directly to a natural gas-price index or quotations instead of linking it to alternative fuels.

To find a reliable natural gas price index requires a liquid market place where prices are set, e.g., at a hub for highly standardized products. In order to sell such a standard product, a seller has to agree with a buyer on a price for such a product. Such an (agreed) price will become the (new) market price, i.e., the principle applied in a 'gas-on-gas' pricing system is that at market places, e.g., exchanges, the price of natural gas

- for the delivery at a defined location, usually at a hub;
- for a certain period; and
- at standardized terms and conditions

is found because of companies placing bids and asks—a polypolistic market structure, including liquidity and transparency being a prerequisite. Natural gas is traded as a commodity.

Typically, the natural gas prices quoted at exchanges for deliveries at hubs are fixed for an agreed period of delivery. Such periods of delivery are, e.g., days or months[65] in the future, i.e., natural gas traded 'now' is agreed to be delivered during one of the above-mentioned defined periods in the future.

While the intraday (if available), the hourly, and the day-ahead markets (sometimes also including the weekend quotations) are called the 'spot market', as natural gas is bought and sold 'on the spot', right now, natural gas deliveries in the future, i.e., up to three years in advance, are traded in the so-called 'forward market'.

[62]See Sect. 6.3.2.2.3.1.

[63]See Appendix F.

[64]In some cases, parties to an existing contract (so far being price-linked to alternative fuels) have tried to alter the pricing to hub-based pricing, e.g., during commercial discussions or during price review negotiations. See also Sect. 6.3.2.7.

[65]For typical terms of standard hub-based contracts, reference is made to Sect. 6.3.4.

A lot size is defined for delivery at each hub, implying that only integer multiples of such a lot are traded.

At exchanges, the prices are quoted throughout the day. At the end of the day for each delivery period, a settlement price is fixed.

As other commodity markets, prices at natural gas markets are inherently volatile, meaning that the prices quoted for the same delivery period will change with time before the beginning of the delivery period. That is, two (or more) buyers buying natural gas for the delivery during a given period will pay different prices, depending on the time they concluded the delivery contract. In fact, natural gas is one of the most volatile commodities currently on the market. After conclusion of the standardized delivery contract for deliveries at a hub, the price is neither negotiable nor adjustable, but fixed for the period of delivery.

The contractual terms and conditions include a firm delivery, as well as a firm off-take obligation. That is, natural gas delivered at hubs is priced with no flexibility—for a firm and continuous delivery during the defined period.

As described, these traded gas prices depend on supply and demand, which themselves depend on many influencing factors as described in Fig. 6.14.

Many different products are traded at the same time. Figure 6.15 provides an impression of the products traded on NBP.

As already described, all these products are highly standardized. Therefore, seller and buyer just need to agree on a price and the quantities for such deliveries.[66] Highly developed markets are not only characterized by highly standardized products and high liquidity in trading, but also by transparency. Prices for exchange-traded products are published by exchanges, and prices for OTC trades might also be published.

If a product delivered under a long-term contract ('LTC'), resembled perfectly a standard product traded at a hub, the parties to this LTC could reasonably agree on a price that is linked to (the published) hub price (development) of such product. By doing so, seller and buyer were able to avoid the regular, e.g., daily or monthly, price-determination process described to find a new market price by placing bids and offers. Similar to price formulae dependent, e.g., on oil products, an exact definition of the gas-price index needed to be agreed between the parties. For example, for deliveries taking place at the German hub NCG, they might agree to use the average of settlement prices published on PEGAS for a certain period as an index for determining the contract price in a specific month. Further, they might agree on a premium to be paid above or below the respective hub price. Such premium might become necessary, if, e.g., flexibility or the delivery point under the LTC is not the same as under the respective hub contracts, the parties decided to use as a price index.

6.3.2.4 Pricing Alterations

Despite all efforts to find a fair price for natural gas delivered under a GSA or LNG SPA, a price under such a contract is always the result of negotiation between seller and buyer. This applies to both price level and price formula. As part of such negotiation result, the parties might also agree on alterations to the indexed-price formula whether gas-price linked or linked to alternative fuels.

Such alterations allow the switch of the 'original' price formula into an 'alternative' price formula if certain conditions are fulfilled. These conditions are defined in the agreement, and they can be both, dependent on the actions of either one of the parties or independent of such actions.

6.3.2.4.1 Alterations Conditional upon Parties' Actions

These alterations depend on actions by either one or both parties to an agreement, i.e., the alteration is not applied automatically, but only if one party or both parties decide to do so during the lifetime of the agreement.

As an example for a switch dependent on the actions of one party, the parties might agree to give the buyer the right to switch an indexed-price formula into a fixed-price formula, either for the complete quantities to be delivered or for only part of those. Such a switch, also referred to as fixed for floating swap, shows characteristics

[66] Assuming a framework agreement between them is in place.

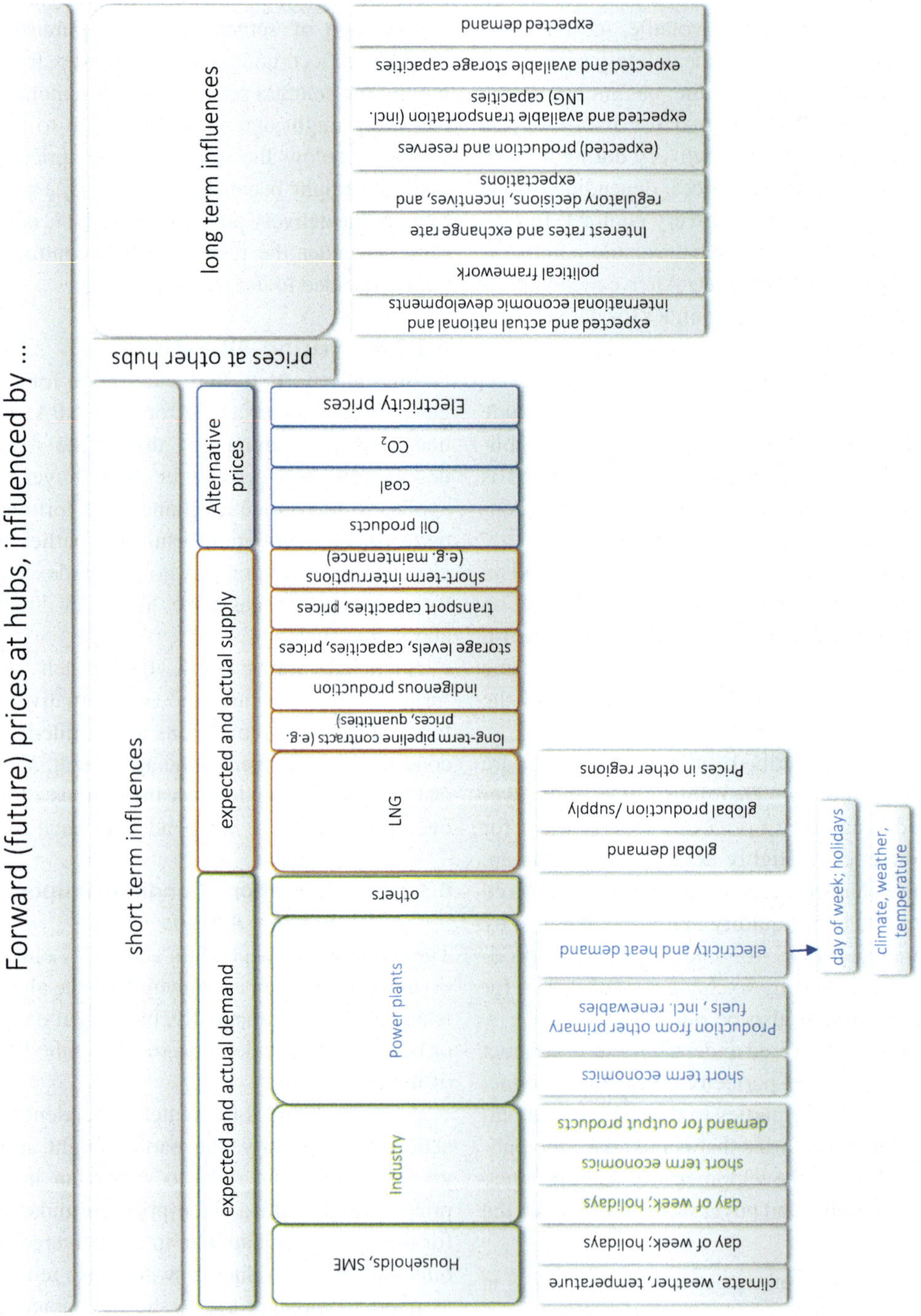

Fig. 6.14 Influencing factors on prices for deliveries at hubs

ICAP UK NBP Gas

Product	Bid	Vol	Ask	Vol	Open	High	Low	Last	Chg	%	VL	VA	Close	Time
NG GAS SPLITS														
BK -Oct-16 (17-31)	:	0	-	0	-	-	-	:	-	-	0	0	39.00	15:42
PROMPT														
Within Day	37.05	0	-	0	37.50	40.00	37.25	37.25	0.75	2.05	0	0	36.50	17:37
Day Ahead	39.75	0	41.40	0	39.00	42.30	39.00	40.28	2.78	7.40	0	0	37.50	17:26
Bal of Week	:	0	-	0	-	-	-	:	-	-	0	0	33.50	04.10.16
Weekend	:	0	-	0	40.50	40.50	40.50	40.50	2.50	6.58	0	0	38.00	17:33
Saturday	:	0	-	0	-	-	-	:	-	-	0	0	34.50	30.09.16
Sunday	:	0	-	0	-	-	-	:	-	-	0	0	35.25	23.09.16
WK/DY NW	39.00	0	-	0	-	-	-	:	-	-	0	0	41.00	16:03
Bal of Month	39.50	0	-	0	-	-	-	:	-	-	0	0	40.75	16:52
Heren Day Ahead	:	0	-	0	-	-	-	:	-	-	0	0	-	12:52
Heren Weekend	:	0	-	0	-	-	-	:	-	-	0	0	0.03	17:25
WEEK														
Wk41-16	:	0	-	0	-	-	-	:	-	-	0	0	37.08	29.09.16
WKDS 42 17/10-21/10	:	0	-	0	-	-	-	:	-	-	0	0	-	16:55
WKDS 43 24/10-28/10	:	0	-	0	-	-	-	:	-	-	0	0	-	16:55
MONTH														
Nov-16	:	0	-	0	43.90	46.25	43.90	45.40	1.10	2.48	0	0	44.30	17:37
Dec-16	:	0	-	0	46.30	47.30	46.30	47.30	1.05	2.27	0	0	46.25	15:16
QUARTER														
Q117	:	0	-	0	46.93	48.50	46.93	47.70	0.60	1.27	0	0	47.10	17:34
Q217	:	0	-	0	-	-	-	:	-	-	0	0	39.85	29.09.16
SEASON														
Sum17	:	0	-	0	41.75	42.84	41.75	42.84	1.44	3.48	0	0	41.40	17:15
Win17	:	0	-	0	47.88	48.47	47.88	48.25	1.22	2.59	0	0	47.03	17:00
Sum18	:	0	-	0	43.10	43.10	42.80	42.80	0.60	1.42	0	0	42.20	17:33
Win18	:	0	-	0	-	-	-	:	-	-	0	0	46.08	04.10.16

October 6, 2016

Fig. 6.15 Screenshot for products traded at NBP on October 6, 2016 (Data provided by 'Montel Nachrichtenservice' [www.montel.de])

of an option insofar as the buyer has the right but not the obligation to ask for a formula switch. Once the option is exercised, the 'original' price formula will be replaced by an 'alternative' fixed price and this new fixed price will be payable independently of whether higher or lower than the 'original' formula price. The 'alternative' fixed price might be determined in two different exclusionary ways:

- the fixed price is pre-determined, i.e., when agreeing upon the switching right, the fixed price itself is already determined, e.g., at 20 €/MWh; or
- the fixed price itself is not pre-determined, i.e., at the moment the buyer chooses to switch the formula the seller makes a proposal for a fixed price based on his/her assumptions and expectations regarding future price development.[67]

Both alternatives lead to a decreased contract price volatility, which might already provide sufficient incentive for the buyer to exercise the switch option. In addition, in the first alternative, the buyer will choose to exercise the switching right, if in his sole view (and based on his expectations) the fixed price provides a better commercial result than the expected formula prices. Contrary to this, the second alternative also reflects the expectations of the seller at the point in time the buyer exercises the option, and the offer will be determined by these expectations. In both cases, the seller might try to hedge the transaction with other trading instruments available.[68]

6.3.2.4.2 Alterations not Conditional upon Parties' Actions

These alterations do not depend on actions by either one or both parties to an agreement, i.e., the alteration is applied automatically.

An example for such alteration might be the automatic switch of an indexed-price formula into a fixed price once certain conditions are fulfilled. These conditions are independent of the parties' actions. An example might be the switch

[67]In many cases, the buyer has the right to reject such a proposal that leads to a prolonged application of the 'original' price formula.

[68]See Sect. 5.5.4.

becoming effective after a certain date or once a certain price level has been reached.

Other common forms of alterations are caps, floors or collars. These are option-like instruments used to limit volatility and level of contract prices.

- Caps work as protection against too high prices. In its simplest form, a cap caps the price at a certain level. As a cap protects against high prices, it is a tool usually wanted by buyers of natural gas. As with all other options, it has a positive value. Hence, a seller of natural gas will not give caps away for free;
- A floor works in the opposite direction, i.e., not allowing the price to fall below a certain level. As it protects against low prices, it is a tool usually wanted by sellers of natural gas. As with all other options, it has a positive value. Hence, a buyer of natural gas will not give floors away for free; and
- Collars are a combination of caps and floors, i.e., the prices are can float only within a certain bandwidth. Collars protect buyers against too high prices and at the same time serve as protection for sellers against too low prices. Dependent on the cap price level and the floor price level, collars can be constructed in a way that the values of the two options outweigh each other. These collars are also named zero-cost collars.

In general, these three instruments can be replicated with a combination of plain-vanilla call options and put options.[69]

In order to reduce volatility, prevent excessively high or low prices, or safeguard that at each point in time the price is not inferior or superior to a defined value, indexed prices can be altered for certain values of the variable. Typically, the inclination of the (linear) equation will be modified piecewise.

A cap limits a price increase otherwise taking place. In contrast to a price discount, a cap is always conditional. That is, if certain price conditions are fulfilled, the cap will be applied. A cap can be defined by either:

- not allowing the price to rise above a fixed level (fixed cap, e.g., the price shall be the lower of the result of an oil product-indexed formula or 30 €/MWh); or
- not allowing the price to rise above another floating-price level (floating cap, e.g., price shall be the lower of the result of an oil product-indexed formula or the result of a hub-price formula).

The conditions to be fulfilled can either relate to the result of the price formula (without cap) itself (i.e., the resulting price shall not be allowed to rise above a certain fixed or floating level) or to parts of the price formula (i.e., the value of a certain variable included in the price formula is not allowed to rise above a certain level). Combinations of the two are possible.

A floor has similar functionality and characteristics except that it always works in the direction of a limitation of a price decrease otherwise taking place.

A collar is a combination of a cap and a floor. Obviously, the collar's structure has to be defined in an unambiguous way, i.e., the conditions triggering the cap must be different from the conditions triggering the floor.

With reference to the general indexed (additive) price formulae[70] (analogously for the multiplicative price formula), a cap and a floor (and, accordingly, a collar) can be described as:

$$P\left(t'(t), \overline{\Psi}_i\left(t'(t)\right)\right) = \begin{cases} P_0^{cap} + \sum_{i=1}^{n}\left[\prod_{j=1}^{m} cap * b_{ij}\left\{\overline{\Psi}_i\left(t'(t)\right) - \Psi_{i0}\right\}\right] \text{for } \overline{\Psi}_i\left(t'(t)\right) > \Psi_{icap}, \\ P_0 + \sum_{i=1}^{n}\left[\prod_{j=1}^{m} b_{ij}\left\{\overline{\Psi}_i\left(t'(t)\right) - \Psi_{i0}\right\}\right] \text{for } \Psi_{ifloor} \leq \overline{\Psi}_i\left(t'(t)\right) \leq \Psi_{icap} \\ P_0^{floor} + \sum_{i=1}^{n}\left[\prod_{j=1}^{m} floor * b_{ij}\left\{\overline{\Psi}_i\left(t'(t)\right) - \Psi_{i0}\right\}\right] \text{for } \overline{\Psi}_i\left(t'(t)\right) < \Psi_{ifloor}, \end{cases}$$

with $cap < 1$ and $floor < 1$ being the factors reducing the inclination. A linear collar is depicted in Fig. 6.16, and a fixed-price collar in Fig. 6.17.

[69]See Sect. 5.5.4.

[70]See Sect. 6.3.2.2.3.1.

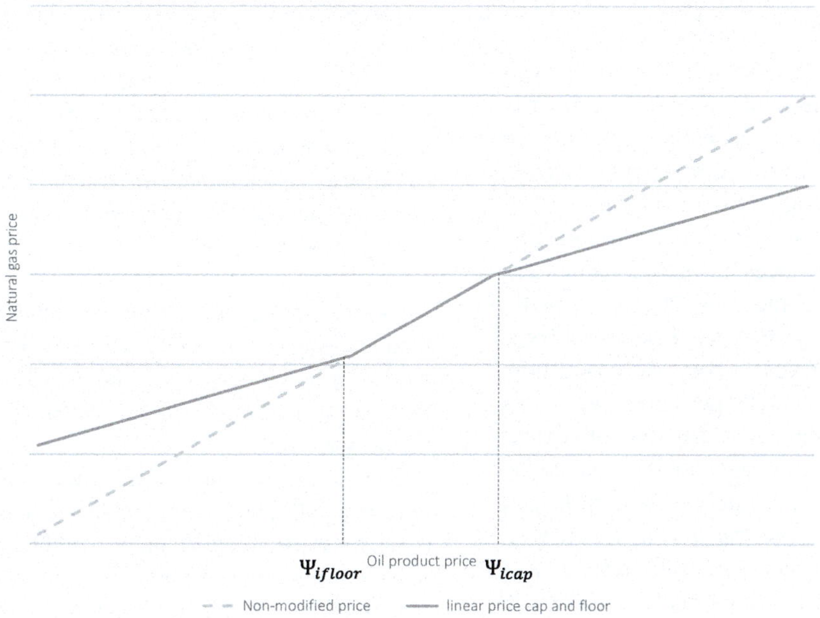

Fig. 6.16 Schematic example of a linear price cap and floor

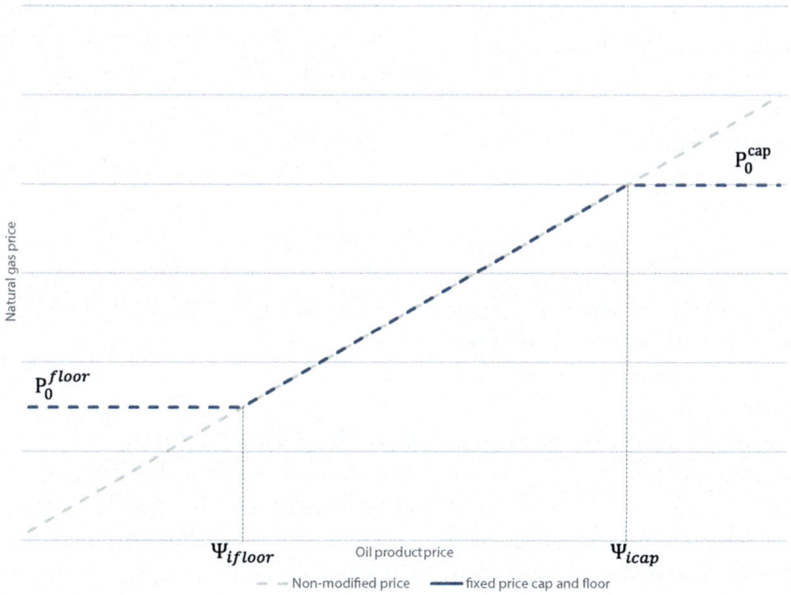

Fig. 6.17 Schematic example of a fixed price cap and floor

Collars

When prices of crude oil dropped considerably at the end of 1997, primarily because of the increase of oil production triggered by OPEC (the largest increase since 1988), and consequently also prices for oil products used for the determination of natural gas prices decreased, natural gas producers selling to Europe partly encountered severe financial problems. This was because costs of production, treatment, and transportation were not always covered (e.g., Russian oil export revenues are estimated to have fallen by approximately 25% during the first half of 1998 compared to the first half of 1997.) Hence, producers requested price floors from importers. Importers, inclined to accept the necessity of such a request, demanded in return price caps. These arrangements proved to be profitable for both parties, yet in different periods. While the floors helped producers to overcome low prices by the end of the millennium, importers profited by the caps primarily during the middle of the first decade.

6.3.2.5 Regulated Pricing

In some countries, the price of natural gas for selected segments or even the whole market is regulated by a governmental or other regulatory authority. Neither supply and demand for natural gas nor supply and demand for alternative fuels but political arguments will determine the price (level) of natural gas to be paid by customers of the regulated segments.

The level of regulated prices can be set following different objectives[71]:

- cost of service: the level of the natural gas price (s) is set to cover the 'cost of service', including capital expenditures, operating expenditures, and a rate of return or margin of the supplier; or
- social and political: the prices of natural gas are set based on a compromise between the perceived necessity to cover the costs of service and provide what is perceived as 'adequately' or 'reasonably' priced natural gas to customers of the relevant segments. In this context, prices are often set below cost of service, i.e., production, transport, and storage of natural gas. This requires (state) subsidy of such prices.

As these prices are set by the authorities at a distinct point in time only, neither volatility nor market developments are necessarily directly observed or included in the determination of the natural gas price.

If the regulated natural gas prices are low or, more generally, lack competitive incentives, such as e.g., in some countries in the Middle East, inefficient consumption of energy may be fostered.

In Europe, regulated pricing was exposed to criticism of the EU.[72] Therefore, it can reasonably be assumed that, if at all, it will persist only in selected segments, as, e.g., households.

Regulated Prices in France

A European example of regulated pricing is the household segment in France. Here the so-called 'Base', 'B0' and 'B1' regulated 'tariffs' (i.e., prices) continue to exist in 2016, while the regulated 'tariffs' for groups of customers with higher natural gas demand were abolished by the end of 2015. The price to be paid consists of an annual base price and a commodity price, fixed over the regulatory determined period of one year.[73]

6.3.2.6 Capacity Price

In addition to a commodity price, often a so-called capacity price has to be paid by a buyer to a seller.

By a 'classical' capacity price, the maximum capacity reserved (or used) during a period for deliveries to a customer is priced. Such prices were and are supposed to cover the seller's transportation and storage costs.

[71] See International Gas Union (2016), p. 11 and p. 59.'

[72] See, e.g., CMS Law (2016) and European Commission (2015).

[73] See tarifgaz.com (2016).

A capacity price is charged either based on a

- capacity reserved,
  In this case, the capacity price is restricted to represent the amount payable for one unit of capacity reserved, yet not necessarily utilized, i.e., irrespective of utilization, during the term of a (commodity) contract.[74] The capacity price represents the fixed costs, while the commodity price represents buyer's specific variable costs; or
- capacity used
  In this case, the capacity price is charged for the maximum flow during the term of a (commodity) contract. Then both, the capacity price and the commodity price represent buyer's specific variable costs.

during a specified period.

If the capacity price is charged on a reserved capacity, e.g., a reserved capacity of 10,000 kWh/h (or 10,000 kW) the buyer has to pay this reserved capacity independently from the actual off-takes. In this case, the capacity price is also referred to as base price or basic charge. A base price can be paid on a monthly, a quarterly, a seasonal, or an annual basis. Shorter-term installments are common. In practice, primarily because of their inherent simplicity, base prices are typically included in contracts with end users.

Capacity Price I

In a natural gas delivery contract, a maximum hourly quantity of 10,000 kWh/h is agreed. The supplier reserves a transportation capacity of 10,000 kWh/h to be able to transport such deliveries to the customer.[75] The capacity price for such capacity reservation shall be 20 €/kWh/h/a.

Accordingly, the customer must pay 10,000 kWh/h * 20 €/kWh/h/a = 200,000 €/a to the supplier for reserving the capacity, i.e., payable in addition to any commodity price.

Consequently, despite of having agreed upon a capacity price but not a base price, the application of such capacity price to a fixed, maximum capacity turns the result for the buyer into the same as if a capacity-induced base price had been agreed.

If, admittedly inconsequently, the capacity price is not applied to the capacity but to the maximum flow during a period, e.g., during a year, then the initial concept of a capacity price being a price payable irrespective of utilization is abandoned. Nonetheless, such pricing exists. It is commonly referred to as 'capacity utilization' pricing. If the capacity price is charged on such capacity utilization, the buyer has to pay a charge based, e.g., on the maximum hourly off-take during a year. Again, such capacity price can be paid on a monthly, a quarterly, a seasonal, or an annual basis. Shorter-term installments are common.

Capacity Price II

In a natural gas delivery contract, a maximum hourly quantity of 10,000 kWh/h is agreed. The supplier reserves a transportation capacity of 10,000 kWh/h to be able to transport such deliveries to the customer.[76] The capacity price shall be 20 €/kWh/h/a. Assuming that the maximum requirements of the customer during a year were not 10,000 kWh/h but 8700 kWh/h only, the customer had to pay 8700 kWh/h * 20 €/kWh/h/a = 174,000 €/a to the supplier for using the capacity. This amount depends on the maximal flow during a period but not the capacity.

While capacity prices have been used at all levels of the natural gas industry, after implementation of unbundling of companies, i.e., after transportation and storage services are offered separately, charging capacity prices has become less common. However, there are still some commodity-delivery contracts with separated commodity and capacity prices.

[74]Following this logic, prices paid for the transportation of natural gas in pipelines or for storage services can also be regarded as (a special case of) capacity prices.

[75]A reservation of storage capacity shall be neglected.

[76]A reservation of storage capacity shall be neglected.

In long-term contracts, capacity prices could be indexed, i.e., expressed as a function of an, e.g., inflation index. The formula used is identical to the multiplicative price formula shown in Sect. 6.3.2.2.3.2. However, the sum of weighting factors does not necessarily equal one (or 100%). To reflect cost developments, most contracts foresee even up to 75% of the price being fixed and only as low as 25% being indexed to, i.e., escalated with, (cost development) indices.

German Indices used as Variables of Long-Term Capacity Prices

- Index of hourly wages paid by utilities: Statistisches Bundesamt: Index tarifliche Stundenlöhne in der Energieversorgung: Fachserie 16; Löhne und Gehälter; Reihe 4.3—Index der tariflichen Stundenverdienste im produzierenden Gewerbe und im Dienstleistungsbereich; früheres Bundesgebiet; WZ 35;
- Index of hourly wages paid by mining companies: Statistisches Bundesamt: Index der tariflichen Stundenverdienste im Bergbau und Gewinnung von Steinen und Erden. Fachserie 16; Löhne und Gehälter; Reihe 4.3—Index der tariflichen Stundenverdienste im produzierenden Gewerbe und im Dienstleistungsbereich; Früheres Bundesgebiet; WZ 2008;
- Index of monthly salaries of utilities: Statistisches Bundesamt: Index tarifliche Monatsverdienste im Wirtschaftszweig Energieversorgung, insgesamt (available under www.destatis.de, GENESIS-Online, Statistics Code 62221-0001, WZ08-35);
- Base salary agreed between the union of mining, chemistry and energy, and the association of crude oil and natural gas producers: Ecklohn des Lohntarifvertrages zwischen der Industriegewerkschaft Bergbau, Chemie, Energie (IG-BCE) und dem Wirtschaftsverband Erdöl- und Erdgasgewinnung e.V. (WEG); or
- Investment index: Statistisches Bundesamt: Index der Erzeugnisse der Investitionsgüterproduzenten (lfd. Nr. 3) Fachserie 17; Preise; Reihe 2;—Index der Erzeugerpreise gewerblicher Produkte (Inlandsabsatz)—1.1 Aktuelle Ergebnisse

6.3.2.7 Price Review

The parties to a long-term contract may not consider the price (formula)[77] defined in a contract to be an adequate tool for the appropriate adjustment of the price to any changes in market environment. The contract price (formula) may not be able to reflect, e.g., general modifications of the market(s) and pricing structure. However, it is generally the parties' intention to ensure competitiveness of natural gas sold and purchased.

Therefore, additional provision(s) to be used to adapt the price (formula) are often included in natural gas contracts. Typically, the objective of such provisions is to provide a contractual 'tool'

- to the buyer to market economically the natural gas purchased; and
- to the seller to support covering its costs of production, processing, and (if applicable) transportation

in the long-term and under circumstances when the price formula is not considered any more the appropriate tool to accomplish such objectives. (Yet, applicability in the long-term does not include applicability at each point in time or period.)

The most important 'tools' to adapt a price (formula) and potentially other provisions of a

[77] (Formula) as, at least in principle, price review clauses could also be applied to fixed prices.

contract are[78] either (extra-contractual) commercial discussions or price review clauses.[79] In addition, hardship clauses may also be used by the parties to adjust the price. Yet, these clauses serve other purposes as well.[80]

6.3.2.7.1 Commercial Discussions

Obviously, price provisions can be modified by agreement between the parties to a contract. Modification of price provisions may include:

- adjustment of the base price;
- modification of price variables and the respective elements as, e.g., the reference periods; adjustment period, and time lag; and
- replacement of (parts of) variables with others.[81]

Such modifications could be structured to be neutral, i.e., giving neither seller nor buyer an advantage, or for the benefit of either seller or buyer.[82]

In any case, modifications of the price might also require the appropriate evaluation, consideration, and potential amendment of economically relevant (non-price) provisions of a contract, as, e.g., volume flexibility or payment provisions.

In commercial discussions, it is also possible for parties to agree to change the underlying pricing concept fundamentally, e.g., by moving from an oil product-indexed pricing to an entirely hub-based pricing.[83]

6.3.2.7.2 Price Review Clauses

In long-term contracts, the pricing provision is often supplemented by a price review clause (also known as price 're-opener' or 'revision' clause), granting both parties the right to ask for an adjustment of the contract price (formula) at defined points in time during the term of the contract.

Yet, price review clauses are usually restricted to modifications of the price (formula). That is, amendments of other contractual provisions, e.g., quantities, quality, delivery point, or flexibility provisions, based on the price review clause are excluded—implicitly or explicitly. Per se, a price review clause does not constitute a right to adjust the price of a long-term contract, but just a right to (re-) negotiate the price (formula).

'Technically', the price review clause is no (second) indexation, it is not supposed to replace the price formula, and the phrasing is necessarily vague. The adjusted contract price is not calculated and re-written from scratch, but only amended to the extent considered necessary to reflect the relative market changes.

A typical (yet generalized) example of a price review clause may be:

> *If circumstances [....], beyond the control of the Parties, change significantly compared to the underlying assumptions in the prevailing price provisions, each Party is entitled to an adjustment of the price provisions reflecting such changes. The price provisions shall in any case allow the gas to be economically marketed based on sound marketing operation.*
>
> *Either party shall be entitled to request review of the price provisions for the first time with effect of dd/mm/yyyy and thereafter every three years.*
>
> *Each party shall provide the necessary information to substantiate its claim.*
>
> *Following a request for a price review the Parties shall meet to examine whether an adjustment of the price provisions is justified. Failing an agreement within 120 days either Party may refer the matter to arbitration in line with the provisions on arbitration of the Contract.*
>
> *As long as no agreement has been reached or no arbitration award has been rendered all rights*

[78] Others, as, e.g., fiscal methods shall not be discussed here.

[79] As price review clauses have gained the most attention in natural gas markets worldwide, they will be discussed subsequently in more detail than the others.

[80] See Sect. 6.3.2.8.

[81] For example, replacing oil product-price quotations in a price formula with, e.g., day-ahead quotations or month-ahead quotations at a defined hub.

[82] Since such consensual price changes are normally forward-looking, they will seldom completely match the parties' expectations, given the inherent degree of uncertainty as to future price developments.

[83] In this particular case, however, the risk sharing and the initial economic bargain between the buyer and the seller might change, as the buyer would be relieved of any

(Footnote 83 continued)
volume risk at least as long as the markets are liquid. Furthermore, the parties to a contract are also free to alter non-price provisions of a contract (e.g., changing quantities or flexibility or delivery points).

and obligations under the agreement—including the price provisions—shall remain unchanged.[84]

In short[85]: If at any time, either of the parties shows that the circumstances in the relevant (natural gas) market(s) have significantly changed in comparison to the ones when the contract was signed (or renegotiated for the last time), i.e., during[86] a so-called review period, and such changes were and are beyond the control of the parties and not reflected in the prevailing price (formula), then this party, by notifying the other and by substantiating the reasons, may request to enter into (re-) negotiations of the price (formula). During these negotiations, the parties are supposed to determine whether such reasons exist and justify a revision of the contract price. In case an agreement cannot be reached during such negotiations, the case is referred to arbitration (or a court).

Further, many price review clauses contain so-called 'in-any-case' clauses, which work as an insurance for the buyer that should allow the natural gas to be economically marketed, assuming the buyer is using sound marketing practices. Such clauses are triggered by many buyers during their price review discussions.

Despite the fact that price review clauses of different contracts are worded heterogeneously, some common features can be identified. The price review process typically consists of two steps: only after identifying that the so-called trigger criteria are met, the parties will determine how the price review shall be affected. This includes the parties' decision to refer the matter to arbitration. To accomplish this, each price review clause contains a

- trigger,
- procedural stipulations, and
- methodological rules.

A precondition for a price review is that the defined trigger criteria are met. The trigger is an event or events allowing the parties to invoke the price review procedure. This includes, e.g.:

- the reference to a specific market in which the conditions (sometimes specified in detail) are deemed to have changed during a specified review period.[87] This specific market may be delineated, e.g., geographically or by segments[88];
- the condition that the market changes must have caused the value of natural gas in this market to change;
- a proviso whereby such change must have been, e.g.,[89]
  - not reflected in the price (formula), e.g., the indices used in the price formula change uncorrelated to the triggering market changes.
    This implies that changes of market prices of alternative fuels, which are used as variables in the price formula, may not be used to trigger a price review[90];
  - significant,
    Determining whether a change in circumstances or in the market is significant[91] may require extensive discussions between the parties and considerable effort. On the other hand, parties may be inclined to include quantified benchmarks to determine the significance of changes. This, in turn, may result in establishing an (unwanted) overlying, second price indexation; and
  - beyond the control of any party to the respective contract; and
- a proviso that each party has acted reasonably and prudently.
  If, e.g., a buyer not acting reasonably and prudently, e.g., by dumping prices or increasing the marketing costs anomalously, were entitled to require a price review and subsequent price adjustment, the seller would have to bear all such costs. As such behavior of the buyer is (at least assumed) not (to be) in line

[84] Hobér (2014), p. 231.

[85] The following only sketches the main principles most frequently included in a price review clause.

[86] Often more specific: at the beginning and at the end.

[87] Which is not contained in the above example.

[88] See below—pricing principle applied.

[89] Often in addition, the unforeseeability is discussed.

[90] An exception to this rule is the imposition of (new or changes to existing) taxes or other duties on these alternative fuels.

[91] Sometimes also referred to as 'material'.

with the intention of the parties concluding the contract, reasonable and prudent market behavior of both parties is a prerequisite for any price review negotiation and arbitration.

The possibility for price review is frequently defined periodically. The price will be automatically reviewed at certain dates during the term of a contract (e.g., every three years). Non-periodic possibilities may also be included. These are 'special events', which may be used by a party to invoke an extraordinary price review negotiation.

Procedurally it is typically stipulated that (re-) negotiation of the price (formula) has to take place or even that alternative forms of dispute resolution, such as mediation, are invoked before referring the case, as ultima ratio, to arbitration. In this context, occasionally also the minimum time requirement of such negotiations is defined in the price review clause.[92] In all cases, parties have to be aware that (re-) negotiation, mediation, and arbitration are sometimes (very) lengthy processes.[93]

Generally, neither of the parties is entitled to terminate the contract as a consequence of a triggered price review. However, this, as well as other rules, will depend on the legal system chosen to dominate the contract.

Methodologically, for both the initial determination of the contract price (including the price formula) and the price review, the parties to a long-term contract used, and use, different approaches. They are either based on a comparison

- with values of natural gas in an end user market, the 'market-value principle' or 'indifference principle'[94];

  or
- with other contracts considered by the parties to be equivalent, the 'benchmark principle'[95];

  or
- on taking the seller's perspective and adding all cost elements, the 'cost-plus' principle.[96]

In case the price review clause uses a 'change concept' (e.g., with wording as described in the example above), the price adjustment shall reflect 'changes', i.e., any adjustment will be equivalent to such changes. Changes of prices can only be identified by comparing prices at two points in time (at least). Therefore, the adjustment that a party can claim is the difference (the 'delta') of the difference (the 'delta') of the reference prices[97] at the beginning and at the end of the review period and the difference (the 'delta') of the contract prices at the same points in time.[98] This is known as the delta-delta method. That is, the price adjustment, Δ, is:

$$\begin{aligned}\Delta = {} & (\mathrm{P}_{\mathrm{Market}}(\mathrm{t}_2) - \mathrm{P}_{\mathrm{Market}}(\mathrm{t}_1)) \\ & - \left(\mathrm{P}\left(\mathrm{t}'(\mathrm{t}_2), \overline{\Psi}_{\mathrm{i}}\left(\mathrm{t}'(\mathrm{t}_2)\right)\right)\right. \\ & \left. - \mathrm{P}\left((\mathrm{t}_1), \overline{\Psi}_{\mathrm{i}}\left(\mathrm{t}'(\mathrm{t}_1)\right)\right)\right)\end{aligned}$$

If such comparison reveals a difference considered significant, the contract price may be adjusted accordingly.[99] That is, if the difference of contract prices is larger than the difference of reference prices, i.e., $\Delta < 0$, the contract price for the time (periods) after to price review date may

[92]For example, six months.

[93]For example, only the process of arbitration can take several years. If the periodicity of price review is three years, the parties may start a new 'round' before having finalized the previous.

[94]See Sect. 6.3.2.2.2.1.

[95]See Sect. 6.3.2.2.2.2.

[96]See Sect. 6.3.2.2, e.g., for adaption of the 'plus'; yet, this approach is dominated by the inherent information asymmetry between seller and buyer.

[97]Determined based on the indifference principle, the benchmark principle, or the cost-plus principle.

[98]A problem, even after having agreed to the general methodology, may arise when trying to define the 'start date' and 'end date'. Whether this is just a day, a month, a quarter, or any other period at the beginning and end of the period may give rise to a potential dispute between the parties. It is usually argued that one day is not appropriate to represent the price at 'the beginning' or 'the end' of the review period. Therefore, frequently the reference prices and the contract price are averaged over, yet arbitrarily chosen, periods.

In more sophisticated forms, the change of circumstances may refer to changes of non-price parameters, such as, e.g., regulatory changes.

[99]Such calculations have to be performed comparing comparable prices, i.e., by taking, e.g., transport and flexibility costs into appropriate consideration.

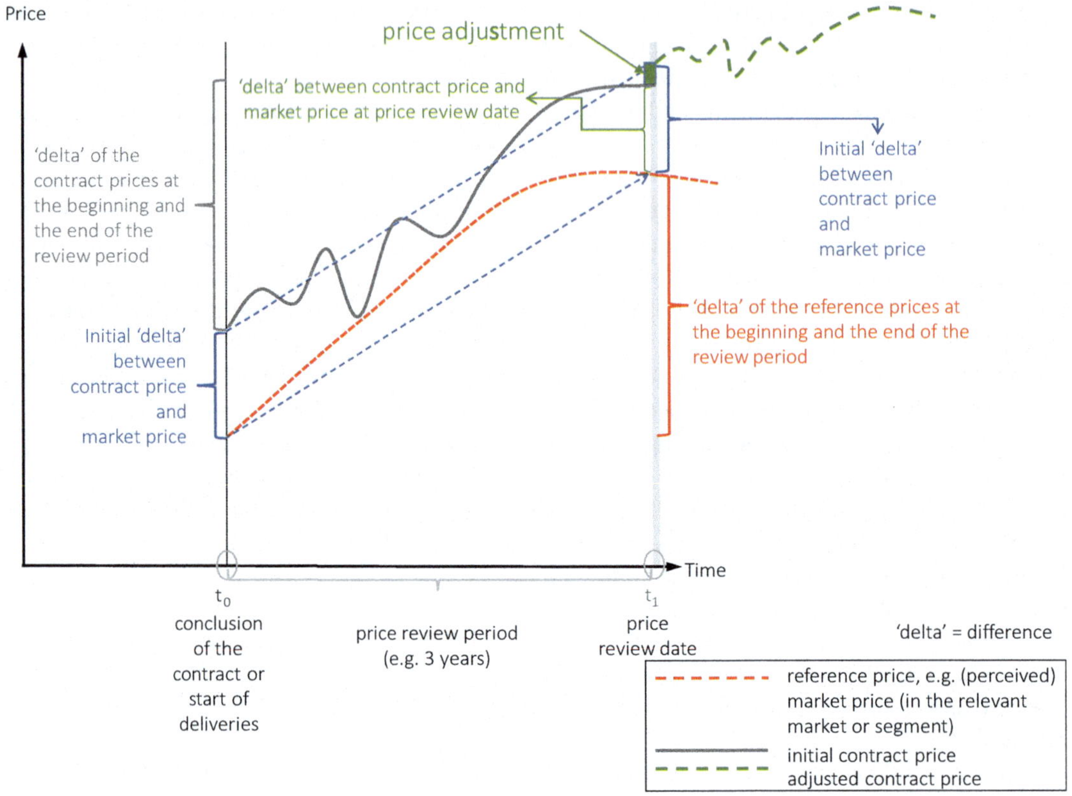

Fig. 6.18 Schematic illustration of the delta-delta method

be decreased and vice versa. Such calculation complies with the inherent requirement that changes but not the absolute values shall be taken into account. This is depicted in Fig. 6.18. Here the difference between the contract price and the market price at the beginning of the review period is larger than at the end of the review period, i.e., $\Delta > 0$. That is, the difference between the contract prices at the beginning and the end of the review period is smaller than the difference of the reference prices at the beginning and the end of the review period. Hence, the contract price has to be increased.

The result of a price review may be an adjustment of the price formula. Such adjustment usually aims at recovering the initial bargain between the parties.

6.3.2.8 Implementation

Gas-on-gas pricing, (oil product) linked pricing, and regulated pricing are worldwide the most commonly used pricing systems. The International Gas Union[100] annually publishes a survey of natural gas price-formation mechanisms and shows therein the dominant pricing methods used in various sectors or regions.

In 2015, 52% of worldwide pipeline imports were based on gas-to-gas pricing and 38% on an oil (product) price link. The remaining 10% had a pricing system referred to as 'bilateral monopoly'.[101] This pricing system is mainly used in the Former Soviet Union and in the Middle East.

[100]See International Gas Union (2016).

[101]*The price is determined by bilateral discussions and agreements between a large seller and a large buyer, with the price being fixed for a period of time—typically one year. There may be a written contract in place but often the arrangement is at the Government or state-owned company level. Typically there would be a single dominant buyer or seller on at least one side of the transaction, to distinguish this category from GOG, where there would be multiple buyers and sellers trading bilaterally.*; See International Gas Union (2016), p. 11.

LNG imports, however, were based 31% on gas-to-gas pricing and 69% on an oil (product) price link.

Also, regional differences can be observed. Whereas in North America nearly all deliveries are based on gas-to-gas pricing, European deliveries show a different picture. The pricing is based to:

- 64% on gas-on-gas competition,
- 30% on oil (product) price indexation,
- 4% different forms of regulated pricing, and
- 2% bilateral monopoly and other.

In Asia, oil (product) price indexation dominates. The detailed results are:

- 14% gas-on-gas competition,
- 59% oil (product) price indexation,
- 26% different forms of regulated pricing, and
- 1% bilateral monopoly.

Less developed markets like the ones in Africa show only 8% oil (product) price indexation and 7% gas-on-gas competition. The remaining parts are based on regulated prices, bilateral monopoly, or other methods.

6.3.3 Hardship

If a party is experiencing substantial hardship, i.e., suffering losses or being exposed to other economically adverse conditions, a hardship clause, if included in the contract, may allow for a review of any of the contractual terms, including but not limited to the price. Such hardship, again, has to be a result of a significant change in circumstance, which, again, are beyond the control of the parties.

The International Chamber of Commerce (ICC) states:

> *A hardship clause ... requests re-negotiation of the contract if the continued performance of one party's contractual duties has become excessively onerous due to an unforeseen event beyond the control of that party.*[102]

Also for hardship clauses, there is no single industry-wide standard, rather their design and use varies between different markets and different legal systems. However, the main differences between hardship and price review clauses is that the former typically do not contain a provision for periodic adjustment of prices.

There is often an argument in the application of a hardship clause as to whether it is a specific legal entity (i.e., company) as a whole that must be suffering hardship or whether hardship is suffered only in relation to the specific contract in which the relevant clause is being invoked. This has to be solved with regard to the wording in the particular contract.

Typically, a hardship clause is more than a clause that allows discussions between the parties to a contract:

> *Furthermore, in order to preclude the argument that a hardship provision represents little more than agreement to agree between the parties and so is unenforceable on that basis such a provision might go on to state that if the parties are unable to agree a basis for revision of the contract price then an independent arbiter such as an expert (...) or an arbitral panel (...) will be appointed to decide the issue.*[103]

6.3.4 Term

The term of a contract is one of the essential clauses, despite its usual (textual) brevity. The term, i.e., the period during which the parties' contractual rights and obligations are agreed, is typically defined as the period between the date of signature of the contract and the termination date, including the period during which natural gas is delivered and off-taken. There is nothing like a 'typical' term of contracts. While LTCs with terms of up to 30 years still exist, shorter-term contracts, i.e., even for daily deliveries, are concluded as well. Introducing contractual preconditions, the parties could also agree that a contract only becomes fully effective if certain conditions precedent are fulfilled. The parties to a natural gas contract, in particular to supply-type contracts, typically do not have extraordinary termination rights.

[102] International Chamber of Commerce (2003).

[103] Roberts (2014), p. 79.

Long-Term Gas Contracts: A Very Peculiar Business

LTCs, being GSAs with terms typically exceeding 5 years, between the producers and the buyers were the backbone of the development of the European gas industry. The number of players on the sellers' and on the buyers' side was limited. Sellers were a couple of very large—often state-owned—producers. In the Norwegian case, the producers founded a state-supervised monopoly GFU (Gassforhandlingsutvalget, Gas Negotiation Committee) that negotiated with the sellers. Buyers were the large European midstream companies like Ruhrgas (Germany), Gaz de France (France), or Eni (Italy). Contract negotiations did not only take years but the teams also met afterwards very frequently to discuss ongoing contractual topics. Such discussions and negotiations were tough and the reviews of the price that usually took place every three years were prepared thoroughly and with studies from outside experts to determine whether price adjustments are justified.

At hubs, forwards (or futures) are traded. That is, seller and buyer assume the right and obligation to deliver and off-take natural gas at a defined hub during an agreed period. The products traded are deliveries for defined future

- days,
- weekends,
- working days next week,
- balance of the month,
- months,
- quarters,
- seasons,
- calendar years, and
- gas years.

6.3.5 Delivery Point

The delivery point is a very clearly defined point where the title and risk of the natural gas sold passes from seller to buyer. In GSAs, a detailed sketch of the pipeline systems and the facilities surrounding the delivery point often complements the definition of the delivery point.

The delivery point specification frequently includes the specification of the delivery pressure, or minimal and maximal values thereof, as well as a reference to the quality specifications, either directly in the GSA or indirectly by reference to the respective transportation agreement(s).

A hub, being physical or virtual,[104] can be chosen as the delivery point as well.

As in GSAs, the definition and agreement on the delivery point is one of the most important topics when negotiating an LNG SPA. The geographical point where the LNG is physically delivered from seller to buyer is usually also the point at which risk of loss, title to, and custody of the LNG is transferred from seller to buyer.[105] A precise (geographic) description is necessary for LNG SPAs, as well as for GSAs.[106]

In practice, LNG SPAs use Incoterms® rules as orientation when defining the delivery point.[107] Most common are[108]:

- FOB (Free on Board):
  'Free on Board' obliges the seller to deliver the goods on board the vessel nominated by the buyer at the defined port of shipment.

[104]See Sect. 5.5.3.1.

[105]As the delivery point is a point in its classical meaning, it cannot be the place where natural gas or LNG is measured. Further, for tax reasons, i.e., for avoiding being responsible for paying import taxes as a seller, it might be agreed between the parties that the delivery point and title transfer point are not at the same location.

[106]Some contractual relationships allow for two or more delivery points. In these cases, the contract also stipulates either who, i.e., seller or buyer, has the right to nominate one of the potential delivery points for a specific delivery or deliveries during a specific period or which parts of the delivery have to be delivered at the various delivery points.

[107]Incoterms® are International Rules for the Interpretation of Trade Terms and were developed by the International Chamber of Commerce (ICC) in order to simplify and standardize international commercial transactions. The most recent version Incoterms® 2010 became effective as of January 1, 2011.

[108]Some LNG SPAs use CFR (cost & freight) or CIF (cost, insurance, and freight) instead.

For LNG, this means in detail that custody, title, and risk of LNG are transferred from seller to buyer at the point of interconnection between (seller's) loading facilities and (buyer's) LNG vessel. As example may serve the definition of 'Delivery Point' under GIIGNL[109]'s Master FOB LNG Sales Agreement:

'Delivery Point' means the point of loading of LNG at the Project's loading terminal as specified in the Confirmation Notice where the inlet flanges of the loading lines of the LNG Vessel connect with the outlet flanges of the loading lines of the terminal's loading facilities.[110]

After delivery, the buyer bears all costs; or

- DAP (Delivered at Place):

  'Delivered at Place' obliges the seller to deliver the goods at the disposal of the buyer at the nominated place.

  Until the end of 2010 when Incoterms® 2000 were in place, the slightly different DES (Delivered Ex-Ship) clause was used instead of DAP.[111] DES obliged the seller to place the goods at the buyer's disposal on board the ship at the port of destination.

  For LNG, this means in detail that custody, title, and risk of the LNG are transferred from seller to buyer at the point of interconnection between (seller's) LNG vessel and (buyer's) LNG discharging facilities. An example may serve the definition of 'Delivery Point' under GIIGNL's Master Ex-Ship LNG Sales Agreement[112]:

  'Delivery Point' means the point of discharging of LNG at the Buyer's discharging terminal as specified in the Confirmation Notice where the outlet flanges of the discharging lines of the LNG Vessel connect with the inlet flanges of the loading lines of the terminal's discharging facilities.

  The seller bears all risks and costs associated with bringing the goods to the delivery point.

The bespoke Incoterms® 2000 and Incoterms® 2010 serve as indication for the parties when agreeing upon the delivery point. However, the precise requirements of the parties are reflected in the LNG SPA. The decision about delivery point also defines who incurs boil-off losses[113] during the transportation of LNG.

Other things being equal, buyers of LNG prefer FOB provisions, while sellers prefer DAP (or DES). The reasoning behind this is geographical flexibility. A buyer with a FOB provision takes custody, title, and risk of LNG at the loading port and has (theoretically) the option to divert the cargo from its originally intended port of destination to another port in a different world region that might offer higher prices. Other than pipelines, which simply connect two distinct geographical points LNG vessels offer the flexibility to be diverted and to deliver LNG at any other port that accommodates a regasification facility and that has sufficient spare discharging capacities at the time of arrival of the vessel.

In case of a FOB-based LNG SPA, the buyer can easily decide to divert a cargo without asking the seller for permissions or assistance. Such a decision right might be limited by so-called destination clauses[114] that prohibit contractually the transportation to, and the discharging at, an alternative port. Under such contractual provisions, however, the buyer bears all (fixed and variable) costs of transportation of the LNG.

Under a DES- (or DAP-) based LNG SPA, the seller is obliged to deliver the LNG at the originally-intended port of destination, even when it seems to be more profitable to divert the cargo to another destination. In such a case, the

[109]Groupe International des Importateurs de Gaz Naturel Liquéfié.

[110]See also Sect. 6.3.9.

[111]Incoterms® 2010 rules eliminated four Incoterms® 2000 rules, *inter alia*, DES (Delivered Ex-Ship), and introduced two new ones: DAT (Delivered at Terminal) and DAP.

[112]See also Sect. 6.3.9.

[113]See also Sect. 7.2.2.2.

[114]Destination clauses might be prohibited under competition law rules in some parts of the world. For example, beginning 2000, the European Commission has started to investigate destination clauses.

only possibility[115] for a seller to generate additional profit is to find an agreement with the buyer, which allows him to divert the cargo whereby such agreement usually stipulates some (additional) profit sharing between seller and buyer. After delivery under a DES- (or DAP-) based LNG SPA, the buyer might decide to re-export the LNG instead of regasifying it, assuming that the regasification facility has re-exporting or better re-loading capacities at its disposal.

GSAs do not offer any sort of geographical flexibility as the delivery point is usually defined to be one specific point within a pipeline (e.g., where a pipeline crosses a border). GSAs offer geographical flexibility only in those (rare) cases where two or more potential delivery points are defined and where either seller or buyer do have the option to choose between these delivery points.

6.3.6 Quality

The quality of natural gas to be delivered at the delivery point is specified in detail.

In case of a GSA, inevitably, the quality will have to correspond to the upstreamcarrier's specifications. That is, the seller cannot deliver natural gas with any of the parameters exceeding the range defined in the applicable rules of the upstream carrier. The seller will also (physically) not be in a position to do so, as its natural gas is transported in a commingled stream. The upstream carrier has to see to it that the parameters of the natural gas transported in such commingled stream at each point in time comply with the quality specifications for the system/network. The same applies to the downstream carrier. At the delivery point, connecting the upstream with the downstream system, only natural gas complying with the quality specifications of the downstream system will be accepted for transportation.[116] The problem arises if the quality specifications in the upstream system are not identical to the quality specifications of the downstream system. Then one of the parties to the GSA, i.e., the seller or the buyer, may be exposed to a (contractual) risk.[117] In this case, the parameters of the natural gas delivered may be within the specifications of the upstream system, i.e., 'on-spec', but exceed the range defined in the downstream system, i.e., 'off-spec'. As the upstream and downstream carriers have no obligation to accept or convert off-spec natural gas, either the seller or the buyer of the commodity contract will be in contractual quality default either under the GSA or under their respective transportation agreements.[118]

In case of an LNG SPA, the contractual quality specification has either to correspond to the specification purported by the operator of the liquefaction plant (in case of a FOB-based contract) or the regasification plant (in case of a DAP- (or DES-) based contract).

Risk Arising from Differing Quality Specifications

Let, at an international delivery point, the range of the gross calorific value of natural gas transported in the upstream system be 10.6–12.3 kWh/m^3. Let further, the range of the gross calorific value of natural gas transported in the downstream system be 10.6–11.9 kWh/m^3. Further assume that the parties to a GSA agreed on a quality specification in line with the specification of the upstream carrier.

[115]This assumes that the seller does not intend to willfully breach the LNG SPA by non-delivery, which leaves him exposed to contractual claims by the buyer as well as reputational damages.

[116]Analogously, the seller has to deliver natural gas at the delivery point with a pressure high enough (but not too high) that allows the natural gas to enter the downstream system in a safe manner.

[117]Dependent on the quality specification in the natural gas delivery contract, the buyer might have a problem, if the quality of the natural gas delivered corresponds to the quality specification of the upstream system or, vice versa, and less probably, the seller if the quality of the natural gas delivered corresponds to the quality specification of the downstream system.

[118]For a discussion on harmonization of parameters in the European Union, see Drasdo et al. (2013).

If the upstream carrier will make available to the seller at a point in time at the interconnection point of the system, being the delivery point, natural gas with a gross calorific value of, e.g., 12.1 kWh/m^3, it fully complies with its obligations under the GSA and under its transportation agreement with the upstream carrier. According to the transportation contract, the seller has to accept this natural gas at the delivery point and sells it at the same point, according to the sales contract, to the buyer. The buyer, endeavoring to bring the natural gas to its market, will pass it at the delivery point to the downstream carrier. The downstream carrier, however, might have the right to reject the natural gas, as it does not comply with the quality specifications of its system. The gross calorific value exceeds the upper boundary by 0.2 kWh/m^3.

Consequently, the buyer will not be able to fulfill its off-take obligations under the contract with the seller. In accordance with the assumption above (i.e., the quality specification of the GSA is in line with the specification of the upstreamcarrier) the buyer might have to pay for natural gas not taken.

6.3.7 Nominations

As described in Sect. 6.3.1, the parties to a GSA might agree on volume flexibilities. In such a case, the contract stipulates either for the buyer and/or the seller the right to choose the quantities to be off-taken or delivered during a certain period. The parties to a contract have to inform each other about future off-takes and deliveries. Such information is usually referred to as nomination.

For GSAs, such information may be non-binding and preliminary or binding. While the non-binding information about planned deliveries or off-takes is frequently agreed upon to be performed annually, seasonally, quarterly, monthly, or weekly before the respective delivery period of the next year, season, quarter, month, or week, the daily information, called 'nomination' will usually be binding upon the parties.[119] Nominations will have to be in accordance with the applicable procedures at the delivery point and the rules applicable for the transportation system(s) connected.

In the case of a so-called 'buyer's nomination'—being typical for supply type contracts—the buyer has to nominate, i.e., to order hourly and daily quantities for delivery by the seller, typically, for (each individual hour of) the following day. In doing so, the buyer will always have to respect the quantity restrictions agreed upon in the contract, as, e.g., 'minimum-flow' requirements,[120] which set daily or hourly minimum flows and are necessary to take account of the seller's (predominantly) technical requirements.[121]

Primarily when selling associated natural gas, sellers frequently reserve the right to nominate themselves, i.e., to prescribe natural gas to be off-taken the following day by the buyer. Again, this has to be performed within relevant limits, i.e., the quantity restrictions, defined in the contract. Obviously, in these cases the buyer usually has no nomination rights. Seller's nomination rights essentially safeguard the seller's security of demand and enable her/him to produce natural gas and other commodities, such as crude oil or condensate, in an economically and technically optimal manner.

Irrespective of whether there is a buyer's or seller's nomination regime in place, intraday re-nomination rights may be included. According to this, quantities nominated on a preceding day could be revised during the day of delivery, again within clearly prescribed limits defined in the delivery contract.

The right to nominate is a consequence of volume flexibility. Under a buyer's nomination regime, such volume flexibility has a positive

[119] Note that some parties prefer to refer to the annual, seasonal, quarterly, monthly, or weekly information also as (non-binding) nomination or 'pre-nomination'.

[120] If allowed also by minimum flow requirements ,zero nominations' are possible.

[121] For example, the reservoir may be harmed if certain minimum production rates are not maintained.

value for the buyer, and the seller, providing volume flexibility, incurs costs.

With associated gas fields, the production costs also include the missed value from the production of crude oil, i.e., flexibility of natural gas deliveries may imply lower or belated oil production. Therefore, often, producers, i.e., sellers, of associated gas do not provide its natural gas buyers with flexibility. Frequently, a seller's nomination regime, which is sometimes called 'gas as produced', is agreed. Under such a seller's nomination regime, the buyer might be obliged to off-take natural gas during some periods. Effectively the buyer receives 'negative' volume flexibility. This has a positive value for the seller.

As described in Sect. 6.3.1.2.3, for LNG SPAs a fixed schedule is agreed between the parties. This gives only limited or no flexibility' to the parties. Usually, the schedule is agreed upon and therefore fixed for a year's period.

6.3.8 Other Commercially Relevant (Contractual) Elements of Commodity Deliveries

In this section, only selected topics relevant for either long-term natural gas delivery contracts, GSAs and LNG SPAs, or short-term contracts, including deliveries at hubs, will be described briefly.[122]

- Definitions
  Especially in international GSAs and LNG SPAs, the first section will define the major terms used. This includes, *inter alia*, units, periods, and quantities;
- Invoicing and payment (also referred to as 'billing')
  The seller will render an invoice to the buyer within a defined period, mostly, following the period of delivery, typically a month. GSAs between companies only rarely foresee prepayments, i.e., payments to be effectuated during the month of delivery. This is unlike deliveries to household (or small customers) where frequently prepayments are set.
  Such an invoice will have to contain details of quantities delivered, quantities not delivered, e.g., due to default, off-spec etc., prices as well as applicable taxes, duties, and levies. The invoice will also detail penalties, e.g., due to default or under-deliveries.
  After receipt of the invoice, the buyer will be obliged to approve the invoice and pay the amount due within a further period. This payment period may be a decisive, commercially relevant part of the contract and, consequently, an integral part of the commercial bargain between the parties. For various, primarily industrial, customers, using large quantities of natural gas, the costs of natural gas supply constitute a major part of their total operating expenses. These expenses have to be covered by revenues from sales of the final products. Therefore, the buyer is interested in a long payment period, while the seller will usually endeavor to agree on a short one;
- Over-deliveries, under-deliveries, over-takes, and under-takes
  Despite all contractual obligations regarding quantities to be delivered and off-taken during a certain period, it occurs during the lifetime of an agreement that either seller or buyer do not fulfil these obligations. The seller might either deliver more (over-delivery) or less (under-delivery, also referred to as shortfall) than nominated, or the buyer might take more (over-take) or less (under-take) than nominated. Each of these cases might lead to negative financial consequences for the other party, and therefore the contract typically stipulates for penalties on the misbehaving party;
- Credit risk (or performance assurance)
  As the natural gas delivered by the seller will be paid by the buyer typically in the month following the month of delivery and as natural gas deliveries frequently involve large payments, the credit risk, depending on the

[122] For an example of a framework agreement accepted by most of the market participants in Europe, the reader is referred to European Federation of Energy Traders (EFET) (2007) or for a general (legal) discussion, including specific issues of LNG contracts, to Roberts (2014).

creditworthiness of the buyer, may have to be mitigated. To this end, the seller may request from the buyer, or more generally, the creditor may request from the debtor, to guarantee the payments for a defined future period. Such collateral support might be achieved by agreeing that the buyer will provide to the seller either a bank guarantee, a letter of credit from any other institution, a parent company guarantee (in case the buyer is an affiliate of a mother company considered to be creditworthy), or even a cash deposit[123];

- Force Majeure
  'Force majeure' is usually defined as any event or circumstance beyond the control of the party affected while acting as a reasonable and prudent operator, which results in or causes failure to perform its obligations and such failure could not have been overcome (and foreseen) by the party. Examples of force majeure events are:
  - strikes, lockouts or other industrial action;
  - laws and acts of government;
  - epidemics, quarantine;
  - insurrections, riots, wars, and civil and military disturbance;
  - earthquakes, fires, storms, floods, landslides, lightning, and other natural calamities;
  - breakage or accident to machinery, oil or gas pipelines, freezing of lines, failure of gas and oil wells and storage facilities;
  - loss, physical inoperability, or failure of the facilities; or
  - physical inoperability or failure of the transportation system.[124]

  Whether these events may affect the seller and the buyer or only one of them to be effective remains to be agreed in the contract. However, some contracts also include 'indirect force majeure' by acknowledging that, e.g., the buyer may claim force majeure if its customer(s) effectively claim force majeure. As accurate and comprehensive information has to be provided immediately in case of force majeure to the party not affected, the parties' obligation for information exchange plays an important role;
- Destination clause
  In traditional GSAs, the buyer agreed not to (re-) sell any natural gas purchased in a country or region outside the contractually defined one. In Europe, such 'destination' clauses were prohibited by the European Union at the beginning of this millennium. In LNG SPAs, destination clauses continued to exist but are currently fading away[125];
- Non-compete clauses
  Similar in effect to destination clauses, sometimes the parties also agreed to 'no-compete' clauses, preventing the seller to sell (additional) natural gas to other buyers or (potential) customers of its buyer in the same country or region. Such clauses have ceased to exist, too;
- Taxes and other public charges
  Taxes, duties, and levies imposed before the delivery point are usually paid by the seller, whereas, after the delivery point (in the direction of flow), such payments are the obligation of the buyer. This is particularly relevant to international GSAs and LNG SPAs; and
- Further clauses
  - natural gas contracts are governed by strict confidentiality clauses, prohibiting the parties to reveal most of the information contained therein to third parties without the permission of the other party, with the exception of, e.g., governmental representatives, insurances, or legal representatives bound by equivalent confidentiality obligations;
  - the parties usually agree on stipulations regarding the liability for indirect (consequential) and direct losses[126];

[123]Additional stipulations in case of material adverse change and will have to be agreed upon separately. As example may serve Sect. 17 (2) of European Federation of Energy Traders (EFET) (2007).

[124]It seems noteworthy that various contracts include 'acts of God' as an applicable event.

[125]See Sheldrick (2015).

[126]Also for this legal issue, the reader is referred to, e.g., European Federation of Energy Traders (EFET) (2007), Sect. 12 or the more general discussion in Roberts (2014), p. 391 et seq.

- applicable law, language, and settlement of disputes are legal issues. The location of dispute settlement, be it by arbitration or court, the language of the contract, and the applicable law must not necessarily coincide;
- the parties agree to exchange information regularly in a defined manner, revealing all information necessary to fulfil the obligations under the contract to the other party; and
- other terms as assignment, severability, entire contract clause, form of modification.

6.3.9 Standard Contracts

To facilitate the negotiation process of a GSA or LNG SPA, many standard contracts have been developed, or as Roberts puts it[127]:

> *When the parties come to prepare the agreements which they require for the sale and transportation of gas (or LNG) they may wish to be guided by the application of any available standard contract terms.*

The further downstream, the more companies work with standard contracts. Nearly all contractual relationships between local distribution companies, i.e., companies that sell natural gas, e.g., to households, use standard contracts. Apart from a quick 'negotiation process', which is indispensable, e.g., for the B2C, standard agreements help sellers to mitigate contractual risks. On top of this, the goal of further digitization seems unachievable without the increased use of standard contracts.

However, also for the wholesalelevel, standardized contracts have been developed to facilitate negotiation processes.

The European Federation of Energy Traders (EFET)[128] introduced a so-called General Agreement Gas. The appendices to this general agreement reflect specifications for delivery of natural gas at different delivery points, in detailed appendices, exist for transactions at[129]:

- Austrian Virtual Trading Point,
- Gaspool Balancing Services in Germany,
- NBP (National Balancing Point) in the UK,
- PEG (Point d'échange de gaz) in France,
- Polish Virtual Point (Punkt Wirtualny),
- PSV (Punto di Scambio Virtuale) in Italy,
- PVB (Punto Virtual de Balance) in Spain,
- TTF (Title Transfer Facility in the Netherlands),
- UK Beach, and
- ZBT (Zeebrugge hub).

In addition to these EFET agreements, for trades in the UK other standard contracts like

- Short Term Flat NBP Trading Terms & Conditions 2015 (NBP 15)[130] or
- Standard Terms for the Sale and Purchase of Natural Gas for UK Short Term Deliveries at the Beach Sub Terminals using 0600 h to 0600 h Gas Day (Beach 2015 (6:6))[131]

are in place.[132]

Further, the Association of International Petroleum Negotiators (AIPN)[133] developed a Gas Sales Agreement in 2006.

For the Australian market and jurisdictions with similar legal systems such as New Zealand, AMPLA[134] introduced a Model Gas Sales Agreement for non-complex wholesale transactions.

The North American Energy Standards Board, Inc. (NAESB)[135] developed a Base Contract for Sale and Purchase of Natural Gas for transactions in North America.

For the sale and purchase of LNG there is variety of standard contracts available:

[127]Roberts (2014), p. 37.

[128]See European Federation of Energy Traders (2016b).

[129]Transactions at NetConnect Germany (NCG), the virtual trading points in the Czech Republic, the Slovak Republic, and the Czech/Slovak border point Lanžhot do not require a specific appendix.

[130]See Joint Office of Gas Transporters (2015a).

[131]See Joint Office of Gas Transporters (2015b).

[132]NBP 15 and Beach 2015 (6:6) replaced NBP 97 and Beach 2000, respectively. The change became necessary mainly due to regulatory changes in the definition of the UK downstream gas day.

[133]See Association of International Petroleum Negotiators (AIPN) (2016).

[134]See AMPLA (2013).

[135]See North American Energy Standards Board (NAESB) (2016).

- The Association of International Petroleum Negotiators (AIPN)[136] developed an LNG Master Sales Agreement in 2009 and prepared an update in 2012;
- The European Federation of Energy Traders (EFET)[137] introduced an LNG Master Agreement in 2010; and
- The GIIGNL[138] published in 2011 two Master LNG Sales Agreements; one for delivery free on board and one for delivery ex ship.

6.3.10 Categorization of Deliveries and Contracts

6.3.10.1 Source: Depletion and Supply-Type Contracts

From a supply perspective, natural gas sales contracts can be categorized as:

- depletion-type contracts, or
- supply-type contracts

The parties to a depletion-type contract agree upon selling and purchasing (parts of) the natural gas economically recoverable produced from a specific reservoir. The quantities to be defined for different periods change over time in accordance with the production profile, i.e., the performance of the reservoir itself. The changing nature of reservoir performance will necessitate that the quantities are continuously adapted. In addition, these quantities will be influenced by the production and transport facilities in place.

This kind of contract provides the seller with a guaranteed sale of the natural gas in the dedicated field, while at the same time giving the buyer the expectation of receiving the reserves of natural gas.

Moreover, the agreement usually lasts for as long as the given field reserves are economically recoverable, i.e., depletion type contracts are often also so-called 'life of field contracts'. Often, yet not always, depletion type contracts include a seller's nomination regime.

Trivially, depletion type contracts are concluded between producers and their buyers (mostly importers or other producers) only. On all other levels of the natural gas industry, supply-type contracts prevail.

In supply type contracts, the quantity of natural gas delivered is defined independently of the production profile of specific reservoir(s). This implies that also the hourly and/or daily quantities are independent of the production rate of a single reservoir. Accordingly, the term of the contract is not influenced by production profiles, but defined in advance by the parties at the time the contract is concluded. The quantities, for each period chosen to be specified, are clearly defined throughout the term of the contract, but may vary for different periods, e.g., for the first years, a smaller daily and annual quantity may be set than for subsequent years.

Usually, the seller can source the natural gas to be sold under a supply-type contract without restrictions. Hence, sellers owning and operating a (large enough) portfolio of natural gas fields will decide to sell the bulk of the production under supply-type contract(s). Concluding a supply-type contract involves the seller taking the risk of availability of the commodity, i.e., the so-called 'reserve risk'. Supply type contracts typically rely upon a buyer's nomination regime.

The seller under a supply-type contract is typically interested in selling natural gas, often over the long-term, to a buyer who is regarded as being reliable and technically and commercially capable to off-take all quantities committed. The buyer's interest in purchasing natural gas under a supply-type contract often includes, in addition to the trivial fulfilment of its demand requirements and the diversification of supplies, a requirement for flexibility and security of supply.

The major distinguishing terms and conditions of depletion and supply type contracts are summarized in Table 6.5.

---

[136] See Association of International Petroleum Negotiators (AIPN) (2016).

[137] See European Federation of Energy Traders (EFET) (2016a).

[138] See Groupe International des Importateurs de Gaz Naturel Liquéfié (GIIGNL) (2011a) and Groupe International des Importateurs de Gaz Naturel Liquéfié (GIIGNL) (2011b).

Table 6.5 Summarized comparison of typical major terms and conditions of depletion- and supply-type contracts

Type of contract main clause	Depletion	Supply
Quantity	(Percentage of) all natural gas quantities produced from a given field	Quantities for different periods defined in the contract
Term	Life of field	Defined in the contract
Delivery point	Various (mostly upstream)	Various (often downstream)
Nomination	Seller's or buyer's nomination	Predominantly buyer's nomination

6.3.10.2 Term: Long-Term and Short-Term Contracts

Historically, the natural gas industry, be it in Europe, Asia or America, relied on LTCs. Throughout the years, particular features were included in these contracts. These features—nowadays frequently associated with LTCs, even if they are not necessarily linked to the term—are[139]:

- a duration, i.e., term, of five to thirty years;
- concluded between sellers and buyers operating (at least frequently) in a monopolistic or at least oligopolistic markets;
- including high quantities, partly covering the total demand of the buyer, yet, also allowing buyer to off-take such quantities flexibly;
- implying the 'traditional' risk-sharing scheme, i.e., that the buyer bears the volume risk, while the seller bears the price risk, implemented, *inter alia*, by
  - containing a 'take-or-pay clause', granting seller a reliable, long-term revenue to compensate for large upfront investments;
  - containing natural gas prices linked to oil-product prices (or sometimes coal or electricity prices), including a price review clause, also providing (at least to some extent) stability of prices;
- (formerly) also containing a so-called destination clause, which prevented the buyer to sell natural gas purchased under such a contract outside the region of destination, defined in the contract;
- reducing transaction costs of the parties to such contract; and
- providing a high security of supply.

As natural gas markets, primarily in the US and in Europe, were regulated in order to increase competitiveness, unbundling of integrated companies was adopted, and competitors entered the markets, which, in turn, became liquid and market participants entered, *nolens volens,*[140] progressively more into short-term contracts.

Typically, these short-term contracts share the following features:

- the term does not exceed one or two years, often contracts for daily, weekly, or monthly deliveries are concluded;
- fixed or indexed prices based on gas-hub prices and not indexed to competitive fuels;
- a hub is often chosen as delivery point;
- if the trade is to occur at a hub, standardized framework contracts are concluded between the parties. Single transactions are included as exhibits to the (general) framework contract;
- (in comparison to LTCs) smaller quantities; and
- no restrictions on subsequent use of the natural gas purchased.

Hence, long-term contracts and short-term contracts are not distinguished by the term only, as the name may suggest. The main features of these contracts are different, because the underlying market models under which these concepts were developed differ. However, LTCs do not necessarily have to be abolished in competitive markets. *Inter alia*, they might, be used for diversification of supplies and prices. If (legally) permitted, long-term contracts will be used to

[139] For a detailed discussion of long-term contracts, see, e.g., Neuhoff and Hirschhausen (2005).

[140] As in some countries, long-term contracts were no longer (legally) permitted, at least in the case where players of particular levels are involved.

Table 6.6 Distinguishing features of selected delivery contracts at various levels of the industry

Parties/contractual stipulation	Seller: producer Buyer: importer	Seller: importer Buyer: regional utility	Seller: regional utility Buyer: local distributor	Seller: local distributor Buyer: household	Seller: various Buyer: industrial
Quantities	Large quantities	→	→	Small quantities	Various
Flexibility	Limited	Enhanced by utilization of storages by importer	Traditionally fully flexibile contracts currently: reduced flexibility in case hub-based pricing is agreed	Traditionally fully flexibile contracts for household and SME customers, limitation apply for industrial customers	Different—according to customer's requirements
Price (level)	Lower prices	Increasing prices →	Increasing prices →	Increasing prices →	Lower prices than households
Price (structure)	Regional differences				
Delivery point	Border or hub	Hub or network	Hub or city gate	Household, plant gate	Plant gate

complement the market participants' portfolio—despite their monopolistic heritage.

6.3.10.3 Segment: Import, Utility and End User Contracts[141]

As depicted in Figs. 5.4 and 5.5 natural gas is sold at various levels (not necessarily all of them existing in every country). As the players, their business model, their competitive situation, and their customer's requirement differ, so do the contracts. Therefore, several stipulations of a typical import contract concluded between a producer and an importer do not equal the respective stipulations of a sales contract between a regional utility and a local distribution company and, even more so, not to a contract with an end user.

This categorization includes contracts between:

- two producers;
- producers and importers;
- importers and (regional) utilities;
- (regional) utilities and local distributors; and
- local distributors and end users, such as
  - households,
  - small and medium enterprises (SMEs),
  - industrial customers,
  - power plants.

In addition, also importers and (regional) utilities deliver to industrial customers or local distributors.

All of these contracts might have different stipulations and frequently also a different structure. Nonetheless and despite all distinguishing features enumerated in the following, several features are common to all natural gas delivery contracts, no matter at which level of the industry they are concluded. In simplified form, at least

- price,
- term,
- quantities, and
- delivery point

have to be agreed upon.

The main distinguishing features of contracts at selected levels are summarized in Table 6.6.

[141]This section consciously omits pure trading agreements.

Table 6.7 Characteristics of GSAs and LNG SPAs

Parameter	GSA	LNG SPA
Duration	Short-term and long-term	Short-term and long-term
Quantity	Small to large	Usually full lots of cargo size
Volume flexibility	On import level limited	Within the agreed schedule: very limited[a], Outside of the schedule[b]: Very high
Geographical flexibility	Limited	Within the agreed schedule: very limited, Outside of the schedule: Very high

[a]Assuming that LNG storage facilities, e.g., at the regasification plant are regarded separately, as they are typically 'behind' the delivery point and therefore not part of the LNG SPA
[b]That is, taking into account potential additional LNG spot deliveries

6.3.10.4 Commitment: Interruptible and Firm Contracts

Typically, natural gas is delivered based on so-called 'firm' delivery contracts. The seller is, at each point in time, obliged to make available the quantities stipulated in the contract.[142]

'Interruptible' delivery contracts are characterized by clauses according to which the seller can at its sole discretion switch off (interrupt) the delivery of natural gas to the buyer on an hourly or daily basis. This means that such interruption is controlled and triggered by the seller. Such contracts are mostly concluded by buyers who are able to switch fuels, i.e., to use other (primary) energy sources in order to accomplish their requirements.

In order to be operational and well defined an interruptible contract must contain the definition of the (with reference to periods of hours or days):

- maximum quantities to be interrupted;
- maximum number of consecutive periods of interruption;
- maximum quantities to be interrupted during the term of the contract (or a delivery period, e.g., one year) or a sub-period (e.g., one month) thereof;
- maximum total number of periods of interruption during the term of the contract (or a delivery period, e.g., one year);
- minimum time lag for the announcement of an interruption; and
- minimum time lag for the announcement of the end of an interruption.

[142]See Sect. 6.2.

From a seller's perspective, Interruptible delivery contracts can be regarded as 'negative sinks', i.e., 'sources' of flexibility imposing at the same time restrictions for buyers.

The price to be paid for interruptible deliveries is usually discounted in comparison to the price of firm deliveries. Even if in theory such discount equals the value of flexibility, current markets do not necessarily confirm such an approach. Sometimes interruptible deliveries are sold at the same price as firm deliveries.

6.3.10.5 Physical State: LNG and Pipeline-Commodity Contracts

Natural gas is traded either in the gaseous state (under GSAs) or in its liquid state (under LNG SPAs). The physical state influences the characteristics of the respective contract type. This is illustrated in Table 6.7.

References

AMPLA (2013) Model gas sales agreement. Available at: https://www.ampla.org/modeldocuments/documents/model-documents62

Association of International Petroleum Negotiators (AIPN) (2016) Homepage. Available at: https://www.aipn.org/default.aspx

Bundesverband der Energie- und Wasserwirtschaft (BDEW); Verband Kommunaler Unternehmen (VKU) and Groupement Européen des Entreprises et Organismes de Distribution d'Energie (GEODE) (2011) Leitfaden, Abwicklung von Standardlastprofilen Gas

CMS Law (2016) ECJ ruling: Poland's system of regulated prices violates EU regulations. Available at: http://www.cms-lawnow.com/ealerts/2015/09/ecj-ruling-polands-system-of-regulated-prices-violates-eu-regulations

Corbeau A-S, Volk D, Sinton J, Jiang J, Ping J, Teng T, Boshu L, Fen Y (2012) Gas pricing and regulation: China's challenges and IEA experience, pp 1–119. Available at: http://www.iea.org/publications/freepublications/publication/ChinaGasReport_Final_WEB.pdf

Drasdo P, Karasz M, Pustišek A (2013) Dis-harmony in European natural gas market(s)—discussion of standards and definitions. Zeitschrift für Energiewirtschaft, 37

Energy Charter Secretariat (2007) Putting a price on energy—international pricing mechanisms for oil and gas, Brussels. Available at: http://www.energycharter.org/fileadmin/DocumentsMedia/Thematic/Oil_and_Gas_Pricing_2007_en.pdf

European Commission (2015) European commission—press release—the EU gas market: the Commission calls on France to revise its system of regulated prices for non-household users. Available at: http://europa.eu/rapid/press-release_IP-12-542_en.htm?locale=en

European Federation of Energy Traders (EFET) (2016a) LNG. Available at: http://www.efet.org/Standardisation/Legal-EFET-Standard-Contracts-and-Documentation/LNG

European Federation of Energy Traders (2016b) Gas & gas annexes. Available at: http://www.efet.org/Standardisation/Legal-EFET-Standard-Contracts-and-Documentation/GasAndGasAnnexes

Groupe International des Importateurs de Gaz Naturel Liquéfié (GIIGNL) (2011a) Master ex-ship LNG sales agreement

Groupe International des Importateurs de Gaz Naturel Liquéfié (GIIGNL) (2011b) Master FOB LNG sales agreement

Hobér K (2014) Recent trends in energy disputes. In: Talus K (ed) Research handbook on international energy law. Northampton

International Chamber of Commerce (2003) ICC force majeure clause 2003—ICC hardship clause 2003. Available at: http://store.iccwbo.org/t/ICC%20Force%20Majeure%20Hardship%20Clause

International Gas Union (IGU) (2015) Wholesale gas price survey—2015 edition—a global review of price formation mechanisms 2005–2014. Available at: http://www.igu.org/sites/default/files/node-page-field_file/IGUWholeSaleGasPriceSurveyReport2015Edition.pdf

International Gas Union (IGU) (2016) Wholesale Gas Price Survey—A Global Review of Price Formation Mechanisms 2005 to 2015 (May), p. 32. Available at: http://www.igu.org/

Joint Office of Gas Transporters (2015a) Short term flat NBP trading terms & conditions. Available at: http://www.gasgovernance.co.uk/sites/default/files/NBP2015-FINAL.pdf

Joint Office of Gas Transporters (2015b) Standard terms and conditions for the sale and purchase of natural gas for U.K. short term deliveries at the beach sub terminals using an 0600 hours to 0600 hours gas day, 2015, pp 1–18. Available at: http://www.gasgovernance.co.uk/sites/default/files/Beach2015(6_6).pdf

Mellbye P (1996) Troll on stream; the story and its perspectives. In ONS Conference 1996. Stavanger

Melling AJ (2010) Natural gas pricing and its future

Neuhoff K, von Hirschhausen C (2005) Long-term vs. short-term contracts: a European perspective on natural gas, Available at: https://www.repository.cam.ac.uk/bitstream/handle/1810/131595/eprg0505.pdf?sequence=1

North American Energy Standards Board (NAESB) (2016) Wholesale gas WGQ contract standards and models. Available at: https://www.naesb.org/wgq/cont.asp

Petroleum Association of Japan (2016) Homepage

Roberts P (2014) Gas and LNG sales and transportation agreements, 4th edn, London

Sheldrick A (2015) Destination clauses on LNG will soon fade away: gas union president. Reuters. Available at: http://www.reuters.com/article/us-asia-lng-contracts-idUSKBN0LT0YY20150225

Stern J, Rogers HV (2014) The dynamics of a liberalised European gas market: key determinants of hub prices, and roles and risks of major players, Oxford

tarifgaz.com (2016) Les tarifs réglementés du gaz en 2016. Available at: http://tarifgaz.com/tarifs/reglemente

7 Transportation

Summary

- Natural gas can be transported either in the gaseous state in a pipeline (system) or in liquefied form (as LNG) by trucks or ships. Truck- and ship-based transportation is also possible for compressed natural gas (CNG).
- Pipeline systems consist of pipelines (of various diameters and serving different purposes), processing plants, compressor stations, metering stations, valves, and SCADA systems.
- The costs of constructing and laying a pipeline are typically proportional to the length of that pipeline for any given pipeline diameter. Offshore pipelines are generally more expensive than onshore pipelines.
- The 'LNG-chain' consists essentially of liquefaction plants, regasification plants—both including storage facilities, and the transportation itself (mainly by ship).
- The main participants in the transportation market are carriers, also called transporters, transport system operators or pipeline system operators and their customers, also referred to as shippers.
- A capacity reservation system is used to determine the basic principles of capacity reservation in transportation systems and contracts. In contrast, the price-determining variables and their functional relationships are defined in a pricing system. Both systems are—implicitly or explicitly—constituting elements of transportation contracts but are, at least in principle, independent of each other, i.e., the capacity reservation system does not have to correspond to or match the pricing system.
- The following types of capacity reservation systems are distinguished: point-to-point capacity reservation system, entry-exit capacity reservation system, and postalized capacity-reservation system.
- Pricing concepts for natural gas transportation are: distance-related, entry-exit, or postage stamp.
- As pipelines are regarded to be a natural monopoly, they are often regulated.

7.1 General

'Transportation' is the movement of objects from one location to another, i.e., the transfer of objects in space.[1] Literally, '*trans*' '*portare*' means 'carrying over', 'conveying' or 'taking across'. In general terms, the elements of any transportation system are an 'infrastructure', a 'vehicle' and an 'operation'. While the infrastructure consists of fixed installations that allow

[1]Compare to storage, Sect. 8.1.

A. Pustišek and M. Karasz, *Natural Gas: A Commercial Perspective*,
DOI 10.1007/978-3-319-53249-3_7

'vehicles' to operate, the 'vehicles' are the devices operated and as such used to move the 'objects'. Yet, this generalized view is only partially applicable to natural gas transportation.

Natural gas can be transported[2]:

- in pipelines, in which the natural gas remains gaseous.
  In this case the infrastructure is the system of pipelines, the vehicle is not defined, and the objects are the (gaseous) molecules of natural gas;
  or
- by truck or ship, the natural gas being cooled to liquid state, i.e., transformed to LNG.
  In this case the infrastructure is either the roads used by trucks or 'sea-routes' (typically not well defined, as no distinct 'routes' exist on the sea), the vehicle is the vessel or truck, and the object are the molecules of (liquid) natural gas;
  or
- by truck or ship, yet not in liquid state but as compressed natural gas (CNG).
  The natural gas is compressed and the relatively smaller quantities of natural gas are then transported by ship over shorter distances than if transported in form of LNG or by pipeline. Again, the infrastructure is not always well-defined, but the vehicle and the object are—as above.

Depending on distance and quantity to be transported, several options may apply and be revealed to be economically viable. A graphical depiction, which is based on certain assumptions, is shown in Fig. 7.1.

In 2015, approximately 68% of worldwide international natural gas trades were performed using pipelines and the remaining 32% were mainly performed by shipping LNG,[3] as, on a worldwide scale, CNG transportation and trade have only limited impact.

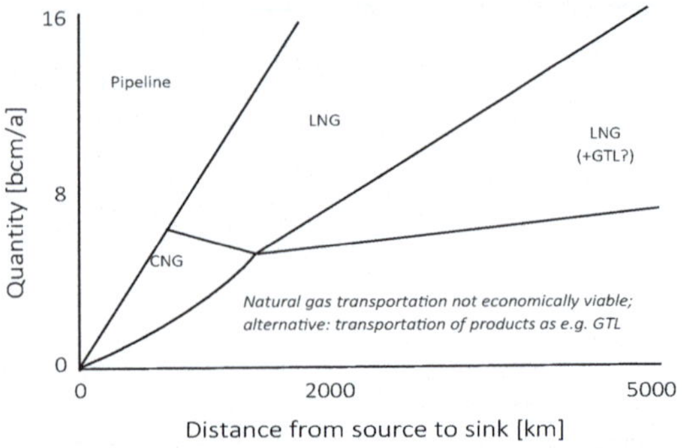

Fig. 7.1 Economically preferable options to transport natural gas from sources in dependence of the quantity and distance After Wood et al. (2008)

For any transportation system, two types of optimization problems are of relevance: design optimization and operation optimization.

In design optimization, the objective is to optimize, e.g., the pipeline (system) design by selecting appropriate layouts, equipment, etc. The optimal design of transportation projects, be it pipeline networks or LNG facilities, is determined based on technical and economic criteria. Such an, often complex, optimization problem is also referred to as 'static' (design) optimization.

Some Relevant Parameters for Pipeline Design

For pipeline design, *inter alia*, the following design variables need to be determined: quantities to be transported, number of compressor stations, compressor station locations, compressor station capacity, compressor station energy demand and fuel, lengths of pipeline segments between compressor stations, diameters of pipeline segments, routing of pipelines, and suction/discharge pressures at each compressor station. The objectives may be minimization of investment cost or

[2] In addition, sometimes 'gas-to-wire', i.e., natural gas transformed to electricity and then transporting the electricity; 'gas-to-liquids' (GTL), i.e., natural gas transformed to liquid hydrocarbon products as, e.g., kerosene or diesel, and then transporting the liquids; or 'gas-to-solids', i.e., natural gas hydrates, transported by ship or truck, are referred to as methods of natural gas transportation. As natural gas (itself) is not transported, such methods shall not be discussed subsequently.

[3] See British Petroleum (2016), p. 29.

maximization of the net present value of the respective project, whereby the latter assumes revenues, i.e., capacities rented and prices achieved, for such rental to be forecasted (or known).

In operation optimization, the network configuration (including compressor and measurement stations) is given, and the goal is to operate the pipeline network 'optimally'. The objective function of such optimization would often be minimization of costs by accepting the flow and all (technical) parameters as restrictions. This is performed in a modern, unbundled environment by the carrier and referred to as 'dynamic' (technical) optimization. In a traditional environment, it constitutes an integral part of the integrated portfolio optimization. It is not performed on a stand-alone basis. Rather, all technical parameters, including, but not limited to, operational costs (used to describe optimal operating conditions), are taken as restrictions for the integrated system's optimization.

7.2 Outline of Physical Transportation and Assets

7.2.1 Pipeline

Effectively, natural gas, being gaseous, is rather 'pushed' or 'pressed' but not 'transported' in the sense that a 'vehicle' is used to move the 'object' of natural gas. Nevertheless, the general term transportation, as well as the specific terms 'transmission' (for long haul, mostly high-pressure) and 'distribution' (for short-haul, mostly low-pressure), are used throughout the industry for moving natural gas from one location to another.

When transporting natural gas via pipelines from the wellhead, i.e., the production site, to the burner tip, i.e., the consumption site of the end user, several assets, and processing steps are involved. The economic and technical challenge is mainly caused by the gaseous state. It necessitates that, once natural gas has left the producing well, a pipeline system has to be used to direct the flow. Usually, the next element is a natural gas processing plant, where the natural gas produced is cleaned before it is directed further to the transportation system.

The infrastructure to transport natural gas from the wellhead to the burner tip consists of the following main (physical) assets:

- pipelines which can be categorized as:
  - flow lines:
    Typically, flow lines are buried 1–2 m underground. 'Wet' natural gas (i.e., natural gas containing significant amounts of hydrocarbon liquids and contaminants) from a wellhead is transported in flow lines to production site facilities. One of the main technical problems in flow lines is corrosion, as the natural gas transported is not yet treated, i.e., amongst others, not yet dehydrated;
  - gathering lines:
    Flow lines from several wells in the producing area are constructed to 'converge'. The natural gas is then transported in gathering lines to the processing facility. These are medium-size steel pipes (usually under 0.5 m diameter) that carry unodorized, raw natural gas. Also gathering lines are typically buried 1–2 m underground. Corrosion is equally problematic as in flow lines. It is primarily due to economies of scale, applicable already at this upstream stage, that the natural gas production from different wells is gathered and then transported to the processing facility. This, of course, will always be accompanied by optimizing, i.e., minimizing, the lengths of the pipelines involved;
  - transportation pipelines:
    Natural gas is transported across long distances in transportation pipelines to and from compressor stations, to metering stations, and to distribution centers or storage facilities. Transportation pipelines are large steel pipes (up to more than 1 m

in diameter). They carry odorized or unodorized natural gas at a pressure of up to 125 bar.[4] In order to alleviate problems of, e.g., seam failures, corrosion, materials failure, or defective welding, the construction is carefully monitored and inspected;

- distribution pipelines (also referred to as takeaway pipelines):
  Distribution pipelines, also known as 'mains', are the middle step between high-pressure transportation lines and low-pressure service lines. They are small to medium sized pipes (0.1–0.7 m in diameter), carrying mostly odorized natural gas at varying pressure levels and typically made from a variety of materials, including steel, cast iron, plastic, and occasionally copper, being flexible and easy to replace; and
- service pipelines:
  Service pipelines connect distribution pipelines to a meter. Usually less than 0.05 m in diameter, they are typically made from plastic, steel, or copper and carry odorized natural gas at low pressures.

Several interconnected pipelines form a pipeline system.

The 'Most Important' Pipeline Deal

In the middle of the cold war, the German natural gas major Ruhrgas and the Soviet foreign-trade organization Sojuznefteexport inked, on February 1, 1970, a deal that became famous as the 'natural gas against pipes' deal. After 15 negotiation sessions —five of which were top-level meetings— the two parties agreed on a gas sales contract over 20 years with an annual quantity of three bcm to be delivered to Ruhrgas. In return, the Soviet Union received 1.2 million tons of large pipes with a 1420-millimeter diameter. The German steel company Mannesmann manufactured the pipes. The value of the deal was around one billion Deutsche Mark (at that time 250 million US $). The deal was inked only three years after an embargo for exports had been lifted. The pipes were used to build a transportation pipeline from Orenburg in Western Siberia to the western border of the Soviet Union at Uzhhorod, in today's Ukraine.

An approximation of the volume flow rate in a pipeline is[5]:

$$\varphi^2 = \text{const} * D_{\text{Pipeline}}^5 * \frac{p_1^2 - p_2^2}{l_{\text{Pipeline}}}$$

with

p_1	pressure at the delivery (inlet) point [Pa = N/m^2],
p_2	pressure at the redelivery (outlet) point [Pa = N/m^2],
l_{Pipeline}	length of the pipeline [m],
D_{Pipeline}	diameter of the pipeline [m],
const	constant [m^6/N^{2}s^2],
φ	volume flow rate [m^3/s]

This capacity, however, is also dependent on the ground temperature, as depicted in Fig. 7.2. This has the implication that usually capacity of pipelines is higher in winter than in summer. Yet, such capacity increase is, because of unpredictability of temperature, not suited to be relied upon by carrier.

[4]Whereby offshore pipelines may have inlet pressures as high as 150 bar, see Chandra (2006).

[5]See, e.g., Keyaerts et al. (2008), p. 2. This equation is an approximation only: compressibility, temperature, and specific gravity are assumed constant (and included in const); the exponent of the diameter is not exactly = 5, but better described by (2.x * 2), with (typically) $0.4 < x < 0.7$ (see also calculation of costs in Sect. 7.3.1.1). For the approximate calculation of flows of (natural) gas in pipeline, *inter alia*, also the Panhandle equation or the Weymouth equation are used. This approximation is also used to determine the capacity of the pipeline.

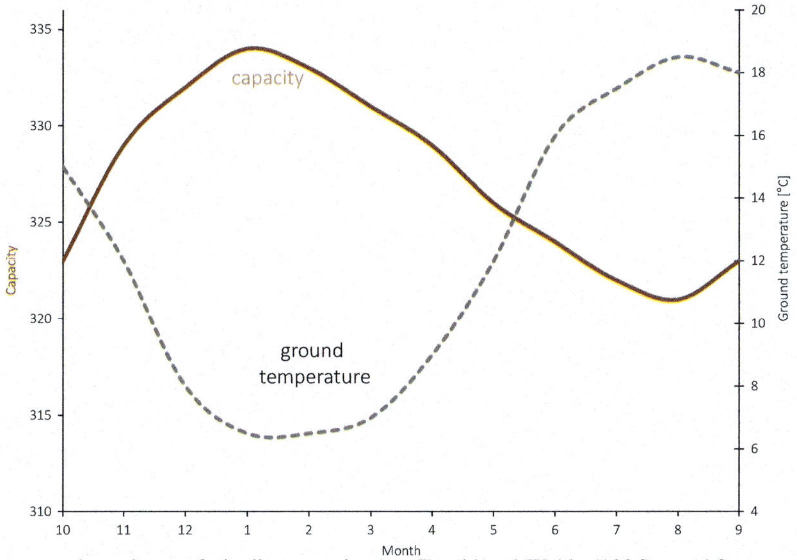

Fig. 7.2 Temperature dependence of pipeline capacity (see Fasold and Wahle (1996), p. 116)

- processing plants:
  In processing plants, natural gas liquids and impurities, liquid, gaseous, or solid, are extracted from the natural gas.

  The (chemical) composition of natural gas differs by region.[6] Nonetheless, as a rule, all natural gas extracted from underground reservoirs usually has to be subjected to an extraction process to remove water (dehydration) and other impurities. Of course, details of such extraction process will depend on the actual composition. The objective, however, is general: in the processing plant, compatibility of all quality parameters with the requirements of the downstream pipeline system is established, i.e., the (so-called) quality of natural gas is adjusted. Such quality parameters are defined (usually) differently in each national pipeline system. However, the (technical) rationale of such rules is the objective to prevent operational problems, pipeline deterioration, or even pipeline rupture.

  In addition, processing plants are facilities designed to recover natural gas liquids from a stream of natural gas. Such liquids are extracted and directed to a separate plant. There, the so-called 'fractionation' process is initiated, removing saturated hydrocarbons from natural gas and separating them into distinct parts, or 'fractions', such as propane, butane, and ethane;
- compressor stations:
  The pressure of natural gas moving through a pipeline decreases along the length of the pipeline due to friction losses. As pressure decreases, velocity of the natural gas in the pipeline increases.[7] Therefore, natural gas has to be compressed in order to move and increase the quantities transported, thus keeping the operation of the pipeline system technically feasible and economically viable. This is performed by compressors: devices boosting the pressure of natural gas in a pipeline and allowing the natural gas to be transported through a pipeline. Using (additional) energy for operation (frequently by 'extracting' small parts of the natural gas in the pipeline, so-called fuel gas[8]), the pressure

[6] See also Sect. 4.1.2.

[7] Neither the decrease nor the increase are linear. For further reference, see, e.g., Hairston and Moshfeghian (2013).

[8] As a rule of thumb, it is often assumed that the fuel gas consumption in long-haul transportation pipelines is typically lower than 0.5%/100 km.

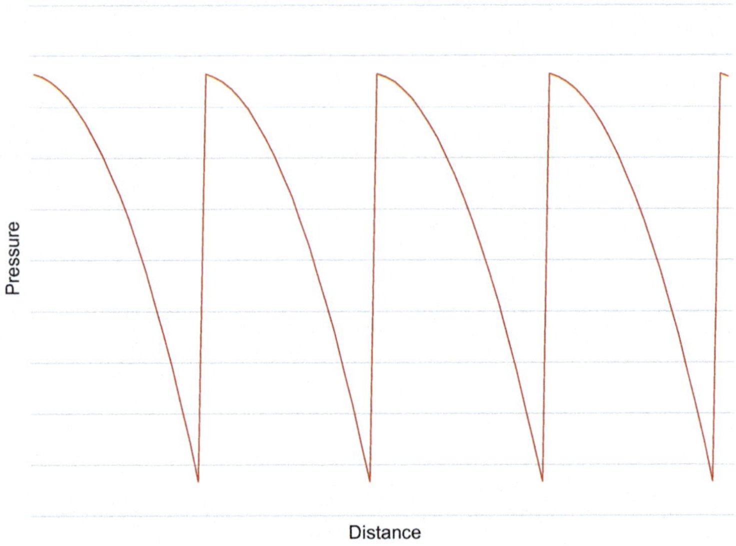

Fig. 7.3 Schematic pressure along a pipeline with equidistant compressor stations

of the natural gas in the pipeline is increased, by simultaneously decreasing the velocity[9] and increasing the temperature. Compressors are also used in the aboveground facilities of storages to pump natural gas back into underground storages.

These compressors are typically either

– reciprocating compressor: (simplified) the pressure of natural gas is increased by linear movement of the driveshaft.
  Operation: natural gas is entering the suction manifold and then flowing into a compression cylinder. A reciprocating moving piston compresses the natural gas. The crankshaft is powered by an internal combustion engine;
  or
– centrifugal compressor: (simplified) the pressure of natural gas is increased by means of mechanical rotating vanes or impellers.
  Operation: Kinetic (rotatory) energy is transformed to pressure energy. To this end, a rotating disk or impeller is used to increase the velocity of the natural gas. This natural gas is directed to a divergent duct section and its pressure is increased.[10]

Such compressors are either fueled by natural gas (referred to as fuel gas and commonly diverted from the pipeline) or electricity.
Primarily to increase performance and enhance security of supply (redundancy), usually several compressors are pooled in compressor stations. In such stations, the natural gas will, before passing through the compressors, typically enter through station valves and then pass through a scrubber or filter (to collect any particles and/or liquids). As the pressure of natural gas decreases as a function of length of a pipeline, compressor stations are usually placed at 50–200 km intervals along the pipeline (mainly in long-haul pipelines). Then pressure, as a function of distance/length, can be depicted as shown in Fig. 7.3.

Given that the objective function of pipeline (network) design is cost minimization,

[9]In general, velocity has to be considered already when constructing the pipeline. Typically, velocities are approximately 20–30 m/s. There is no uniform opinion on the maximum velocity of natural gas flowing in a pipeline.

[10]Such compressors may either be equipped with either a wet- or dry-seal configuration to prevent natural gas leakage. While wet seals use oil around the rotating shaft, dry-seal compressors use the opposing forces created by hydrodynamic grooves and springs.

the selection and determination of the number of compressor stations, their locations and design, the pipelines' diameters, the lengths of the pipelines, and the suction and discharge pressures at each compressor station are the main variables of network design optimization. That is, the placement, design, and number of compressor stations will always have to be determined in view of the objective function with due consideration for technical and economic constraints;

- metering stations:
  Measuring the flow of natural gas moving through pipelines, metering stations are placed along the pipelines and at end points. Usually the volume is measured. As the energy content and not the volume of natural gas is typically traded, respectively, defined in a GSA, measuring the volume implies that the gross calorific value of the natural gas at the respective station has to be pre-set, defined, or also measured. Only then, the energy flowing through the pipeline in a period or at a point in time can be determined.
- valves:
  Movement of natural gas along a pipeline is controlled in part by a series of valves, which can be opened to allow the gas to move freely or closed to stop gas flow along a section of pipe. Large valves may be placed every 5–30 km along the pipeline;
  and
- SCADA systems:
  Pipeline operators use supervisory control and data acquisition (SCADA) systems, to track and manage the natural gas as it travels through the pipelines, ensuring that all shippers receive timely delivery of their portion of natural gas. These (real-time) systems are essentially sophisticated communications systems that take measurements and collect data along the pipeline (usually in metering or compressor stations and valves) and transmit the data to a centralized control station. Flow rate through the pipeline, operational status, pressure, and temperature readings may all be used to assess the status of the pipeline at any one time.

7.2.2 LNG

When natural gas is cooled to approximately −161.5 °C at atmospheric pressure, the volume shrinks to approximately 1/610 of the volume in gaseous state at norm temperature. By keeping the low temperature, LNG can be transported by ship or truck from the liquefaction to the regasification plant. At the outlet of the later, it is—again in gaseous state—usually injected into a pipeline system and transported nationally, regionally, or locally to the end user. Occasionally, LNG is also stored, either before shipping at the liquefaction facilities or at the regasification facilities.[11]

Already in the 1940s liquefied natural gas was used for peak shaving.[12] Today, liquefied natural gas ('LNG') is predominantly used as an alternative method to transport natural gas over long distances.[13]

The 'LNG-chain' consists essentially of:

- a liquefaction plant, incl. storage facilities;
- a transportation, i.e., mainly shipping; and
- a regasification plant, incl. storage facilities.

7.2.2.1 Liquefaction Plants

As the name suggests, in liquefaction plants natural gas is liquefied. Cooled down to a temperature of approximately −161.5 °C, natural gas changes its state from gaseous to liquid. Most liquefaction plants consist of more than one so-called train, working in parallel to liquefy natural gas. To make use of economies of scale, some supporting processes (e.g., electricity generation) are centralized within one plant. A further advantage of having available more than one train is optimization of maintenance. For instance, if more than one train is available, maintenance of one train does not lead to a complete shutdown of the plant.

A train process usually starts with metering the incoming natural gas stream. To avoid plant

[11]For a detailed discussion of all technical aspects of LNG, see, e.g., Mokhatab et al. (2014).

[12]See, e.g., Chandra (2006), p. 49.

[13]However, some industry players regard LNG as a separate 'product', traded on its own 'commodity' market.

damages during the refrigeration process, natural gas has to be as 'clean' as possible, i.e., as near as possible to chemically pure methane. Impurities like sulphur or CO_2 have to be removed completely, and the natural gas needs to be—at best—water-free, as cooling water will inevitably lead to icing, which again might damage or plug the units used to liquefy the natural gas stream. After the dehydration process, NGLs such as propane and butane are separated.

Then, the actual liquefaction process sets in. The main processes used in industry are the cascade process or the mixed refrigerant MCR process. Both processes use propane and heat exchangers to precool the natural gas to approximately −30 °C. The cascade process then requires two further heat exchangers using ethylene and methane as refrigerants. The MCR process uses instead a mixture of refrigerants consisting of methane, ethane, propane, and nitrogen.

After liquefaction, LNG has to be stored onsite until it is loaded to a ship (or a truck). LNG storages are aboveground or semi-buried.[14]

All the processes described require the use of energy. As a primary energy source, natural gas is used as feed. As a rule of thumb, approximately 8% of natural gas measured at the inlet of the plant is used as fuel, i.e., only 92% of the natural gas arriving at the liquefaction plant are transported in form of LNG.

The nominal capacity of liquefaction plants is denominated in million tonnes of LNG per year [mmtpa] whereby one tonne of LNG corresponds to approximately 1400 Nm^3 of natural gas. At the end of 2015, 90 liquefaction trains worldwide with a total nameplate capacity of 288.5 mmtpa were in operation.

7.2.2.2 Transportation of LNG

Usually, LNG is transported by ships.[15] These vessels are specially designed to keep the LNG in a liquid state. Sometimes (with newer vessels), double-hull passive insulation is supported by active refrigeration aboard. However, many older ships do not use active refrigeration but only passive insulation. These ships usually use boil-off gas,[16] i.e., LNG that has changed state from liquid to gaseous during the voyage, as propulsion fuel.

Most LNG vessels sailing today carry LNG either in spherical tanks (Moss sphere design) or in geometric membrane tanks (membrane design).

Speed of Modern Vessels

A typical a modern vessel can reach in open waters a speed of 20 knots, i.e., it is able run 480 nautical miles or approximately 890 km per day.[17] Assuming an average speed of 20 knots, a one-way voyage from Ras Laffan in Qatar to Pyeong-Taek in the Republic of Korea with a distance of 6232 nautical miles takes approximately 13 days (omitting loading and discharging).

The nominal transportation capacity of an LNG vessel is measured in m^3 of LNG. The density of LNG is between 430 and 470 kg/m^3. Such m^3 of LNG corresponds to approximately 630 Nm^3 of natural gas.[18] At the end of 2015, 449 LNG vessels with a total nameplate capacity of 64.6 million m^3 of LNG were in operation.

A majority of liquefaction plant operators run their own fleet of LNG vessels. In order to avoid unnecessary (and operating cost increasing) partial or temporary shutdowns of liquefaction trains, the size of the vessel fleet is customized with the liquefaction capacity. Thereby, the number of vessels necessary depends on several factors, such as:

[14] See further description in Sect. 8.3.1.4.

[15] In some cases, when transportation distances are relatively short and the quantities to be transported are small, specially designed trucks might be used transportation. Recently, also tests with trains commenced [LNG World News (2016)].

[16] Typical boil-off rates are at approximately 0.15% per voyage day. Obviously, this factor depends on several factors such as ship's and tank's design, outside temperatures, etc. As already described in Sect. 4.1.2, LNG changes its chemical composition due to boil-off.

[17] One nautical mile corresponds to 1852 m.

[18] Calculated for a density of 450 kg/m^3. Then, as shown above, 1 tonne of LNG corresponds to 1400 Nm^3 of natural gas.

- capacity of liquefaction plant, incl. on-site storage capacities;
- capacity and speed of loading facilities at the liquefaction plant and thereby duration of a loading process;
- vessels' capacities;
- typical heel[19];
- average distance to regasification plants and vessels' speed, thereby defining the average duration of a voyage (taking into account two ways);
- capacity and speed of offloading facilities at the regasification plant and thereby duration of an unloading process;
- capacity of regasification plant, incl. on-site storage capacities; and
- port restrictions.

Considering these main technical and commercial factors, the number of vessels employed is optimized.

7.2.2.3 Regasification Plants

In regasification plants LNG is regasified, i.e., transformed from the liquid state back to the gaseous state. To fulfil this task, LNG vessels are received and offloaded. As do liquefaction plants, regasification plants also have on-site storages available. Again, these LNG storages are aboveground or semi-buried.[20]

In a regasification plant vaporizers are used to increase the temperature of LNG and thereby bring it into the gaseous state. In principle, vaporizers are heat exchangers fueled with seawater. Dependent on the geographical location of the regasification plant, the seawater used may have to be heated. As natural gas is used for this heating, variable costs increase.[21] After this process, the natural gas is sent into the pipeline grid where it commingles with other natural gas already available. Regasification does not necessarily take place onshore; floating storage and regasification units (FSRUs) exist as well.

The nominal capacity of regasification plants depends on several factors:

- offloading restrictions,
- LNG storage capacities,
- vaporizing capacities, and
- send-out capacities.

Out of these factors, the send-out capacity (measured in Nm^3/h) is the most restrictive. It describes the pipeline capacity available to bring regasified natural gas into the pipeline grid.

At the end of 2015, 117 regasification plants worldwide were in operation.

While historically LNG regasification plants have been built and operated by importers, third party access rules, may require LNG terminals to be owned or operated by companies independent from traders.

7.3 Costs

As the distances between sources and sinks of natural gas, i.e., production sites and consumption sites (the burner tip) are often very large, the costs of transportation heavily influence the commercialization of natural gas resources. They even may prevent natural gas from various sources to be marketed.

7.3.1 Pipeline

In brief, the costs of constructing and laying a pipeline are typically proportional to the length of that pipeline for any given pipeline diameter. Offshore pipelines are generally more expensive than onshore pipelines.

As well as in other areas, the cost of natural gas transport can be divided into fixed and variable costs or, into capital and operating costs.

[19]Heel defines a minimum volume of LNG that remains in the vessel after discharging. It remains there to keep the tanks cold for the ballast voyage back to the loading port.

[20]Some regasification capacities do have reloading capacities for LNG. These facilities allow the reloading of LNG to other (smaller) vessels or to trucks.

[21]Costs could increase by 1.5–3%. See Chandra (2006), p. 64.

7.3.1.1 Capital Costs

The capital costs of pipeline system construction are determined by the costs for the construction of:

- pipelines themselves, which, in turn, are influenced by the costs of:
  - the materials (essentially the pipes),
  - planning (incl. rights of way), and
  - the construction;
- compressors; and
- measurement and control stations.

These capital costs include depreciation and interest on debt and equity. Depreciation is calculated based on the purchase price or replacement value and taking into account the asset-depreciation range. Different depreciation periods can be assumed for different components (pipes, compressors, etc.). In general, depreciation methodology and depreciation periods applied will determine capital costs.

Such capital costs will depend on:

- the capacity of the pipeline,
- the length of the line,
- the interrelation between compressor stations and pipeline diameter,
- the topography and type of terrain traversed,
- applicable legal and regulatory conditions, and
- environmental and safety requirements.

The cost of the construction of natural gas pipelines are almost entirely fix. They represent nearly 95% of the total costs of natural gas transportation.

Since the cost per unit length are (approximately) proportional to the diameter of the pipe and the capacity increases over-proportionally with diameter, (capacity-related) specific costs decrease when pipeline capacity is increased.[22] This can be shown as follows:

Let α_1 and α_2 be two independent parameters describing the pipeline, further, $D_{Pipeline}$ shall be the diameter of the pipeline to be constructed, then the specific (pipeline-) length related costs, $c_{Pipeline}$, can be described by $c_{Pipeline} = \alpha_1 * D_{Pipeline} + \alpha_2$.[23]

For $c_{Pipeline}$ in [€/m] the parameters α_1 and α_2 are estimated to be in the range of $0.46 \leq \alpha_1 \leq 0.79$ and $92 \leq \alpha_2 \leq 120$, the lower value being applicable for distribution networks and the higher for national transportation networks.[24]

Simplifying, the capacity $\kappa_{Pipeline}$ of a pipeline is considered to be proportional to its diameter to the power of γ, with $\gamma \geq 2.58$.[25] If the factor of proportionality is τ, then the capacity can be described by[26]: $\kappa_{Pipeline} = \tau * D^{\gamma}_{Pipeline}$ and thus $\left(\frac{\kappa_{Pipeline}}{\tau}\right)^{\frac{1}{\gamma}} = D_{Pipeline}$.

Because the specific (pipeline-) length-related costs are $c_{Pipeline} = \frac{C_{Pipeline}}{l_{Pipeline}}$, i.e., fixed costs of pipeline construction, $C_{Pipeline}$, divided by $l_{Pipeline}$, being the length of the pipeline, and $C_{Pipeline} = \alpha_1 * \left(\frac{\kappa_{Pipeline}}{\tau}\right)^{\frac{1}{\gamma}} * l_{Pipeline} + \alpha_2 * l_{Pipeline}$, or, by setting $\alpha'_1 = \alpha_1 * \left(\frac{1}{\tau}\right)^{\frac{1}{\gamma}}$: $C_{Pipeline} = \alpha'_1 * \kappa^{\frac{1}{\gamma}}_{Pipeline} * l_{Pipeline} + \alpha_2 * l_{Pipeline}$.

This shows that the fixed costs of pipeline construction, $C_{Pipeline}$, are, under the simplifying assumptions, directly proportional to the pipeline length, $l_{Pipeline}$, and increase disproportionally

[22]See, e.g., International Energy Agency (IEA) (1994), p. 48 or Fasold and Wahle (1996), p. 115 and p. 118.

[23]To be exact, this equation applies to isolated pipelines only. For pipeline systems, i.e., networks, the equations are more complex. Nonetheless, for the prevailing explanation, the simplification is regarded as sufficient.

[24]See Cerbe et al. (2004), p. 234, Fig. 5.50 for pipelines in distribution networks:

$$c_{Pipeline-1} = 0.46 * D_{Pipeline} + 92 \text{ [€/m]},$$

see Füg (1999), p. 9, 10 for national transportation: $c_{Pipeline-2} = 0.73 * D_{Pipeline} + 111$ [€/m] (p. 9) and: $c_{Pipeline-3} = 0.79 * D_{Pipeline} + 120$ [€/m] (p. 10), with $D_{Pipeline}$ = diameter of pipeline [mm].

[25]See Fasold and Wahle (1996) p. 118. See also Sect. 7.2.1, where the flow, dependent on the diameter of the pipeline and the pressure, was approximated. For simplicity, $\gamma = 2.5$ was assumed. In this section, the pressure at the delivery and the redelivery point is assumed to be given. Hence, not the flow, φ, but the capacity, $\kappa_{Pipeline}$, is the quantity used.

[26]For purposes of the calculation of costs this, further simplified, approach for the pipeline capacity is chosen.

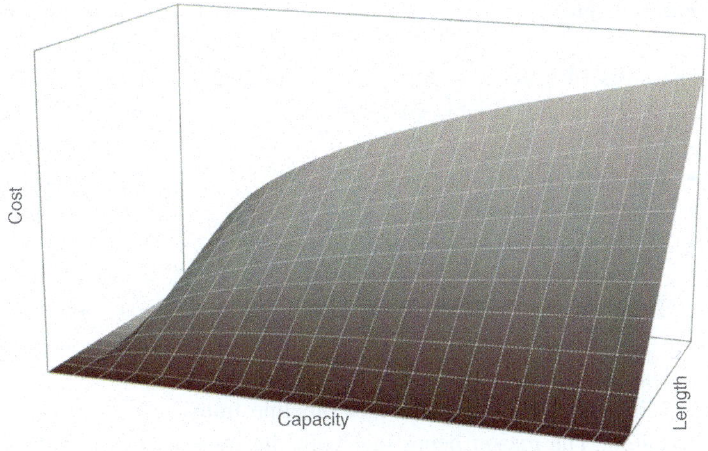

Fig. 7.4 Pipeline construction costs as a function of capacity and length (schematic)

with increased pipeline capacity.[27] This is depicted in Fig. 7.4.

Given the above, the (capacity related) specific total costs, c_{specific}, are:

$$
\begin{aligned}
c_{\text{specific}} &= \frac{C_{\text{Pipeline}}}{\kappa_{\text{Pipeline}}} \\
&= \frac{\alpha_1' * \kappa_{\text{Pipeline}}^{\frac{1}{\gamma}} * l_{\text{Pipeline}} + \alpha_2 * l_{\text{Pipeline}}}{\kappa_{\text{Pipeline}}} \\
&= \alpha_1' * \kappa_{\text{Pipeline}}^{\frac{1-\gamma}{\gamma}} * l_{\text{Pipeline}} + \frac{\alpha_2 * l_{\text{Pipeline}}}{\kappa_{\text{Pipeline}}}
\end{aligned}
$$

This is depicted in Fig. 7.5.

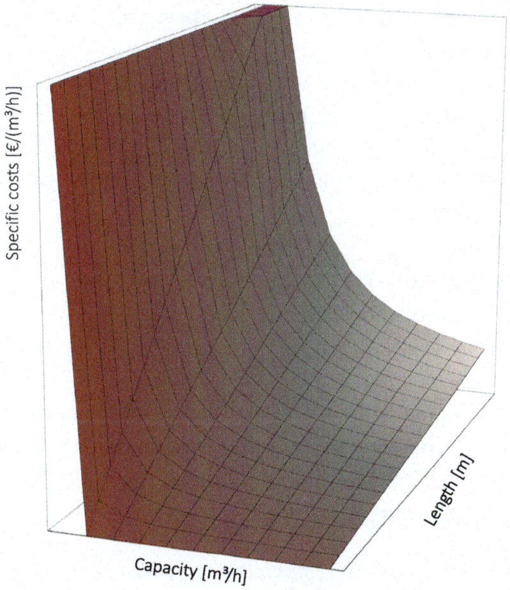

Fig. 7.5 Capacity related specific pipeline construction costs as a function of capacity and length (schematic)

7.3.1.2 Operating Costs

Variable costs of pipeline operation are primarily influenced by costs of:

- operation;
- maintenance and repair;
- tax and insurance; and, predominantly,
- fuel costs (for electricity or gas in kind), providing the energy for compressors.

Given today's technology, leakage of natural gas transportation pipelines is mostly negligible.

As a rule of thumb, variable costs of pipeline operation represent approximately 5% of the total costs of natural gas transportation, the major part being fuel costs for compressors.

[27]The costs of compressors, as well as the costs of planning, can be neglected in the prevailing context; for further explanation, see Pustišek (2005), p. 45 et seq.

7.3.2 LNG

The costs of LNG transportation are influenced by the fixed and variable costs of:

- liquefaction, incl. storage;
- transportation; and
- regasification, incl. storage.

The main difference between LNG and pipeline transportation is that the distance-related capital or fixed costs of LNG transportation are considerably lower than those of pipeline transportation. The reason being that both the liquefaction and regasification facilities are independent of the distance between the source and the sink, i.e., the sending and receiving terminals, while the costs to construct a pipeline between two points are primarily distance related.[28]

Obviously, for all parts of the LNG chain, capital costs vary heavily. Costs are driven by many factors, *inter alia*, for liquefaction or regasification plants, these are of major importance:

- location of plant,
- size of the plant and storages, and
- investments into harbor facilities.

Typical cost range, for a completely new-built liquefaction plant, between 1000 and 1800 US$/tonne of annual capacity of LNG.[29]

According to GIIGNL[30]:

> *The cost of LNG ships today is between $225–250 million for a 135,000 m³ carrier up to approximately $300 million for the larger ships.*

[28]See Sect. 7.3.1. Again, as a rule of thumb, it is generally assumed that the break-even point of LNG and pipeline transportation is to be found at a distance between 4000 and 7000 km. That is, up to this distance pipeline transportation may be more economic than transportation in form of LNG. Yet, such 'estimation' does not consider political, technical, and other economic restrictions and is, therefore, not suited to be applied for practical purposes.

[29]See Songhurst (2014), p. 29.

[30]Groupe International des Importateurs de Gaz Naturel Liquéfié (GIIGNL) (n.d.), p. 3.

Compared to liquefaction plants, regasification plants require relatively low investments. The Dunkirk LNG terminal in Northern France, which went into operation in 2016 and which has a regasification capacity of 13 bcm/a, required an investment of approximately 1 billion €.[31]

Compared to these capital costs, for the whole LNG value chain operating costs are relatively low. An exception to this are the costs for feed gas, boil-off gas, and heating gas already addressed in Sect. 7.2.2.

7.4 Players

The main players in the (pipeline) transportation market are (simplified):

- Owners of the systems: The owners of a transportation system are often companies also pursuing interest in other sectors.[32] In some countries, the transportation system is owned, in whole or in part, by the state;
- The operation of the transportation system, i.e., all services in connection with the sales of transportation capacity to the shipper (i.e., the customer), is performed by the carrier. The carrier is often also referred to as transport(ation) system operator (TSO) or transmission system operator (also TSO) or transporter or on a local and municipal level distribution system operator (DSO) . These companies are usually subject to regulatory supervision; and
- The transportation capacities are rented by shippers. Such shippers could be market participants from all segments, *inter alia*, producers, traders, local distributors, or (primarily industrial) end users or storage system operators (though, to a limited extent, e.g., for offering virtual storage products at hubs). It is the customer of the carrier, which is acting on other markets for natural gas, i.e., the storage and the commodity market.

[31]See Dunkerque LNG (2011).

[32]As, e.g., pension funds or hedge funds or insurances.

What is commonly referred to as 'transportation market' is described by agreements between the carriers and the shippers.

In most countries the transportation market is regulated. Hence, regulators, not being market participants per se, are to be considered as 'background' players in the market.

The main players in the LNG market are (simplified):

- In many cases, the owners and operators of liquefaction plants are affiliated to the producers that deliver the natural gas to be liquefied. These producers are also those market participants that act as seller of the then liquefied natural gas. Other liquefaction plants are owned and operated by producer-independent companies;
- As described in Sect. 6.3.5, sellers might be —depending on the delivery point under the LNG SPA chosen—responsible for transporting the LNG. In order to do so, some producers use their own vessels (again, typically in an affiliated company), or they charter a vessel from a third party. Buyers of LNG act in a similar fashion when they are obliged to transport LNG under a FOB-based contract: They might use their own vessels or they might charter an LNG tanker from a third party; and
- Regasification plants are either owned and operated by independent companies or by companies affiliated to sellers of LNG or companies affiliated to buyers of LNG.

7.5 Pipeline-Capacity Reservation Systems and Pricing Systems

7.5.1 Capacity Reservation

Prior to discussing 'products' and 'prices' for pipeline transportation services offered, the distinction between commonly confounded terms is considered appropriate.

A capacity reservation system is used to determine the basic principles of capacity reservation in transportation systems and contracts. In contrast, the price determining variables and their functional relationships are defined in a pricing system. Both systems are—implicitly or explicitly—constituting elements of transportation contracts but are, at least in principle, independent of each other,[33] i.e., the capacity reservation system does not have to correspond to or match the pricing system.

The following types of capacity reservation systems are distinguished:[34]

- Point-to-point capacity reservation system: the capacity is defined and reserved at distinct delivery and redelivery point (s) as well as along a clearly identified (and contractually described) transportation route or path connecting the delivery and redelivery point(s).[35] Typically, a shipper holding a contract for the reservation of transportation capacity between two points would not be able to switch the entry or exit point, unless it obtains another transportation contract;
- Postalized capacity reservation system: the capacity is defined and reserved, in a network'. Neither the transportation route nor specific entry or exit points are (pre-) defined. The shipper has the right to enter (or exit) natural gas at each entry or exit point; and
- Entry-exit capacity reservation system: the capacity is defined and reserved for different

[33]Capacity-reservation systems are also referred to as 'capacity type' and pricing systems as 'tariff type'; see Lapuerta and Moselle (2002), p. 56.

[34]The following definition refers to Lapuerta and Moselle (2002), p. 56 et seq.

[35]However, as described, e.g., by Lapuerta and Moselle (2002), p. 62, there are modifications of the point-to-point capacity reservation system, applied, e.g., in the US: *"Many contracts in North America allow shippers to designate several 'primary' and 'secondary' entry and exit points. In the above example, a shipper might be allowed to designate A and B as its primary entry points. If so, then it would have a firm right to switch from A to B. The same shipper may be permitted to choose either C or D as a primary exit point, but the other one would be a 'secondary' exit point. If such a shipper wanted to switch deliveries from C to D, then the TSO would allow the switch if feasible. Typically the secondary exit point has priority over simple interruptible service."*

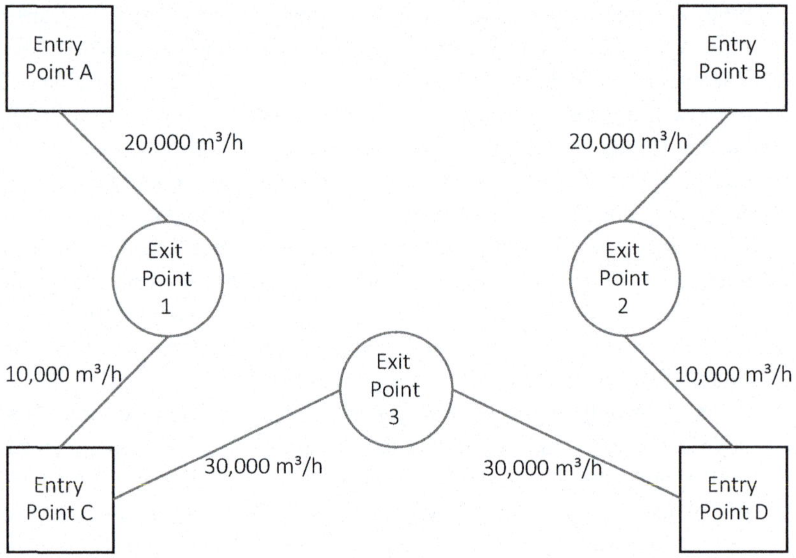

Fig. 7.6 Simplified network for the illustration of total capacity availability (equidistant entry and exit points assumed)

entry and exit points; the transportation path is not. A possible extension is that reservations of entry and exit capacities do not necessarily have to be performed by the same shipper. That is, one shipper may book entry capacities only and use exit capacities reserved by other shippers (which may be end users).

The choice of capacity reservation system has technical and economic implications:

- The geographical flexibility offered to shippers in networks using postalized capacity reservation systems will foster competition in natural gas trading. In such a system, even shippers reserving smaller capacities will be able to serve different customers. Shippers with a large portfolio of customers are likely to gain competitive advantages over shippers with smaller portfolios in point-to-point systems, as they will be able to increase utilization of capacities reserved or reduce the total capacity reserved.[36]
- The flexibility provided to shippers can affect the amount of firm capacity available. That is, in case shippers are offered the option to use different exit points, the total capacity available for sale in the transportation system is lower than in case such option was not offered.

Implications of Capacity Reservation Systems[37]

The starting point is a simplified network with four entry and three exit points. The physical transportation capacities between entry and exit points are depicted in Fig. 7.6.

If the carrier applies a postalized capacity-reservation system, it can sell up to 10,000 m^3/h of firm capacity only. Otherwise, it would not be able to fulfil its commitments. If, e.g., it would sell 20,000 m^3/h and the shipper would endeavor to use entry point C and transport natural gas (20,000 m^3/h) to exit point 1, then the carrier would not be able to accept such a request. Nonetheless, the carrier can sell additional capacity on an interruptible basis.

If the carrier applies an entry-exit capacity reservation system, it can sell up to 60,000 m^3/h of firm capacity, i.e.,

[36]The more customers a shipper serves, the more likely it is that such shipper will profit from equilibration of deliveries.

[37]See another example at, e.g., Lapuerta and Moselle (2002), p. 58 et seq.

20,000 m^3/h at entry point A, 20,000 m^3/h at entry point B, 10,000 m^3/h at entry point C (being the minimum, i.e., the bottleneck; accepting a higher capacity reservation, it would not be able to fulfil it between entry point C and exit point 1), and 10,000 m^3/h at entry point D (accepting a higher capacity reservation it, would not be able to fulfil it between entry point D and exit point 2).

If the carrier applies a point-to-point capacity reservation system, it can sell up to 120,000 m^3/h, i.e., the sum of all capacities between the entry and exit points.

This example can be generalized[38]: the total firm capacity, which can be offered and sold by the carrier, depends on the capacity reservation system applied. While the point-to-point capacity reservation system, typically, will help maximizing the total capacity sellable, it may help large players in the market to gain competitive advantages over smaller ones. The contrary is true for postalized capacity reservation systems, whereas the entry-exit capacity reservation systems could serve as a compromise between the two.[39]

7.5.2 Pricing

7.5.2.1 Price Level

The market value[40] of transportation capacity between two markets may be determined based on the difference between the market values of natural gas in the markets linked by the respective pipeline(s). Under this approach the pipeline is regarded to be an option with the actual difference of market prices for natural gas being the 'intrinsic value' of the option; the possibility of a change of this actual market value provides the option with an additional 'time value'.

However, the actual prices (often referred to as 'tariffs' or 'fee'[41]) to be paid by the shipper might be determined using other methods. As pipeline transportation is commonly regarded to be a natural monopoly, prices or price structures might be regulated. Whether cost-oriented or incentive oriented[42] a regulator, i.e., a governmental authority, determines the price of transportation. This price-determination process requires providing detailed information by the carrier to the regulatory authority. The regulatory authority will, based on the information received mainly by the carrier and in line with national and international legal requirements, and the general objectives, determine prices for the reservation of transportation capacity or approve prices proposed by the carriers. Other than in traditional market environments, where the carrier (in many cases being part of an incumbent vertically integrated company) set the transportation prices in an arbitrary way (e.g., to prohibit competition in the trading sector), regulated pricing is seen as a means to foster competition because the regulator is regarded as neutral.

Regulators might also promote auctions.[43] Either in cases where there are plans to build a new pipeline or where transportation systems already exist, auctions are regarded as a method to determine the market value of transportation capacity. With auctions, it is possible to reflect not only the supply but also the demand side for transportation capacity reservation. That is, the objective is to find a market price for such capacity reservation.

Of course, if a transportation capacity is acquired in an auction, the general pricing concepts described in this section are not necessarily used. Then, a lump-sum fee for the reservation of a capacity between or at two points for one or

[38]The proof may be omitted.

[39]This was confirmed in reality: after changing from the point-to-point system to the entry-exit system in Germany, the total firm capacity available was reduced.

[40]This 'market value' is not necessarily reflected in the actual market prices.

[41]As 'tariff' is often referred to as taxes and duties, the term shall not be used in this book.

[42]See Sect. 5.3.

[43]The types of auctions implemented vary.Regulators and carriers, in addition, frequently also promote and conduct so-called 'open-season' procedures, e.g., for new pipelines to be constructed.

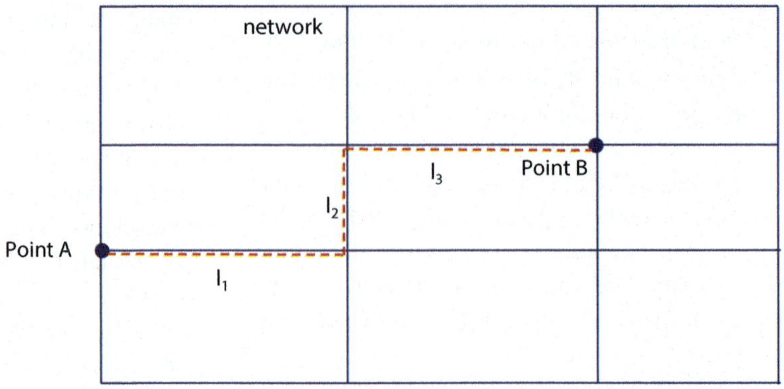

Fig. 7.7 Schematic illustration of the distance related pricing system

more periods has to be paid.[44] Such a lump-sum fee is a result of the bidding process, which, again, is the reflection of market price developments and market price expectations.

7.5.2.2 Pricing Structures

The structure of prices for the reservation of transportation capacity does not necessarily reflect changes of the market value of transportation capacity in time but essentially the denominator of transportation prices. The pricing concepts used are described in the following Sects. 7.5.2.2.1, 7.5.2.2.2 and 7.5.2.2.3 and compared in Sect. 7.5.2.2.4.

7.5.2.2.1 Distance Related

The price for reservation of natural gas transportation capacity is set as a function of the distance between the delivery and redelivery points,[45] the capacity reserved, and the duration of the contract. Such prices are expressed in units of, e.g., [€/(m³/h)/km/a] or of [€/(kWh/h)/km/a]. This pricing system is often applied to international transit or other long-haul transportation, i.e., primarily for transportation in a clearly defined (single) pipeline but rarely in branched networks (see Fig. 7.7).

Distance-Related Transportation Price

Let $l_1 = 100$ km, $l_2 = 50$ km, and $l_3 = 100$ km, let the (volumetric) capacity to be reserved $\kappa^{\text{Volume}}_{\text{Pipeline}} = 10{,}000\ \text{m}^3/\text{h}$, the gross calorific value $GCV_n = 11\ \text{kWh/m}^3$, and the price $P_{\text{Distance}} = 0.01$ €/(kWh/h)/km/a, then the total amount, A_{Distance} to be paid for a one year's capacity reservation will be:

$$l_{\text{Pipeline}} = l_1 + l_2 + l_3 = 100\,\text{km} + 50\,\text{km} + 100\,\text{km} = 250\,\text{km}$$

$$\kappa^{\text{Energy}}_{\text{Pipeline}} = \kappa^{\text{Volume}}_{\text{Pipeline}} * GCV_n = 10{,}000\,\text{m}^3/\text{h} * 11\,\text{kWh}/\text{m}^3 = 110{,}000\,\text{kWh}/\text{h}$$

$$A_{\text{Distance}} = \kappa^{\text{Energy}}_{\text{Pipeline}} * l_{\text{Pipeline}} * P_{\text{Distance}} = 110{,}000\,\text{kWh}/\text{h} * 250\,\text{km} * 0.01\,€/(\text{kWh}/\text{h})/\text{km}/\text{a} = 275{,}000\,€/\text{a}.$$

7.5.2.2.2 Entry-Exit

The price for the natural gas transportation capacities reserved at the delivery (entry) and redelivery (exit) points are set separately and independently. The distance between such points is—at least explicitly—not taken into consideration. Such prices are expressed, e.g., in units of [€/(m³/h)/a] or of [€/(kWh/h)/a]. This pricing

[44] Expressed in units of, e.g., [€/a] or [€/month].

[45] However, it should noted that the distance between a delivery and a redelivery point can be defined and/or measured in different ways. For details, see Pustišek (2005), p. 76 et seq.

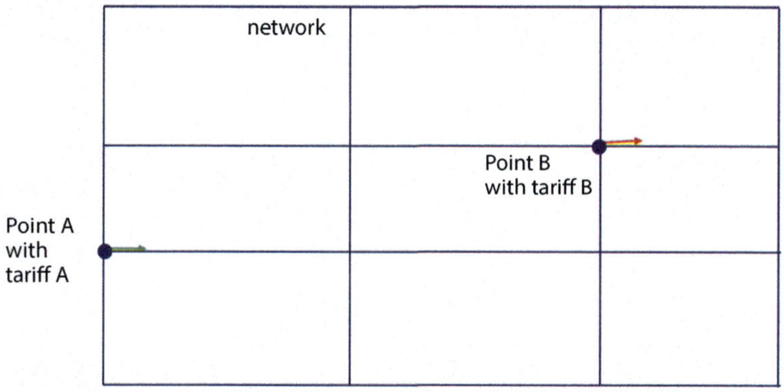

Fig. 7.8 Schematic illustration of the entry-exit pricing system

system is predominantly applied for national high-pressure pipelines and illustrated in Fig. 7.8.

Entry-Exit Transportation Price

Let the (volumetric) capacity to be reserved $\kappa_{Entry}^{Volume} = \kappa_{Exit}^{Volume} = 10{,}000\ m^3/h$,[46] the gross calorific value $GCV_n = 11\ kWh/m^3$, and the price $P_{Entry} = 1.7$ €/kWh/h/a, $P_{Exit} = 0.00219178$ €/kWh/h/d,then the total amount, $A_{Entry/Exit}$ to be paid for a one-year's capacity reservation will be:

$$\kappa_{Entry}^{Energy} = \kappa_{Entry}^{Volume} * GCV_n = \kappa_{Exit}^{Energy} = \kappa_{Exit}^{Volume} * GCV_n$$
$$= 10{,}000\,m^3/h * 11\,kWh/m^3 = 110{,}000\,kWh/h$$

$$A_{Entry/Exit} = \kappa_{Entry}^{Energy} * P_{Entry} + \kappa_{Exit}^{Energy} * P_{Exit} * 365$$

110,000 kWh/h * 1.7 €/kWh/h/a + 110,000 kWh/h * 365 d/a * 0.00219178 €/kWh/h/d = 275,000 €/a.[47]

7.5.2.2.3 Postage Stamp

Being '*The simplest ... type of transmission pricing ... A postage stamp rate is a fixed charge per unit of energy transmitted within a particular zone, regardless of the distance that the energy travels".*[48] This pricing system is predominantly applied for transportation in (local) distribution networks and illustrated in Fig. 7.9.

Postage-Stamp Transportation Price

Let the (volumetric) capacity to be reserved $\kappa_{Postage}^{Volume} = 10{,}000\ m^3/h$, the gross calorific value $GCV_n = 11\ kWh/m^3$, and the price $P_{Postage} = 2.5$ €/kWh/h/a, then the total amount, $A_{Postage}$ to be paid for a one-year's capacity reservation will be:

$\kappa_{Postage}^{Energy} = \kappa_{Postage}^{Volume} * GCV_n = = 10{,}000\ m^3/h * 11\ kWh/m^3 = 110{,}000\ kWh/h$

$A_{Postage} = \kappa_{Postage}^{Energy} * P_{Postage}$

110,000 kWh/h * 2.5 €/kWh/h/a
= 275,000 €/a.[49]

Despite such simplicity, postage-stamp prices can be designed differently:

> *Postage stamp rates are based on average system costs and may have a variety of rate designs ... Rates often include separate charges for peak and off-peak periods, may vary by season, and, in some cases, set different charges for weekday versus weekend and holiday usage.*[50]

In some natural gas transportation contracts, the underlying pricing concept is neither explicitly stated nor revealed. The shipper is just requested to pay a fee for the reservation of a defined capacity during a certain period.

[46]It is assumed that the entry-capacity equals the exit capacity.

[47]Rounded to zero decimal places.

[48]Prete (1998), p. 7.

[49]Rounded to zero decimal places.

[50]Prete (1998), p. 7; also these prices are expressed, e.g., in units of [€/(m³/h)/ a] or of [€/(kWh/h)/a].

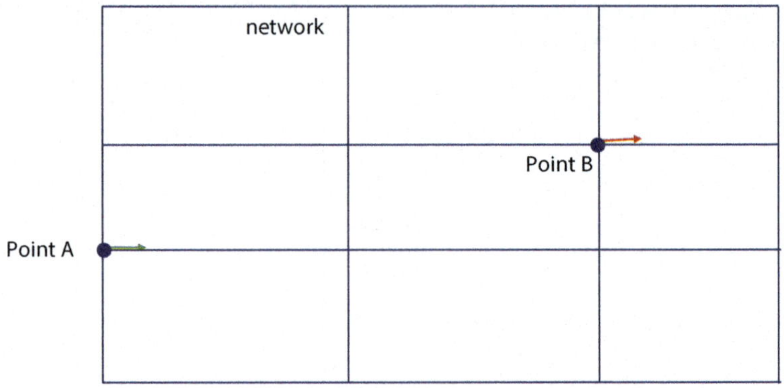

Fig. 7.9 Schematic illustration of the 'postage stamp' pricing system

Table 7.1 Summary of advantages and disadvantages of transportation pricing systems

Pricing system	Advantages	Disadvantages
Distance related	Cost related and cost reflective Equivalent to transportation price structure of competing fuels Well-established system in other branches	Complex structure Not fostering competition In a network distances not unambiguously and transparently determinable
Entry-Exit	Simple Fostering competition (Potentially) market reflective (Potentially) cost reflective	'Pancaking'[a] for international or inter-regional transports (Potentially) cost reflective[b]
Postage stamp	Simple Cross-subsidization (if wanted)	Cross-subsidization (short-haul shippers to long-haul shippers) 'Pancaking' for international or inter-regional transports Not market reflective Not cost reflective

[a]'Pancaking' is the accumulation of non-distance-related transportation prices (or costs) to be paid by customers having reserved transportation capacity in at least two adjacent networks
[b]Cost reflectiveness may be seen as an advantage or disadvantage depending on the actual prices set

However, as carriers in the regulated environments of the European Union will have to observe the requirements of non-discrimination, such fees are calculated based on an underlying pricing concept, which corresponds to one of the above-mentioned ones.

7.5.2.2.4 Comparison of Pricing Structures

Market participants value the advantages and disadvantages of the pricing systems differently.

Some of the frequently discussed, yet necessarily subjective, advantages and disadvantages, as seen by market participants of the pricing systems, are summarized in Table 7.1.

7.6 Contracts

Given the differences in the product, it is evident that natural gas pipeline or LNG transportation contracts have to be distinguished from natural gas or LNG purchase or sales contracts. While the latter set the rules for the purchase or sale, including the transfer of risk and ownership of natural gas as a commodity, the parties to the former agree on the terms and conditions for the reservation of transportation capacity as described above.

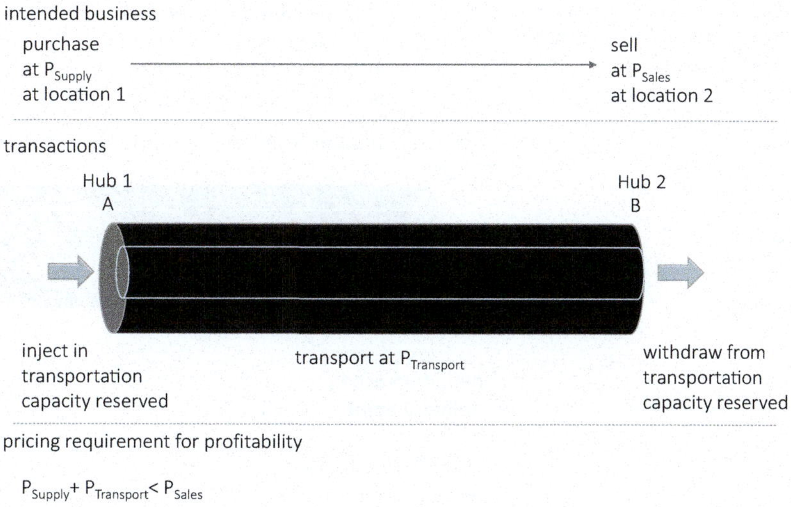

Fig. 7.10 Transport capacity reservation—business model, transactions, and pricing requirements

The contractual elements described subsequently, i.e., primarily the products and prices, are the ones agreed upon between carriers and its shippers, as this constitutes what is commonly referred to as the transportation market. Other contracts, e.g., for operation of pipeline systems, will not be discussed.

7.6.1 Pipeline Contracts

As depicted in Fig. 7.10, a shipper will typically require and reserve transportation capacity between the delivery and the redelivery point[51] if it has natural gas available at the former and intends to sell it at the latter. Simplified, the reservation of transportation capacity between the delivery and the redelivery point is economically viable only if, *ceteris paribus*, supply and transportation costs do not exceed sales revenues.[52]

The traditional model for a natural gas transportation contract[53] through a third party's pipeline is one whereby the shipper delivers natural gas to the carrier at a defined input point (delivery point) and the carrier then transports and redelivers natural gas to the shipper at a defined redelivery point. The carrier receives a payment by the shipper for the capacity made available and the supplementary services provided.

Under such a regime, the shipper has the right but not the obligation to utilize the capacity reserved. Such a natural gas transportation contract is, from an economic point of view, at least comparable to an option, an option to dispose natural gas at one point (the delivery point) and to receive the same quantity of natural gas at the same time[54] at the redelivery point.

7.6.1.1 Products

The product (or good) traded in a 'transportation market', for which the shipper pays a fee (or tariff or price) to the carrier, is the (empty) capacity, typically defined as the maximal hourly (or daily) flow rate or throughput rate between (or at) a delivery point and a redelivery point during a defined period, together with the carrier's service to convey natural gas within such capacity.[55] Metaphorically, it is the 'pipe in the pipe' made available by the carrier to the shipper.

[51]For the definition of these terms, see Fig. 7.11.

[52]Whereby other costs, as, e.g., storage costs, are neglected for the simplicity of the argument.

[53]Whereby the term 'transportation contract' is used as equivalent to 'network-access contract' or 'grid-access contract'.

[54]Whereby this is a 'contractual assumption'.

[55]The so-called 'quantity-transportation agreements', in which the carrier and the shipper agree upon the transportation of a defined (annual or total) quantity to be nominated, and the shipper will pay the price to the carrier according to the quantity of natural gas transported, are less important in today's markets.

Fig. 7.11 Basic concept of transportation and nomenclatures of points

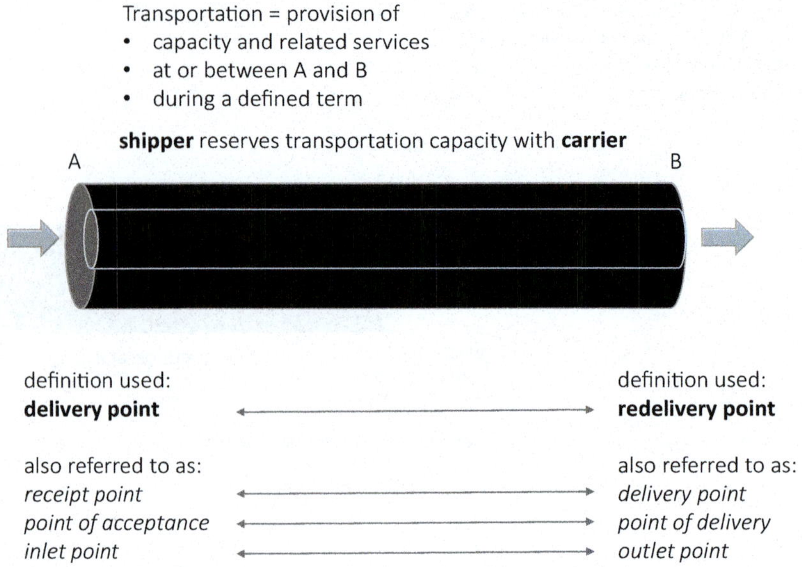

This, together with an illustration of the wording used, is illustrated in Fig. 7.11. That is, the shipper, is put in a position as if it has at its disposal, a pipeline or part of a pipeline system. It transfers possession (but not ownership) of any natural gas delivered at the delivery point (or inlet point, etc., see Fig. 7.11) to the carrier.[56] Custody of such natural gas will be re-transferred to the shipper at the redelivery point (or outlet point etc., see Fig. 7.11).

The product sold is thus: [capacity] defined as [volume/time] or [energy/time] = [power] to be reserved over a defined period at or between two defined points. It is measured in (SI-) units of [m^3/h/a] or [m^3/d/a] or [kWh/h/a] or [kWh/d/a] or [kWh/h/d].

Consequently, the carrier will (at least usually) not own, but only possess the natural gas transported, or even any natural gas.

The shipper will book only those capacities needed to fulfil his other contractual obligations, e.g., from purchase and sales agreements.[57] As a result, the risk of having built pipeline over-capacities, i.e., capacities not sellable, is with the carrier, respectively, its owner.[58]

A potential conflict of interest may arise between carrier and shipper. The former designs the pipeline system (and consequently transportation capacity) based on the volume to be transported in a given unit of time. The latter's intention is to receive the exact amount of energy delivered at the delivery point at the redelivery point. If, however, the capacity is defined as [energy/time] (as usually done in national European transportation systems), the risk of variations of the gross calorific value is borne by the carrier, as the pipeline's physical dimensions, and thus the ability to transport natural gas is restricted by volume, not by the energy content of the natural gas. On the other hand, an energy-based capacity definition applied in a system with gross calorific values for the natural gas transported varying over time and between different locations, may imply the necessity that the carrier will have to accept or release quantities of natural gas (for balancing purposes).

[56]The same (physical) result could be achieved by the parties using a 'buyback' model, whereby the carrier buys natural from a shipper at the delivery point at a defined price (e.g., p) and the carrier re-sells the natural gas to the shipper at the redelivery point at a (higher) defined price (e.g., P + X).

[57]See Sect. 5.1.1.

[58]In several regulated systems, carriers are permitted to 'socialize', i.e., include in the regulated asset base, such losses from over-capacities.

Booking Energy or Volume

Let the gross calorific value of natural gas to be transported in a pipeline be 10 kWh/m^3. Transportation of 100,000 kWh/h will require a (volume-related) pipeline capacity of 10,000 m^3/h. If the gross calorific value changes to 11 kWh/m^3, the required (volume-related) pipeline capacity will be reduced to 100,000/11 m^3/h = 9091 m^3/h.

7.6.1.2 Pricing

Of course, the price to be paid by the shipper to the carrier is one of the most important elements of a transportation contract. The determination of price levels, as well as different pricing systems, was already comprehensively discussed in Sect. 7.5.2.

Under long-term transportation contracts, capacity charges are in most cases escalated by a certain percentage with (in most cases) publicly available indices to reflect market or cost developments during the term of the contract.

Hence, a typical, yet slightly simplified (so-called multiplicative) price formula for transportation services is of the form:

$$P_{Transport}\left(t, \overline{\Psi}_i(t)\right) = P_{Transport0} * \left(\sum_{i=1}^{n}\left(d_i * \overline{\Psi}_i(t)/\Psi_{i0}\right)\right)$$

with:

$P_{Transport0}$	base value at time t_0 = starting point of indexation
Ψ_{i0}	base value of the index i at time t_0 = starting point of indexation
$\overline{\Psi}_i(t)$	average[59] value of the index i at time t
d_i	weighting factor of index i,
n	number of indices
t	time of price (re-) calculation (typically: month)

[59]Many transportation-price formulae are simpler, as no average values but single distinct values are used. In case an average value is calculated, see Sect. 6.3.2.2.3 for further details.

$P_{Transport0}$ is determined either by application of a distance related, entry-exit, or postage stamp pricing system.

Elements to be paid, in addition to the prices for reservation transportation capacities by the shipper to the carrier, may be one or more of the following fees or prices:

- penalties, i.e., charges for exceeding the capacities reserved, for attempting to inject quality-deficient natural gas, or any other non-compliance with contractual stipulations;
- interruptibility discounts, i.e., prices can be discounted if the carrier is allowed to interrupt the provision of services under the contract, i.e., to reject or reduce nominations of the shipper; or
- variable costs, i.e., costs related to the operation and utilization of the transportation system. These are either paid in kind (i.e., as a certain percentage of transport volumes—'fuel gas' for the compressors) or expressed as an indexed-price formula.[60]

Typically, irrespective of the utilization of the capacity reserved, the shipper has to pay the reservation fee to the carrier, i.e., the shipper has a 100% ship-or-pay obligation. In addition, carriers might require their shippers to maintain a minimum flow. Such minimum flow requirements might be used by the carrier to guarantee the operability of the transportation system at each point in time.

7.6.1.3 Main Tasks, Obligations and Rights of the Parties to a Transportation Contract

Natural gas transportation contracts will require the carrier to sell and reserve for the shipper a transportation capacity between (or at a) delivery point and (a) redelivery point(s) of a pipeline (system) operated by the carrier, together with the corresponding services.

[60]See Sect. 6.3.2.2.3.

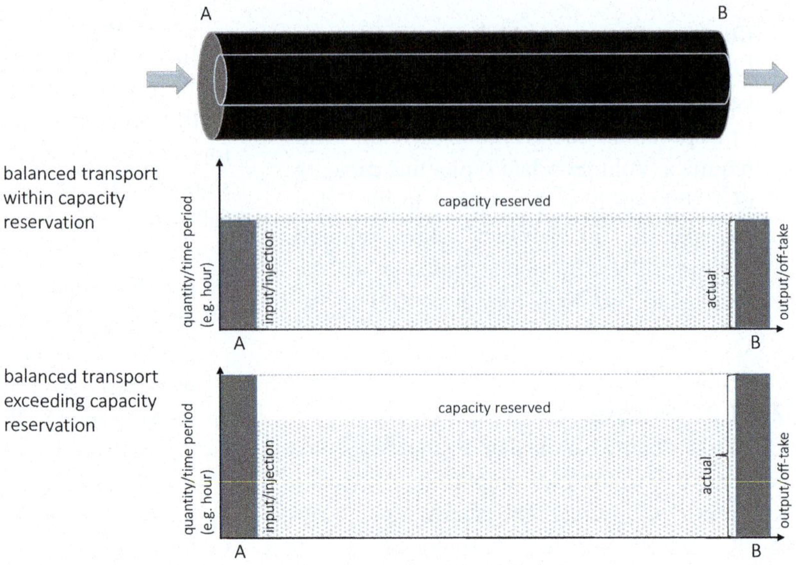

Fig. 7.12 Exceeding capacity reservation

The main tasks of a carrier are the following:

- receipt of the natural gas made available by the shipper at the delivery point (within the restrictions stipulated in the contract);
- return of the natural gas made available by the shipper at the delivery point at the redelivery point considering the restrictions and rules applicable at such point (*inter alia*, quality parameters);
- management and steering of the pipeline (system);
- operate and maintain the pipeline and ancillary facilities (in some cases also to design, construct, install, and commission new pipeline (sections) of the network); and
- other services such as:
  - measurement (including metering the quantity and analyzing and metering the quality of the natural gas at the delivery and redelivery point);
  - receipt and confirmation of (re-) nominations;
  - data management, collection and transfer;
  - allocation and attribution of quantities (to various shippers);
  - evaluation of measurements;
  - documentation; and
  - billing.

Correspondingly, the main tasks of the shipper are:

- payment of the contractually agreed transportation fees (charge);
- nomination of the quantities to be transported;
- delivery of the quantities to be transported at the delivery point in accordance with the terms and conditions of the contract. This includes, *inter alia*:
  - observing the compatibility of the maximum hourly (or daily) quantity to be delivered at the redelivery point with the capacity reserved,
    As depicted in Fig. 7.12, the quantities delivered by the shipper at the delivery point may not exceed the capacity reserved. The same quantity has to be off-taken at the redelivery point.
    The shipper might be obliged to pay overrun charges if shipper causes the transportation of a quantity of natural gas exceeding the reserved capacity;
  - observing the contractual, regulatory, and/or legal restrictions of the natural gas quality parameters;
  - observing the contractual pressure specifications at the delivery point; and

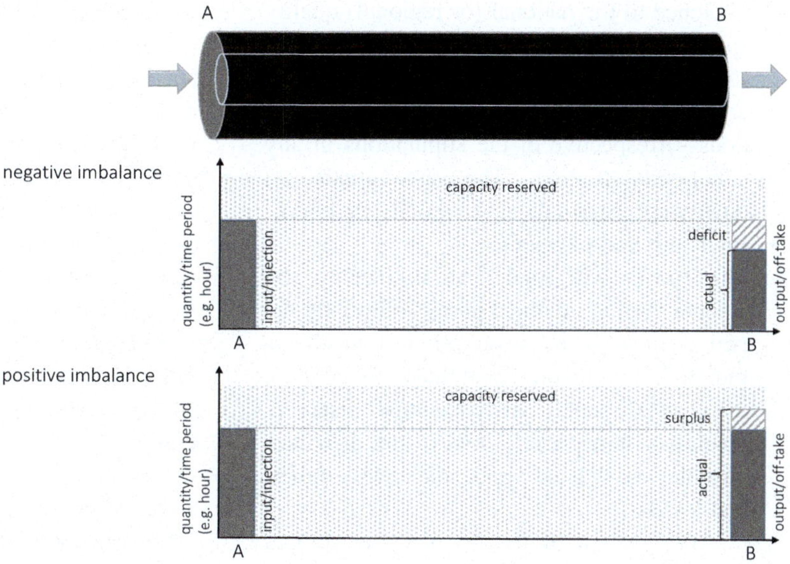

Fig. 7.13 Illustration of imbalances within the capacity reserved

- (if applicable) observing the minimum flow requirements,
- observing the balancing rules (Fig. 7.13) and
- off-take of the quantities transported at the redelivery point.

As mentioned, the shipper is usually not obliged to utilize the capacity reserved. However, vertically integrated companies were supposed to 'contractually congest'[61] pipelines by hoarding capacities neither needed nor used. Thus, other shippers were prevented from gaining access to such capacity. As a consequence, 'use-it-or-lose-it' or 'use-it-or-sell-it' rules, requiring transport capacity booked but not used by a shipper to be made available to other potential shippers, are increasingly implemented at various (congested) points.[62]

If 'use-it-or-lose-it' rules are implemented, they frequently refer to short-term, e.g., day-ahead, bookings only. The main argument against the implementation of long-term 'use-it-or-lose-it' rules is that, given the temperature dependence of natural gas consumption, the unpredictability of (mid- and long-term) temperatures in a region, and the requirements for safeguarding the security of supply, shippers, being traders, utilities, or local distributors have to book capacity for the peak load. Consequently, at all other times during the term of a transportation contract, the capacity will remain idle.

7.6.1.4 Other Commercially Relevant Elements of Transportation Contracts

As commodity contracts, transportation contracts contain various rules that may influence the economic viability of the contract and, therefore, shall be described briefly.[63]

- Definitions
  Again, the major terms used are defined. This includes, *inter alia*, units, periods, and quantities;
- Quality
  The quality of natural gas to be delivered at the delivery point is specified usually with

[61]Contractual congestion should be distinguished from physical congestion. While the latter describes a situation in which the physical pipeline capacity does not suffice to meet the requirements of users, the former describes a situation where capacity in a pipeline is fully booked but (physically) not utilized.

[62]See Agency for the Cooperation of Energy Regulators (2015).

[63]See also Sect. 6.3.4. For a legal discussion of (primarily international) transportation contracts, see Roberts (2014).

reference to the national (or regional) quality specifications. The natural gas made available for transportation by the shipper at the delivery point has to comply with such specifications—irrespective of the stipulations in any commodity contract. The same applies to quality specifications at the redelivery point;

- Invoicing and payment
  The carrier will render an invoice to the shipper within a defined period following the period of capacity reservation, typically a month. Such invoice will have to contain, *inter alia*, details of capacities reserved, quantities transported, prices, and penalties payable, as well as applicable taxes, duties, and levies.
  After receipt of the invoice, the shipper will be obliged to approve the invoice and pay the amount due within a further period;
- Nominations
  The shipper has to inform the carrier about (intended) future utilization capacities reserved. Such information may be non-binding and preliminary or binding. Typically, the non-binding information about planned capacity utilization is frequently agreed to be performed annually, seasonally, quarterly, monthly, or weekly before the respective period of the next year, season, quarter, month or week. Daily, or hourly information, the so-called 'nomination', will usually be binding upon the parties, i.e., the shipper will have to use, and the carrier will have to make available, the capacity nominated, as long as it is complying to the stipulations of the contract.
  Also in transportation contracts, intra-day re-nomination rights may be included. According to this, quantities nominated on a preceding day could be revised during the day of utilization, again not exceeding the capacity reserved and, if applicable, observing the minimum flow requirements;
- Measurement and facilities
  The location, details of installations, access rules, and procedures applicable—*inter alia*, for the determination of quantities transported—rules for testing and calibration and the accuracy required are specified. The carrier evaluates measurement records and uses them as the basis to issue invoices to the shipper;
- Credit risk (or performance assurance)
  Often, the carrier requests from shipper a guarantee of the payments for the capacity reserved for a defined future period. Similar to commodity contracts, this is done either by a bank guarantee, a letter of credit from any other institution, a parent-company guarantee (in case the shipper is an affiliate of a mother company considered to be creditworthy), or even a cash deposit;
- Force Majeure
  Also transportation contracts typically contain a force majeure clause, similar to the one explained in Sect. 6.3.8;
- Taxes and other public charges
  Taxes, duties, and levies imposed are usually paid by the shipper in addition to the payment for the capacity reserved; and
- Other clauses[64]
  - also the parties to a transportation contract usually do not assume liability for indirect (consequential) or even direct losses;
  - the term of a contract is one of the essential clauses;
  - applicable law, language, and settlement of disputes are legal issues;
  - the parties agree to exchange information regularly in a defined manner, revealing all information necessary to fulfil the obligations under the contract to the other party;
  - the assignment typically requires consent of the other party;
  - consents and approvals are typically given in writing;
  - severability, entire contract clause, amendments, i.e., form of modification;
  - warranties and representations; and
  - insurances.

[64]For these, predominantly, legal clauses see, e.g., Roberts (2014).

7.6.2 LNG Transportation Contracts

As with some, primarily international, pipeline transportation contracts LNG transportation contracts are usually tailor-made, in a sense, to allow the shipper to fulfil its obligations under a (or a set of) sales and purchase contracts.[65] Thus, under a FOB-based LNG SPA, the buyer of LNG has to organize sufficient and timely access to LNG vessels in order to fulfil its take obligation at the port of loading, and, under a DAP- (or DES-) based LNG SPA, the seller of LNG has to organize sufficient and timely access to LNG vessels in order to fulfil its delivery obligation at the port of discharging.[66]

Organizing transport capacity in LNG vessels is either possible by using own ships, or, more typically, by securing capacity under charter-party arrangements. A charterparty is an agreement for the hire of a ship, in the case of LNG, of an LNG vessel. Parties to such charterparty are usually the ship-owner and the charterer, i.e., dependent on their agreement on delivery point, the buyer or seller of LNG.

In general, charterparties are either time charter where the vessel is hired for a specific period of time or voyage charter where the vessel is hired for a specific voyage or voyages.[67]

Charterparties, *inter alia*, stipulate the following:

- vessel's name and IMO number,[68] including a detailed description of the capacities, performance (e.g., speed) and condition of the ship;
- details regarding the cargo to be loaded (e.g., quantity and temperature);
- loading and discharging rates;
- lay time;
- loading port and discharging port and respective terminals (especially for voyage charters);
- duration of charter (especially for time charters);
- freight and hire;
- demurrage rates;
- ownership and maximum quantity of heel[69];
- LNG price for paying any changes to the level of heel during the charter; and
- billing and payment.

7.6.3 Standard Contracts

Natural gas transportation being predominantly a regulated business[70] the carrier is usually required to offer pipeline transportation capacity on a non-discriminatory basis to other market participants. This leads necessarily to the development of standardized transportation contracts or third-party access to transportation networks. Such standard contracts are usually in the public domain and published by the carriers on their respective websites.

In addition to these general terms and conditions, the European Federation of Energy Traders (EFET)[71] introduced a so-called General Agreement Gas, which permits appendices. Two of these appendices reflect transport transactions. They might be used for transactions at PRISMA and/or Fluxys/GRTgaz.

Further, the Association of International Petroleum Negotiators (AIPN)[72] developed a Gas Transportation Agreement in 2009.

Further to contracts regarding the sale and purchase of LNG, also standard contracts for the transportation of LNG are available. In 2016, GIIGNL jointly with BIMCO[73] developed a

[65]See Sect. 5.1.1.

[66]See Sect. 6.3.4.

[67]In both cases, the crew is usually provided by the ship-owner. Only a bareboat charter, a subtype of the time charter, permits the charterer to provide the crew.

[68]The IMO number is a unique ship-identification number assigned by the International Maritime Organization (IMO).

[69]See Sect. 7.2.2.2.

[70]See Sects. 5.3 and 7.5.2.1.

[71]See European Federation of Energy Traders (2016); see also Sect. 6.3.9.

[72]See Association of International Petroleum Negotiators (AIPN) (2016).

[73]See Baltic and International Maritime Council (BIMCO) (2016).

LNG voyage-charter party designed for the growing spot market for LNG. Shell issued in 2005 the ShellLNGTime1 for LNG time charters whereby this standard also works well for voyage charters.

7.6.4 Categorization of Natural Gas Transportation Contracts

To facilitate the description, natural gas transportation contracts are categorized. The most important criteria, applied for such categorization are:

- regional scope:
  - international off-shore transportation,
  - international transit,
  - national (high pressure system) transportation,[74] and
  - local distribution;
- type of service:
  - firm transportation,
  - interruptible transportation, and
  - other forms[75]; and
- duration (or term):
  - long-term,
  - short-term, and
  - day-to-day.

The first two categories will be discussed subsequently.

7.6.4.1 Region: International, National, and Local Contracts

In (international) offshore pipelines and transit pipelines, natural gas is transported over long distances. These pipelines are often constructed with the purpose to bring the natural gas from the production site(s) to the region of consumption.

Offshore and Transit Pipelines Used for Imports to Europe

Essentially, four main systems, developed or planned, are used to import natural gas to Europe:

- Northeast
  Russian natural gas is imported via the Nordstream pipeline connecting Russia with Germany via the Baltic Sea; the Yamal pipeline, crossing Belarus and Poland ending near Frankfurt/Oder, Germany; and the Brotherhood pipeline crossing the Ukraine, Slovakia, and the Czech Republic.
- Northwest
  Norwegian natural gas is imported to Continental Europe, *inter alia*: via Europipe and Norpipe to the Emden area, Germany; via Zeepipe to the Zeebrugge area, Belgium; and via Franpipe to the Dunkerque area in France. Exports to the UK are transported via the Langeled pipeline to the Easington area and the Vesterled pipeline to the St. Fergus area.
  The Norwegian offshore-pipeline system has a length of more than 8000 km.
- Southwest
  Natural gas from (primarily) Algeria is transported via the Transmed pipeline to Italy and the Maghreb pipeline to Spain (additional exports to Europe are transported in the form of LNG).
- Southeast
  To import natural gas from Central Asia to Europe several projects, *inter alia*, Nabucco and Southstream have been proposed and investigated. These projects have not been realized. Other projects in this region are still investigated.

National (not necessarily nationalized) carriers operate pipelines within national boundaries. As different regulatory systems apply in different

[74] To be more complete, a further distinction between high and medium pressure, equivalent to national and regional systems, may be introduced, but usually the features of these systems are similar enough to summarize them in one group.

[75] As, e.g., hybrid forms of partly flexible transportation services (offered, e.g., in the US). These shall not be discussed subsequently.

countries, transportation contracts are offered predominantly for transportation within a country. That is, when borders must be crossed, shippers will have to conclude additional contracts.

Local distribution is offered by distribution system operators (DSOs) for natural gas transportation within a clearly defined, often municipal, region. At this stage, the natural gas is transported to the end user.

Depending on the regional scope of natural gas transportation, various types of capacity reservation and pricing systems are used:

- for both international off-shore transportation and international transit distance-related pricing systems, in conjunction with a point-to-point capacity reservation system;
- for most of the (regulated) national (high-pressure system) transportation, an entry-exit pricing and an entry-exit capacity reservation system; and
- for local distribution systems, a postalized pricing system and either a postalized capacity or entry-exit capacity reservation system.

7.6.4.2 Commitment: Interruptible and Firm Contracts

7.6.4.2.1 Firm Transportation Contracts

In a firm or uninterruptible transportation contract, the capacity has to be made available by the carrier to the shipper with priority and—in general—unconditionally. However, despite the 'firm' nature of such capacity, the capacity may be interrupted under specific circumstances.[76] These circumstances include:

- force majeure;
- maintenance, revamping works, repairs, or construction, inspection, tests; or
- other events, which, *inter alia*, may include
  - emergencies not qualified under the applicable law or under a specific contract as force majeure;
  - orders of authorities not qualified under the applicable law or under a specific contract as force majeure; and
  - non-compliance with the rules stipulated in the transportation contract.

[76]Such circumstances are usually neither controlled nor triggered.

> *No pipeline can provide an absolute guarantee of physical delivery .... Physical firmness is inherently a probabilistic concept: what a pipeline defines as "physically firm" service is in reality service with a very low probability of interruption.*

Lapuerta and Moselle (2002), p. 4; and similar p. 24, 25.

An example for the distinction of such circumstances was specified by the Bulgarian carrier in its standard gas transportation contract:

(Footnote 76 continued)

> *15.1. The Transmission Operator shall have the right to interrupt the performance of contracted interruptible capacity transmission services, as well as to terminate or limit natural gas transmission in the following cases:*
>
> *15.1.1 in case of immediate threat to the life, health or property of people and during prevention of such circumstances;*
> *15.1.2 in case of Force Majeure;*
> *15.1.3 at the time of planned reconstructions and planned repairs of the gas transmission system equipment;*
> *15.1.4 in case of failures or other technological reasons beyond the operator's control;*
> 15.1.5 in case of unscheduled repairs of the gas transmission system equipment;
> *15.1.6 in case of enforcement of the sanctioning measures under item ...;*
> *15.1.7 in case of an enforced limiting regime;*
> *15.1.8 in case that the Client fails to comply with the requirements for natural gas quality and pressure at the entry points, set out in ...;*
> *15.1.9 in cases threatening the integrity of the transmission network;*
> *15.1.10 in case that the exit point is no longer operating (the Client guarantor has no valid contract for balancing);*
> *15.1.11 in case that the Client has no valid contract for the purchase/sale of balancing energy;*
> *15.1.12 in case of fulfillment of the obligations imposed to the Operator by an Emergency Plan in accordance with Regulation (EU) No 994/2010 of the European Parliament and of the Council, approved by an Order of the Minister of Economy, Energy and Tourism.,* Bulgartransgaz (2013), p. 17

Similar rules apply for other pipelines, see, e.g., GazSystem (2014), p. 63. The Yamal pipeline connects the Russian natural gas pipeline system with the Polish, via the Belarusian.

Yet, despite such possible interruptions, these transportation contracts are considered by the parties as firm ones.

7.6.4.2.2 Interruptible Transportation Contracts

A carrier, having sold an 'interruptible' transportation capacity, has the right not to make available this capacity—at its sole discretion. That is, while any interruption of transportation capacity availability in firm transportation contracts is not (and cannot) be controlled and triggered by the carrier, the contrary applies to interruptible transportation contracts. Typically, the conditions or reasons for interruptions do not have to be specified. Any specification of such conditions or reasons of interruption would put more of a restriction on the carrier than help increasing the predictability and thus the ability of the shipper to schedule such interruptions.[77]

The incentive to conclude an interruptible transportation contract for the carrier is that it might be able to sell more capacity than physically available in its system at any point in time. Based on his assessment of the use of the (physical) transportation capacity in the system, it will take the risk of 'congestion' or 'overbooking'. The shipper may be incentivized to conclude such interruptible contract by the price, i.e., a low transportation fee, which in turn depends, *inter alia*, on his expectation of the probability of being interrupted.

In any contract, interruptibility should be described by the following parameters (usually during a contract year[78]):

- maximum extent of interruption, i.e., percentage of the capacity interruptible or absolute capacity interruptible, e.g., not more than 40% of the capacity reserved can be interrupted;
- maximum duration of single interruptions, e.g., no interruption shall last longer than two consecutive days;
- maximum total cumulative duration of interruptions, e.g., the sum of all days of interruption during a contract year may not exceed six days;
- maximum number of interruptions, e.g., the number of interruptions during a contract year may not exceed four;
- minimum time lag for the announcement of an interruption, e.g., any interruption has to be announced by the carrier at least 36 h before the start of such interruption; and
- minimum time lag for the announcement of the end of an interruption, e.g., any end of an interruption has to be announced by the carrier at least 12 h before the end of such interruption.

Even if interruptible transportation contracts are often offered at a discounted price for the capacity,[79] some carriers offer these services for the same price as firm ones[80] and give *ex post* discounts.[81]

[77]An example of such restriction may be:

> *if the average daily temperature at XXX falls below Y °C on a day between October 1 and March 31 of any year, then the transportation capacity made available by the transporter to the shipper may be interrupted for ZZZ consecutive days (hours) starting on TTT, given that the transporter informs the shipper with HHH prior notice.*

Generally, however, such a rule imposes a restriction for interruption on the carrier but does not increase (mid- or long- term) predictability for the shipper. The temperature has to be regarded as being unpredictable in the long-term and, therefore, a statistical variable.

[78]Yet, it has to be noted that, in reality, not all of the listed parameters are included in each interruptible transportation contract.

[79]See KEMA (2009), p. 11.

[80]

> *For example, in Austria transit tariffs for interruptible capacity are indeed the same as firm tariffs, with network users being compensated in case of interruptions. Similar mechanisms are also widely used in Germany.*
>
> KEMA (2009), p. 12.

[81]

> *RWE Energy offers interruptible contracts at the same price as firm contracts, if all firm capacity has been sold, and there is a rebate if the customer is curtailed.*
>
> Arthur D. Little (2004), p. 19.

References

Agency for the Cooperation of Energy Regulators (2015) ACER annual report on contractual congestion at interconnection points. Available at: http://www.acer.europa.eu/official_documents/acts_of_the_agency/publication/20150529_acer 2015 report on congestion at ips in 2014.pdf

Association of International Petroleum Negotiators (AIPN) (2016) Homepage. Available at: https://www.aipn.org/default.aspx

Baltic and International Maritime Council (BIMCO) (2016) Homepage. Available at: https://www.bimco.org

British Petroleum (2016) BP statistical review of world energy

Bulgartransgaz (2013) General terms and conditions of the transport contracts for natural gas transmission through Bulgartransgaz EAD gas transmission networks. (April). Available at: http://www.bulgartransgaz.bg/files/useruploads/files/general_terms_and_conditions_ngtn_gtntt.pdf

Cerbe G, Dehli M, Kätelhön JE, Kleiber T, Lehmann J, Lendt B, Mischner J, Mundus B, Pietsch H, Spohn D, Thielen W (2004) Grundlagen der Gastechnik – Gasbeschaffung, Gasverteilung, Gasverwendung, 6th edition, Munich

Chandra V (2006) Fundamentals of natural gas—an international perspective, Tulsa

Dunkerque LNG (2011) LNG terminal, Key points

European Federation of Energy Traders (2016) Gas & Gas Annexes. Available at: http://www.efet.org/Standardisation/Legal-EFET-Standard-Contracts-and-Documentation/GasAndGasAnnexes

Fasold H-G, Wahle H-N (1996) Einfluss der Rohrrauhigkeit und der Rohrreibungszahl auf die Transportkapazität und die spezifischen Kosten von Gasrohrleitungen. gwf-Gas/Erdgas, 137(3): 109–118

Füg T (1999) Zur Preisbildung in Erdgastransportsystemen, Idstein

GazSystem (2014) Transmission network code of the Polish section of the transit gas pipeline system JAMAL – EUROPE, (January). Available at: http://en.gaz-system.pl/fileadmin/pliki/sgt/en/TGPS_TNC_EN.pdf

Hairston D, Moshfeghian M (2013) Low pressure vs high pressure dense phase natural gas pipeline transportation, pp 2–8. Available at: http://www.jmcampbell.com/tip-of-the-month/2013/01/low-pressure-vs-high-pressure-dense-phase-natural-gas-pipeline-transportation/

International Energy Agency (IEA) (1994) Natural gas transportation—organisation and regulation, Paris

KEMA (2009) Study on methodologies for gas transmission network tariffs and gas balancing fees in Europe, submitted to: The European Commission, Directorate-General Energy and Transport, Arnhem. Available at: https://ec.europa.eu/energy/sites/ener/files/documents/2009_12_gas_transmission_and_balancing.pdf

Keyaerts N, Meeus L, D'haeseleer W (2008) Natural gas balancing: appropriate framework and terminology natural gas balancing: appropriate framework and terminology working paper, Leuven. Available at: http://www.mech.kuleuven.be/tme/research/

Lapuerta C, Moselle B (2002) Convergence of non-discriminatory tariff and congestion management systems in the European gas sector

Arthur D. Little (2004) Gastransport services—West European gas transmission tariff comparisons, Appendix to the report to Gastransport Services, Reference 17928/005rep/Apps

Groupe International des Importateurs de Gaz Naturel Liquéfié (GIIGNL), The LNG Process Chain, (LNG Information Paper 2)

Mokhatab S, Mak JY, Valappil JV, Wood DA (2014) Handbook of liquefied natural gas, Oxford

LNG World News (2016) Alaska railroad to begin LNG transport tests. LNG World News, pp 8–9. Available at: http://www.lngworldnews.com/alaska-railroad-to-begin-lng-transport-tests/?utm_source=emark&utm_medium=email&utm_campaign=daily-update-lng-world-news-2016-09-21&uid=19664

Prete L (1998) Transmission pricing issues for electricity generation from renewable resources

Pustišek A (2005) Untersuchungen der Struktur von Preissystemen für Erdgastransportkapazitäten. University of Münster, Germany

Roberts P (2014) Gas and LNG sales and transportation agreements, 4th edn, London

Songhurst B (2014) LNG plant cost escalation. Oxford Institute for Energy Studies (February), pp 8–10. Available at: https://www.oxfordenergy.org/wpcms/wp-content/uploads/2014/02/NG-83.pdf

Wood DA, Mokhatab S, Economides MJ (2008) Technology options for securing markets for remote gas. Proceedings of 87th Annual Convention of the Gas Processors Association (GPA)

Storage

8

Summary

- Storage is used to transfer natural gas in time.
- The functions of storage are:
  - balancing of demand fluctuations,
  - providing security of supply, and
  - serving as a tool to optimize the trader's business, i.e., to maximize (expected) profit.
- Storages can be constructed in:
  - depleted oil or natural gas reservoirs,
  - aquifers,
  - salt caverns,
  - rock caverns,
  - LNG tanks,
  - Pipelines.
- Every storage is described by at least the following interdependent parameters:
  - injection capacity,
  - withdrawal capacity,
  - working gas capacity.
- Storage prices are defined using these three parameters.
- In traded liquid markets, storage functions can, within limits, be replicated by time swaps, i.e., selling and purchasing forwards (or futures).

8.1 General

In general, 'storage' designates the space where objects are put at times when they are not directly or immediately being used or required.[1] As such, storage is (quasi-)permanent containment of goods or objects including an (easily accessible) inlet and outlet. This definition can be applied to natural gas.

Whereas transportation refers to the transfer of objects in space, storage can be (abstractly) described as a transfer of objects in time. For transportation, an object is 'injected at the delivery point' and 'withdrawn at the redelivery point'—(contractually) at the same time[2]; for (physical) storage, an object is 'injected at one point in time' and 'withdrawn at a later point in time'—(typically) at the same location.

Storages are (predominantly) owned and operated by storage system operators ('SSO'), being unbundled companies, (partially) owned by producers, utilities, or local distribution companies or by companies otherwise not active in the natural gas industry. According to most unbundling requirements, storage system operators do not own natural gas (except for purposes of storage operation) and, therefore, do not trade natural gas.

[1] In addition, the term can also refer to the process or act of putting an object into a place where it can be kept or even the state itself, see Merriam-Webster (2016).

[2] In case 'instantaneous transportation', in accordance with regulatory stipulations, is assumed.

A. Pustišek and M. Karasz, *Natural Gas: A Commercial Perspective*,
DOI 10.1007/978-3-319-53249-3_8

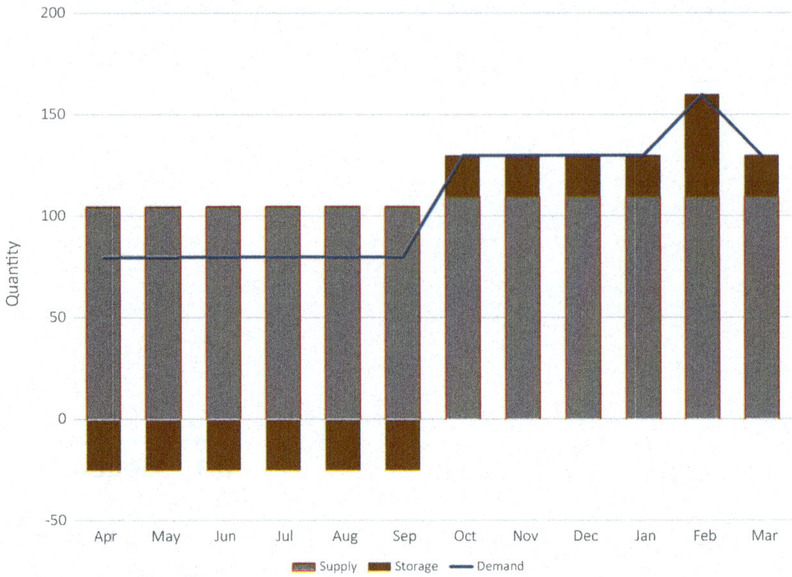

Fig. 8.1 Seasonal demand variation

8.2 Functions

The demand for natural gas is neither constant nor does it correlate to variations of production.[3] Moreover, provisions for interruptions of supplies have to be established. In modern markets, tools for hedging and speculation are required as well. These tasks can be fulfilled by storages. That is, natural gas storages have the following functions[4]:

- balancing of demand fluctuations;
  - short-term, i.e., hourly and daily demand fluctuations;
  - mid- and long-term, i.e., up to annual demand fluctuations;
- provision of security of supply; and
- (economic) optimization, here, supply cost minimization.

Essentially, storages are used to provide volume flexibility, which, in turn, is used for the above-mentioned purposes.

[3]See Sect. 6.2.

[4]Multi-annual balancing of demand fluctuations or any other multi-annual use of storage are only rarely regarded as a function of storages.

Generally, storages should be:

- reasonably close to transportation infrastructure, and
- reasonably close to the main areas of consumption.

8.2.1 Balancing of Demand Fluctuations

The primary function of storages is to eliminate differences between supply and demand during a period. A simplistic example is shown in Fig. 8.1. Demand (blue line) in the winter months is higher than in summer. The supply (grey bars) cannot cover demand in winter and exceeds it in summer. Therefore, natural gas is injected to storage in summer (negative red bars) and withdrawn in winter (positive red bars).

Typically, the high-frequency demand fluctuations are superimposed on the seasonal demand. Contractual flexibility, i.e., volume flexibility provided under a commodity supply contract,[5] may be used to provide part of the flexibility of demand but exceeding fluctuations of demand have to be absorbed by storages. This is depicted

[5]See Sect. 6.3.1.

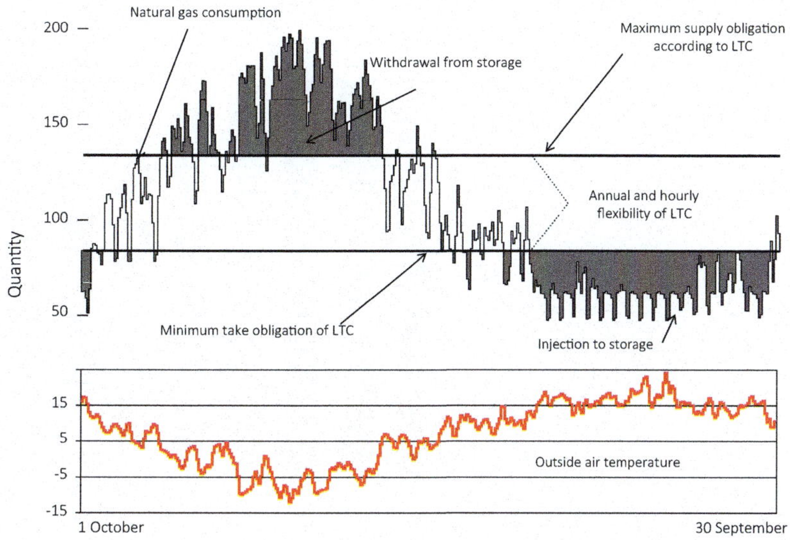

Fig. 8.2 Use of (seasonal) natural gas storage to balance supply and demand

in Fig. 8.2 (note that, contrary to Fig. 8.1, the time scale starts in October not in April).

8.2.2 Provision of Security of Supply

In case of supply interruptions, stored natural gas can be withdrawn to fulfil the demand requirements. This is depicted in Fig. 8.3. In this example, supplies are reduced in February to 80 (arbitrary units), while the demand (line) remains at 130. Withdrawals from storage (red and yellow [for the compensation of interruptions]) are increased to 50 in order to satisfy demand.

8.2.3 Optimization

Even in the case of no supply interruptions or demand fluctuations, storage capacity is used. Such a situation is depicted in Fig. 8.4. In this example, a price increase is anticipated starting in September (yellow, dotted). Accordingly, storage injections are increased when prices are

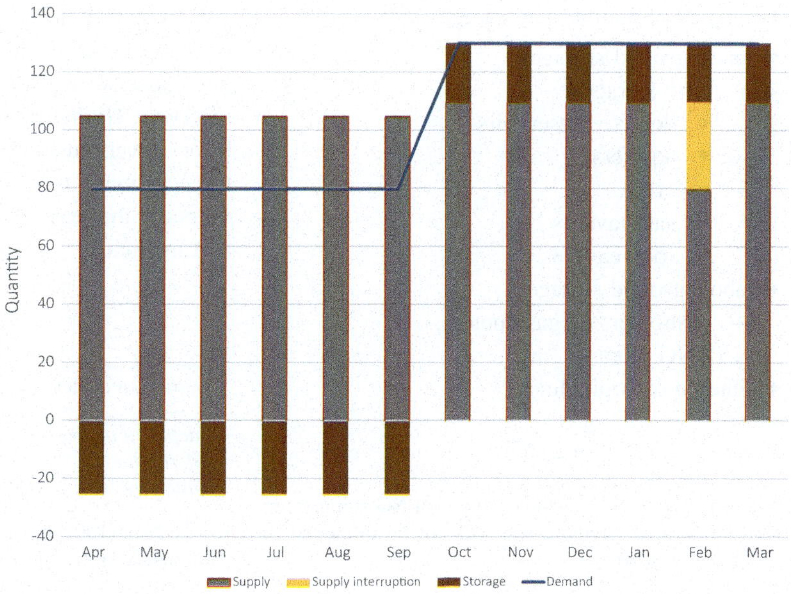

Fig. 8.3 Supply interruption

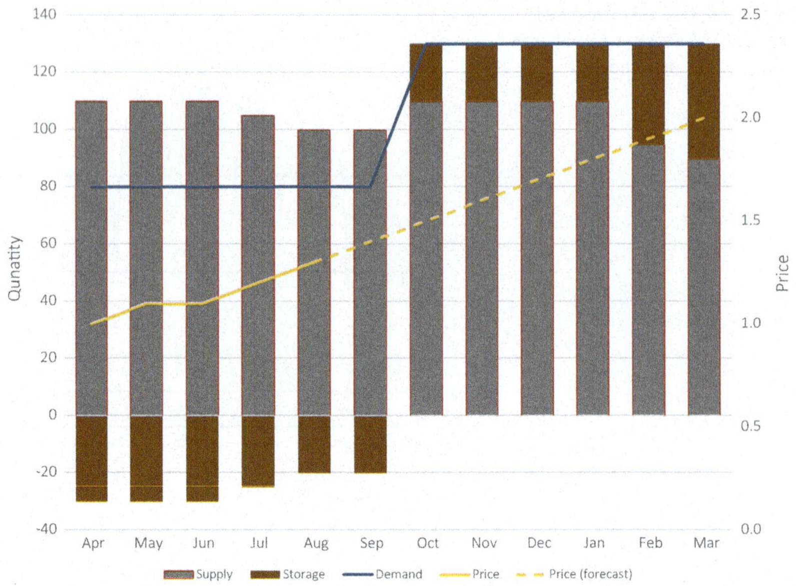

Fig. 8.4 Supply cost minimization

(relatively) low (red, in April to June) and decreased when prices increase (red, in August to September). Then, supplies at higher prices could be replaced by increasing withdrawals from storage (red, in February and March).

8.3 Outline of Physical Storage and Assets

8.3.1 Types of Storages

The types of natural gas storages are:

- underground storages:
  - pore storages,
    - depleted reservoirs,
    - aquifers
  - caverns
    - salt caverns,
    - rock caverns
- aboveground storages:
  - gasholders or gasometers,
  - LNG storages, and
- line pack in pipelines.[6]

Geologic and technological conditions will determine which types and sizes of storage can be constructed.

8.3.1.1 Pore Storage

Porous media in the underground are frequently used as natural gas storage.

8.3.1.1.1 Depleted Reservoirs

Depleted reservoirs are underground formations that originally contained oil or natural gas reserves.[7] When converted to a storage, cushion gas required to maintain the pressure (for depleted fields in the range of 50% of the total gas capacity) has to be injected (or left) in the reservoir. Higher quality depleted reservoirs also have lower cushion gas requirements. Of the three primary types of storage, depleted reservoirs are the least expensive to develop, operate, and maintain.

[6] Aboveground storage facilities (except LNG storage) are not widely used anymore and will not be discussed in detail in this section.

[7] Contrary to depleted gas fields,

> *depleted oil reservoirs are not always satisfactory if residue oil is produced in sufficient quantity to interfere with gas withdrawal operations,*

see Peebles (1992), p. 92. See also Energy Information Administration (2004).

Geologically, the following conditions should be fulfilled for depleted reservoirs:

- To ensure containment within a depleted reservoir (a prerequisite for any gas storage facility), there must be an impermeable cap rock and either structural or stratigraphic containment on the flanks of the reservoir;
- The reservoir rock itself must have sufficient porosity and permeability to allow the natural gas to migrate into the reservoir and to allow withdrawal within an economically sufficient period of time. The overall size of the reservoir and the porosity of the formation determine the amount of natural gas that can be stored. The permeability determines the rate at which natural gas flows through the formation. In turn, jointly with the capacities of surface facilities,[8] permeability determines the injection rate and withdrawal rate of natural gas in or from the storage; and
- The reservoirs must be situated within an economically deep stratum, i.e., not too deep to increase drilling costs, but not too shallow to allow for a safe operation, that must not contain old leaky well bores or have complicated geological structures.[9]

8.3.1.1.2 Aquifers

Aquifers are underground porous, permeable rock formations that act as natural water reservoirs.[10]

These water-containing formations may be reconditioned and used as natural gas storages. Being more expensive to develop than depleted reservoirs, aquifers are usually used only in areas where there are no depleted reservoirs nearby. They are expensive for several reasons:

- Because of the lack of production wells and production history, the geological characteristics of aquifer formations are not as thoroughly understood as those of depleted reservoirs. Hence, *inter alia*, seismic testing must be performed, much like it is done for the exploration of potential natural gas formations;
- The area of the formation, the composition, and porosity of the formation itself, and the existing formation pressure must all be explored and analyzed prior to development of the formation. In addition, the capacity of the reservoir is unknown and may only be determined once the formation is further developed;
- More extensive infrastructure must also be developed including: installation of wells, extraction equipment, pipelines, dehydration facilities, and compression equipment; or
- In aquifer formations, the required cushion gas can amount to 90% of the total gas storage capacity. This volume has to be injected into the aquifer and increases the CAPEX. While it is possible to extract the cushion gas from depleted reservoirs, doing so from aquifer formations could be impossible. As such, most of the cushion gas that is injected into any aquifer formation may remain unrecoverable, even after the storage facility is shut down.

[8]See Sect. 8.3.2.

[9]Which are considered to add 'geological' risk to the storage development.

[10]Naturalgas.org (2013a); for more, see also Energy Information Administration (2004).

8.3.1.2 Salt Caverns

Salt caverns are located in underground salt formations either in salt domes or in salt beds.

Salt domes are thick formations created by halokinesis, i.e., natural salt deposits that, over time, rose up through overlying sedimentary layers to form large dome-type structures. They can have sizes of up to 2 km in diameter and of up to 10 km in height. Typically, salt domes used for natural gas storage are at depths of 300–800 m.

Salt beds are shallower and thinner formations than salt domes. Usually, these formations are up to 300 m in height and composed of multiple, thin layers.

The caverns are typically solution-mined by injecting fresh water through a well drilled into the salt, dissolving the salt into brine with the fresh water and withdrawing the resulting brine for disposal or, e.g., for use as feedstock in the chemical industry.

In comparison to depleted reservoirs or aquifers, salt caverns, being free space and therefore not restricted by porosity and permeability, offer higher injection and withdrawal rates (in relation to the working gas capacity, which is lower). The result is that the working gas capacity in salt caverns can be cycled more often than in both, depleted reservoirs and aquifers, typically up to 12 times per year. Salt caverns also require less cushion gas (usually 20–40% of the working gas capacity).

Salt cavern storages are best suited for peak-load 'shaving', i.e., compensation of short-term demand variations. The working gas capacity is typically much smaller than the one of depleted gas reservoirs and aquifers. In fact, underground salt caverns usually take up only 1% of the area taken up by a depleted gas reservoir and are, therefore, only to a certain degree suited to hold the volume of natural gas necessary to meet base-load storage requirements.

A remarkable and observable issue of salt caverns is convergence.[11] This is the reduction of the cavern volume (induced by the pressure of surrounding rocks), typically measured by the convergence rate, i.e., the percentage reduction per year. Usually convergence rates of approximately 1%/a are assumed and observed, but this is just an approximation as both temperature and pressure will increase with depth and trigger an increase of convergence rates.[12] Further, the storage usage of the cavern in itself influences the convergence rate, because caverns that are nearly empty tend to have higher convergence rates.[13] In addition, the pressure and, consequently, the volume of natural gas to be stored increases with depth, and so do costs. Consequently, optimization with regard to the restrictions will have to be performed while designing a new salt cavern.[14]

8.3.1.3 Rock Caverns

These facilities are constructed in (man-made) excavated caverns in basement rock. Storing natural gas in rock caverns is mainly used only in those regions that do not offer the geological prerequisites necessary to build pore storages or salt caverns.

There are two main concepts:

> *The unlined rock cavern concept is based on the principle that the ground water pressure in the surrounding rock balances the gas pressure inside the cavern. Drilled* water *curtains can be used to improve the function and to allow storage pressures above the natural ground water pressure level. The lined rock cavern (LRC) concept is a technology for underground storage of natural gas at high pressure in lined rock caverns at relatively shallow depths. The main idea is to store the compressed gas in rock caverns lined with a thin impermeable liner and to let the surrounding rock mass carry the* load.[15]

8.3.1.4 LNG

Even if, in comparison with underground storages, LNG storages can provide volume flexibility only to a limited extent, the versatility of LNG storage is not to be neglected. Such volume flexibility can be made available either by:

- storage of LNG at the liquefaction plant,
- storage of LNG at the regasification plant, or
- by the LNG cargo-delivery schedule.

Whereas the first two primarily are suitable as tools used to provide short-term flexibility components, the latter may be used for long-term flexibility components.

[11]For details, see, e.g., Oelizowski and Walaszcyk (2008), p. 14 et seq.

[12]For further details, see, e.g., Plaat (2009).

[13]For this reason, some storage contracts stipulate certain minimum working gas levels.

[14]For a description, see also, e.g., Energy Information Administration (2011).

[15]Johansson (2003), p. 13.

Usually, LNG is transferred via cryogenic pipelines to insulated storage tanks. Also, such tanks are designed to keep the liquid at low temperatures and minimize the amount of evaporation. In addition, a

> *storage facility is designed with a venting feature as an ultimate protection against risk of overpressure due to a 'roll-over' condition in the LNG tank. LNG 'rollover' refers to the rapid release of LNG vapours from a storage tank, resulting from stratification. The potential for rollover arises when two stratified layers of different densities (due to different LNG compositions) exist in a tank.*[16]

Typically, a regasification terminal is equipped with at least two LNG storage tanks.

There are several kinds of storage tanks as, e.g., single, double, full containment, and in-ground membrane-type storage. The main difference between these is the secondary containment.

Also, for LNG storage economies of scale apply. Consequently, as technology has matured storage tank sizes have grown.

Recently, also small-scale LNG facilities, providing increased flexibility, are under investigation and construction.

8.3.1.5 Line Pack

Natural gas networks can be used to store natural gas in limited quantities. Storage capacity is created in pipelines by increasing the pressure above the level required for transportation, i.e., by injecting more natural gas into the pipeline than is being withdrawn, but still within the limit of safe pipeline operation. Such storing of natural gas in pipelines is called line-pack storage. Dedicated line-pack storages[17] include (local) pipeline systems consisting of pipelines with a larger diameter than necessary for pure transportation or pipelines longer than necessary. In this context 'line-pack' has to be distinguished from 'line-pack flexibility'. Whereas the first shall be

> *the total volume of gas present in a pipeline section,*[18]

the latter is

> *the amount of gas that can be managed flexibly by controlling the operation pressure levels between a minimal and a maximal level*

(see Footnote 18). Compared with other storages the volume of such facilities is small. Therefore, line-pack storage can be used for short-term storage only. The economics of line-pack storage are determined by the investment in the larger or longer pipeline and operational expenditures required to increase the pressure.

8.3.2 Surface Facilities

For underground storages, surface facilities include compressor stations for the injection, gas treatment installations, gas measuring stations, communication facilities, a control center, heaters, coolers, filters, etc.

Natural gas originating from the network enters the surface facilities of the storage, is filtered in order to separate solids and liquids, compressed to overcome pressure in the storage, cooled (as the compression process heats the natural gas), and injected via high-pressure pipelines and wells into the storage.

The same wells are used to withdraw natural gas from the storage. As natural gas in the storage absorbs water, initially, free droplets of water are absorbed. To prevent gas hydrates from being formed in the processing plant, the natural gas withdrawn is heated, de-pressurized and then dehydrated in a glycol absorption plant. This natural gas is then injected into the network. At the end of the withdrawal period, it might be necessary to compress the natural gas withdrawn again in order to comply with the pressure specifications in the network (Fig. 8.5).

[16] Groupe International des Importateurs de Gaz Naturel Liquéfié (GIIGNL) (n.d.), p. 6.

[17] Of course, line pack is used in all pipeline systems including those not deliberately built for storage purposes for balancing short-term variations between supply and demand within the respective pipeline.

[18] Keyaertsa et al. (2010), p. 3; see also Keyaertsa et al. (2011), p. 866.

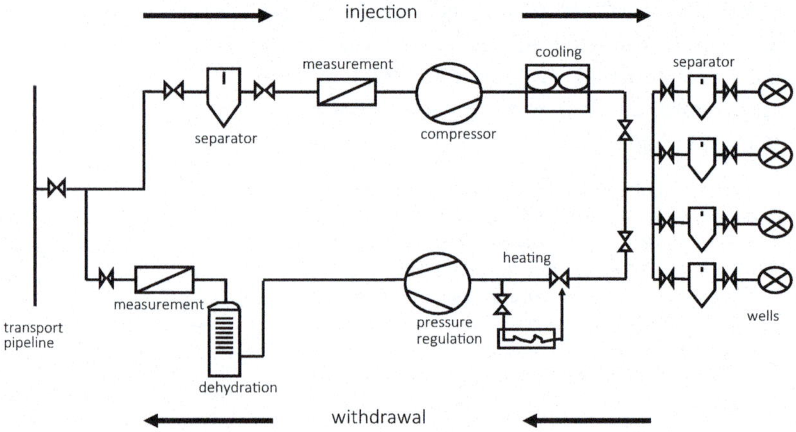

Fig. 8.5 Depiction of storage facilities

8.4 Storage Parameters

Different volumetric measures and parameters are used to characterize a natural gas storage and natural gas contained in such storage. The main parameters are shown in Fig. 8.6.

These parameters are also defined in Table 8.1.

The withdrawal rate and injection rate of a given storage are, *inter alia*, a function of total gas in storage, the pressure in the storage, and the design of surface facilities. While the withdrawal rate increases as a function of the working gas volume, the injection rate decreases as a function of the working gas volume. That is, the maximum injection rate can only be provided when the storage is empty and vice versa for the withdrawal rate. Such functions are schematically illustrated in Fig. 8.7.

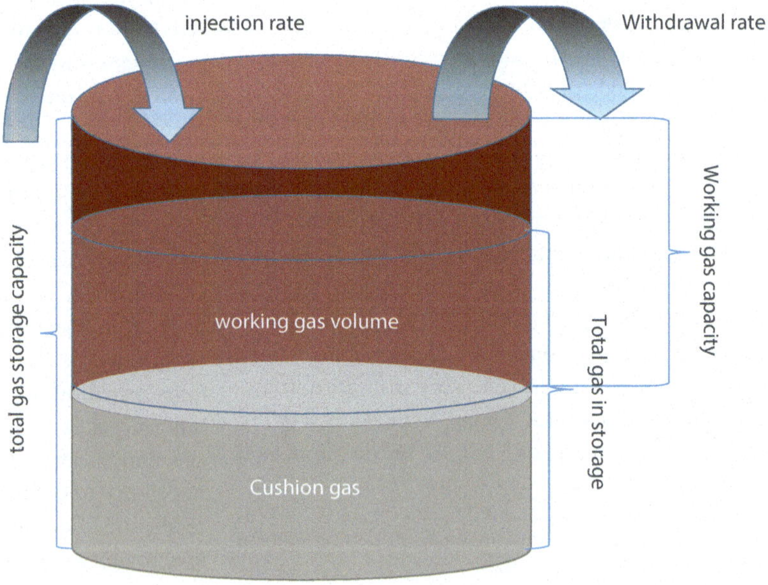

Fig. 8.6 Volumetric parameters of natural gas storages

Table 8.1 Volumetric storage parameters

Parameter	Explanation	Physical quantity	Typical unit
Total gas storage capacity	Maximum volume of natural gas that can be stored in a certain storage facility. As such it is the sum of cushion gas and working gas capacity	Volume or energy	[m^3] or [MWh]
Total gas in storage	Volume of natural gas remaining in the storage at a particular time	Volume or energy	[m^3] or [MWh]
Cushion gas (or: base gas)	Volume of natural gas intended as permanent inventory in a storage required to maintain adequate pressure and deliverability rates throughout the withdrawal season	Volume or energy	[m^3] or [MWh]
Working gas capacity	Capacity that can be used in the storage to inject and withdraw natural gas	Volume or energy	[m^3] or [MWh]
Working gas volume	Volume of natural gas in the storage exceeding cushion gas. The working gas volume can be injected and withdrawn. It is at the disposal of the users	Volume or energy	[m^3] or [MWh]
Withdrawal rate (or: deliverability)	Amount of gas that can be withdrawn from a storage on an hourly or daily basis	Volume/time or energy/time	[m^3/h] or [MW]
Injection rate	Amount of gas that can be injected into a storage on an hourly or daily basis	Volume/time or energy/time	[m^3/h] or [MW]
Withdrawal time	Time necessary to withdraw working gas capacity	Time	[h]
Injection time	Time necessary to inject working gas capacity	Time	[h]
Cycling time	Withdrawal time plus injection time	Time	[h]
Cycling rate	8760 h/a/(cycling time)	1/time	[1/a]

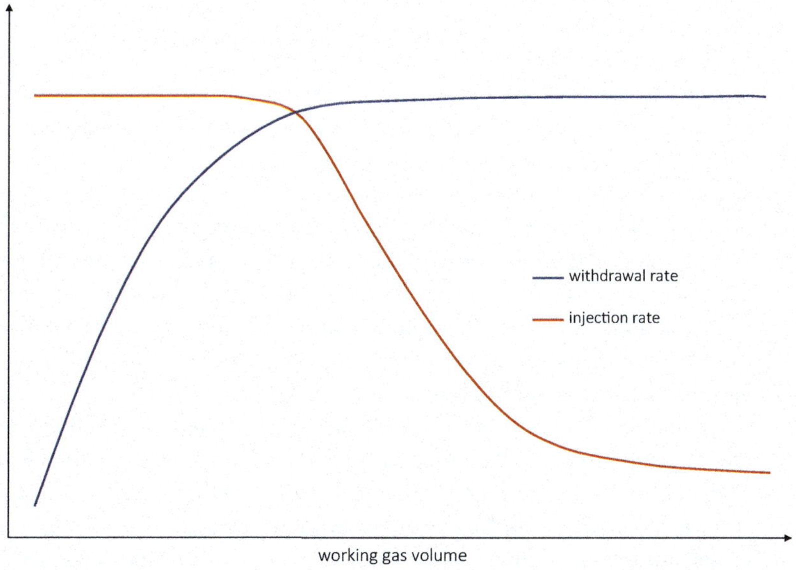

Fig. 8.7 Schematic illustration of relationship between main storage parameters

8.5 Players

Essentially, a storage facility consists of the (underground) capacity and the aboveground facilities, including the connection to the transportation network. Owners of such facilities are not necessarily operators, and operators are not necessarily operating the entire installation as they might use external service provider to fulfil certain tasks.

Hence, in simplified form, the main players in the storage market are:

- Owners of the (underground) storage facility: These companies may or may not have other activities in the energy sector. Increasingly, so-called 'financial investors' invest in storages worldwide. In some countries, storages are owned, in whole or in part, by the state. The operation and maintenance of the facility may be transferred to contractors;
- The owner of aboveground facilities is not necessarily identical to the owner of the storage. These facilities are often constructed according to the requirements of the operators. Hence, the owner of the ground may rent this ground to the storage system operator. The storage system operator constructs, owns, and operates the aboveground facilities;
- The commercial operation of the storage, i.e., all services in connection with the sales of storage capacity to the customer, is performed by the storage system operator. These companies are also subject to regulatory supervision; and
- The storage capacities are rented by customers. Such customer could be market participants from all segments, *inter alia*, producers, traders, local distributors, or (primarily industrial) end users or carriers (though, to a limited extent, e.g., for balancing purposes). It is the customer of the storage system operator who is acting in the other markets for natural gas, i.e., the transportation and the commodity market. These customers acquire storage capacity (and services) to be able to make use of one or more of the storage functions.

What is commonly referred to as the 'storage market' is described by contractual relations between the storage system operator and its customers.

The storage market is regulated in many countries. Hence, regulators, not being market participants per se, are to be considered as 'background' players in the market.

Gas Storage Europe
In Europe, the interests of 30 storage system operators in 17 European countries are represented by 'Gas Storage Europe' (GSE). GSE is part of 'Gas Infrastructure Europe' (GIE), the European association of the transmission system, storage, and LNG terminal operators.[19]

8.6 Contracts

The contractual elements described subsequently, i.e., first and foremost the products and prices, are the ones agreed upon between the storage system operator and its customers, as this constitutes what is commonly referred to as the storage market. If any, the contracts between the storage system operator and the owners of underground or aboveground facilities focus on technical issues.

8.6.1 Products

A reservation of storage capacity is typically required by a buyer, i.e., a customer having a surplus of natural gas in one period or at one point in time and expecting a shortage in a later period or at a later point in time. This is (simplistically) depicted in Fig. 8.8. The customer may purchase natural gas at time 1 at P_{Supply} and endeavor to sell it at time 2 at P_{Sales}. The required storage capacity to complement the transaction will be purchased as long as it is expected that $P_{Sales} - P_{Supply} > P_{Storage}$.

Yet the 'product' is more complex. At least the three essential parameters, i.e., the working

[19]See Gas Infrastructure Europe (2016).

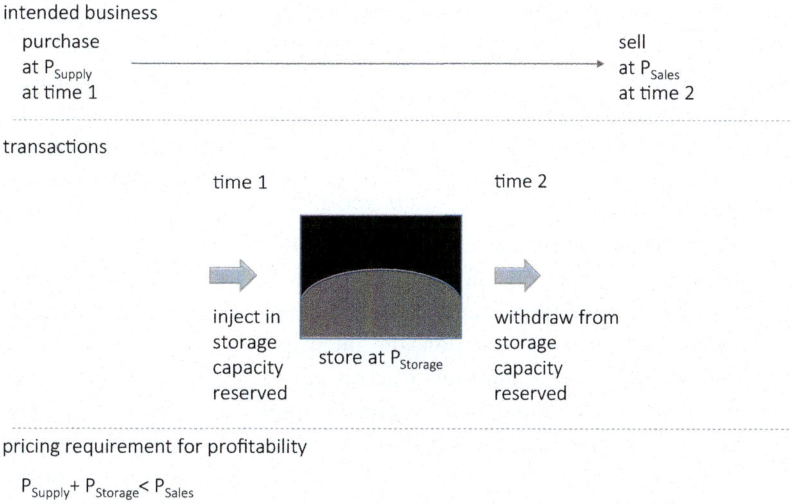

Fig. 8.8 Storage capacity reservation—business model, transactions, and pricing requirements

gas capacity, the injection capacity, and the withdrawal capacity have to be defined and reserved in order to enable the customer to utilize storage. In more detail, the product traded in a 'storage market', i.e., sold by a seller, in this case the storage system operator, and purchased, i.e., paid for by the customer, is:

- a reservation of injection, working gas and withdrawal capacity, i.e., a right but not an obligation
  - to make available natural gas to be injected by the storage system operator into a (defined or not defined) storage during a period at the delivery point[20];
  - to store the natural gas injected in the (defined or not defined) storage for a defined period; and
  - to receive natural gas withdrawn by the storage system operator from a (defined or not defined) storage during a period at the redelivery point (which is typically identical to the delivery point).

Typically, injection and withdrawal rates are a function of the working gas volume in place;

- in addition, other parameters such as minimum flow (for either the injection rate or the withdrawal rate) or the cycling rate may be defined, i.e., restricted;
- long-term contracts for storing natural gas in salt caverns might also foresee clauses for revision of working gas capacity due to salt creep (convergence);
- provision (and receipt) of supplementary services, as, e.g.[21]:
  - measurement (including metering the quantity and analyzing and metering the quality of the natural gas at the delivery and redelivery points);
  - receipt and confirmation of (re-) nominations;
  - data management, data collection and data transfer;
  - allocation and attribution of quantities (to various storage customers);
  - evaluation of measurements;
  - documentation; and
  - billing.

That is, just as for transportation, the customer is put in a position as if it has at its disposal a storage facility, respectively, the capacities. It transfers possession (but not ownership) of any natural gas delivered at the delivery point to the storage system operator. Custody of such natural

[20]Here the identical terms as in Sect. 7 are used.

[21]Compare with the services for transportation listed in Sect. 7.6.1.2.

gas will be re-transferred to the customer at the redelivery point at another point in time, defined by the customer.

The storage 'products'[22] offered by storage system operators are:

- firm or interruptible
  While a firm product has to be provided by the storage system operator at each point in time during the term of the respective storage contract,[23] i.e., the reservation, the interruptible product can be interrupted according to the contractual stipulations agreed upon between storage system operator and the customer[24]; and
- unbundled or bundled
  When purchasing unbundled capacities, the customer defines each of the three components, i.e., injection capacity, withdrawal capacity, or working gas capacity (within the technical limits of the storage facility and on offer by the storage system operator) independently. Each of the components is priced separately.

  Bundled capacities, on the other hand, consist of 'bundles', exactly specified by the storage system operator, of

  - maximum injection rates,
  - maximum withdrawal rates, and
  - maximum working gas capacity.[25]

  The pre-defined combination of injection, withdrawal, and working gas capacities is usually called a 'standard bundled unit' ('SBU'). As these SBUs do not necessarily meet customers' requirements, additional firm or interruptible unbundled capacities are on offer to fit these needs.

[22]Sometimes also referred to as 'services'.

[23]Essentially the same rules and restrictions apply as for firm transportation contracts; see Sect. 7.6.4.1.2.1.

[24]Essentially the same rules and restrictions apply as for interruptible transportation contracts; see Section 7.6.4.1.2.2.

[25]In addition, some storage system operators require for operational reasons minimum rates, e.g., during certain seasons or a certain minimum utilization of the working gas capacity.

Combinations of the above characteristics, as e.g., firm bundled products, are necessary to describe the product completely.

Derived from the above and primarily the requirements of unbundling, the principles of storage utilization (also transferred to contracts) are:

- Only natural gas injected can be withdrawn. Consequently, the (physical) working gas volume can never be negative.
  Exceptions are so called 'loaning contracts' or 'pseudo storage contracts': these contracts allow the customer to withdraw natural gas before (re-)injection;
- The physical restrictions of the individual storage have to be observed. This refers, e.g., to a declining injection or withdrawal rate at the end of the respective period;
- Compatibility of storage injection and withdrawal capacities with capacities of the transportation system used has to be ensured by the user of the storage capacity; and
- At the end of the agreed storage period, the storage has to be emptied. (This is, *inter alia*, also a consequence of the unbundling requirements, according to which the storage system operator is not allowed to participate in trading and, therefore, cannot purchase natural gas.)

8.6.2 Pricing

While natural gas commodity prices as well as transportation capacity[26] prices are one-dimensional, i.e., consisting of one price component only, storage prices are necessarily three-dimensional, consisting primarily of three capacity components (even if the working gas capacity, is priced in commodity-like units).

Hence, storage prices are typically defined using the components, i.e.:

[26]Neglecting fuel gas; see Sect. 7.6.1.2.

- Injection rate price, $P_{Injection}$, quoted in [monetary units/capacity/period], e.g., [€/(kWh/h)/a = €/kW/a] or [€/(m^3/h)/a];
- withdrawal rate price, $P_{Withdrawal}$, quoted in [monetary units/capacity/period], e.g., [€/(kWh/h)/a = €/kW/a] or [€/(m^3/h)/a]; and
- working gas capacity price, $P_{Working\ gas}$, quoted in [monetary units/quantity/period], e.g., [€/kWh/a] or [€/m^3/a].[27]

According to this (still simplified) pricing system, the total amount to be paid by the customer to the storage system operator, $A_{Storage}$, for the reservation of a storage product consisting of:

$\kappa_{Injection}$	injection rate,
$\kappa_{Withdrawal}$	withdrawal rate, and
$V_{Working\ gas}$	working gas capacity

is calculated by:

$$A_{Storage} = P_{Injection} * \kappa_{Injection} + P_{Withdrawal} * \kappa_{Withdrawal} + P_{Working\ gas} * V_{Working\ gas}$$

In order to incentivize storage capacity reservation corresponding to the physical parameters of each storage, several storage system operators offer storage bundles. These are products with the three components, and consequently the relationships between them are fixed for predefined absolute prices. Such bundles are priced in [monetary units/bundle/period], e.g., [€/bundle/a].

Typically, the amount to be paid for a defined storage product will be lower, if such storage product is purchased based on the bundled price rather than based on prices for single components, i.e., so-called unbundled prices. This is a consequence of commercial incentives given by the storage system operator to the market and driven by the fact that storage system operators intend to market their 'whole storage', i.e., the maximum injection, withdrawal, and working gas capacity, and not unbundled services only. (In the extreme, a customer could block the whole storage facility by booking the whole working gas capacity even without booking injection and withdrawal capacity.) In addition to commercial incentives given by the pricing structure described, storage system operators try to prohibit such customer comportment by rules included in the general terms and conditions for utilization of storages.

Standard Bundled Units

A standard bundled unit shall be defined by:

working gas capacity = 100,000 m^3/a
injection rate = 25 m^3/h/a
withdrawal rate = 25 m^3/h/a

The price of the bundle shall be 7500 €/a.

If, the prices for the unbundled services are defined by:

$P_{Injection}$ = 55 €/m^3/h/a,
$P_{Withdrawal}$ = 55 €/m^3/h/a, and
$P_{WorkingGas}$ = 0.05 €/m^3/a,

the same product would cost 7750 €/a, when replicated with unbundled services.

As storage prices are three-dimensional, the major problem is comparability. As long as not all prices of single components of one 'price bundle' are lower than the prices of single components of another 'price bundle', the simple statement 'capacity reservation in storage 1 is cheaper than in storage 2' will not be possible without additional reference to a specific storage product (and vice versa for higher prices).

This can be shown by the simple example summarized in Table 8.2. Despite the same product and different prices of all components, the total costs, i.e., total amount to be paid by the customer to the storage system operator, is identical in both cases. This is a result of $P^1_{Injection} < P^2_{Injection}$ and $P^1_{Withdrawal} < P^2_{Withdrawal}$ but $P^1_{Working\ gas} > P^2_{Working\ gas}$.[28]

[27]Note that the 'working gas capacity' is, strictly speaking, not a 'capacity' but an (empty) 'volume'. While a capacity is measured as [quantity/time], the working gas capacity is a quantity; see Sect. 8.6.1.

[28]The prices of storage 1 are denoted by the superscript 1 and the price of storage 2 by the superscript 2.

Table 8.2 Identical storage costs with different prices of storage components

	Product 1		Price 1		Product 2		Price 2	
Working gas capacity	1,000,000	m^3	0.05	€/m^3/a	1,000,000	m^3	0.0475	€/m^3/a
Withdrawal rate	250	m^3/h	55	€/m^3/h/a	250	m^3/h	60	€/m^3/h/a
Injection rate	250	m^3/h	55	€/m^3/h/a	250	m^3/h	60	€/m^3/h/a
Cycling time	8000	h			8000	h		
Total costs			77,500	€/a			77,500	€/a

Table 8.3 Example of two different storage products with different component prices reduced to one-dimensional prices

	Product 1		Price 1		Product 2		Price 2	
Working gas capacity	1,000,000	m^3	0.05	€/m^3/a	1,000,000	m^3	0.025	€/m^3/a
Withdrawal rate	250	m^3/h	55.00	€/m^3/h/a	500	m^3/h	52.50	€/m^3/h/a
Injection rate	250	m^3/h	55.00	€/m^3/h/a	500	m^3/h	52.50	€/m^3/h/a
Cycling time	8000	h			4000	h		
Gross calorific value	11.0	kWh/m^3			11.0	kWh/m^3		
Total costs			77,500	€/a			77,500	€/a
Price related to working gas capacity			7.05	€/MWh/a			7.05	€/MWh/a
Price relating to withdrawal rate		or	310 28.18	€/m^3/h/a €/kW/a		or	155 14.09	€/m^3/h/a €/kW/a

Often three-dimensional storage prices are 'reduced' to one dimension. This may cause misinterpretation and ambiguous numbers, as shown in Table 8.3. Different products (as, e.g., product 1 and product 2) having different prices for the single components (injection rate, withdrawal rate, and working gas capacity) will have identical prices when reduced to a (one-dimensional) working gas capacity related price but differ by 100% when reduced to a (again one-dimensional) withdrawal rate related price. In this example, product 2 is the faster storage, with cheaper prices for each single component. Nonetheless, the working gas capacity related price is identical to the one for product 1. Even though several market participants tend to use working gas capacity related storage prices (with the argument of alleged simplicity), one-dimensional prices are not suited for comparison of storage products. 'Shrinking' three dimensions to one will always cause loss of information and ambiguity.

Under long-term storage contracts, capacity charges are in most cases escalated to a certain percentage with (in most cases) publicly available indices to reflect market and/or cost developments during the term of the contract.

Hence, a typical, yet slightly simplified (so-called multiplicative), price formula for storage services is of the form:

$$P_{Storage}(t, \bar{\Psi}_i(t)) = P_{Storage0} * \left(\sum_{i=1}^{n} (d_i * \bar{\Psi}_i(t)/\Psi_{i0}) \right)$$

with:

$P_{Storage0}$ base value at time t_0 = starting point of indexation

Ψ_{i0}	base value of the variable/index Ψ_i at time t_0 = starting point of indexation
$\bar{\Psi}_i(t)$	average[29] value of the variable/index Ψ_i at time t
d_i	weighting factor of variable/index Ψ_i,
n	number of variables/indices
t	time of price (re-) calculation (typically: month).

This formula is applied either to:

- a defined bundle; or
- withdrawal rate; or
- injection rate; or
- working gas capacity.

In the European Union, the Member States can, according to Article 19 of the European Union Gas Directive, implement negotiated or regulated access to storage capacity. The Guidelines for Good TPA Practice for Storage System Operators (GGPSSO)[30] have been developed by the Madrid Forum and published by the European Regulators Group for Electricity and Gas (ERGEG). They, while not legally binding, define a minimum set of rules for third-party access to the storage and price structure. Despite such guidelines, published information and, in particular, prices for storage capacity reservation vary considerably throughout the various natural gas markets in the European Union. Nonetheless, non-discriminatory pricing of storage capacity and third party access is implemented in all member states of the European Union.

Similar to transportation,[31] the price level for storage capacities can be determined using various approaches. In the case that not only the access to the storage but also price setting are regulated, a regulatory authority will determine the prices. Alternatively, the storage system operator might set the price, negotiate the price, or perform auctions.

The prices for reservation of a bundle or individually defined injection rate(s), withdrawal rate(s), and working gas capacities, may be adjusted or supplemented by one or more of the following elements:

- service fees, i.e., fees that are charged for additional services provided in connection with the storage contract such as accounting, billing, measurement, etc.;
- penalties, i.e., charges for exceeding the capacities reserved, for attempting to inject quality-deficient natural gas, or any other non-compliance with contractual stipulations;
- multi-annual, seasonal, quarterly, or monthly discounts or adjustment factors, i.e., typically, storage system operators quote the prices for reservation of storage capacity for a year. Any other period might have a different value in the market and, therefore, will be priced differently. Storage system operators frequently grant discounts for multi-annual capacity reservations. That is, the price paid for one year will be reduced by x% if a multi-annual contract is concluded. On the other hand, the price paid for a monthly storage capacity reservation might be not equal to 1/12 of the annual price but 1/X, with $X < 12$. The analogous applies for seasonal or quarterly capacity reservations;
- cycle charges, i.e., some storage system operators limit the number of cycles permitted during the term of a contract. Others include the number of cycles in the pricing system. That is, if the customer intends to fill and empty the storage more frequently than foreseen by the storage system operator, a surcharge or percentage increase of the price will be applicable;
- interruptibility discounts, i.e., in strict analogy to transportation capacity prices, storage prices can be discounted if the storage system operator is allowed to interrupt the provision of services under the contract, i.e., to reject or reduce nominations of the customer for either injection or withdrawal; and

[29] Many storage price formulae are simpler as not average values but single distinct values are used. In case an average value is required to be calculated, see Sect. 6.3.2.2.3 for further details.

[30] See European Regulators' Group for Electricity and Gas (ERGEG) (2011).

[31] See Sect. 7.5.2.1.

- variable costs, i.e., costs related to the operation and utilization of the storage. These are either paid in kind (i.e., as a certain percentage of quantities injected or withdrawn —'fuel gas' for the compressors installed at storage facilities) or (more commonly) expressed as an indexed-price formula.[32]

The diversity and number of these elements further complicates or even hinders comparability of prices.[33]

8.6.3 Other Contractual Elements

Several clauses in storage contracts correspond widely or are analogous to the respective clauses in commodity or transportation contracts.

- Delivery and redelivery point
  Since (typically) the storage customer is responsible for organizing transport services to and from the storage facility, the delivery point for quantities meant to be injected, as well as the redelivery point for quantities meant to be withdrawn from the storage, have to be exactly defined. This applies to delivery and redelivery at a virtual point, too.
  In the case that delivery and redelivery points do not coincide, the contract is also referred to as 'wheeling' contract;
- Natural gas quality and pressure
  At the delivery and redelivery point(s), the gas has to fulfil certain quality requirements that are usually in line with the requirements of the transport system operator upstream (or downstream) of the storage facility. Minimum and maximum pressure requirements have also to be fulfilled in order to allow for smooth technical operation of the systems involved;
- Nominations
  Natural gas quantities to be injected or withdrawn from a storage have to be nominated by the customer. These nominations have to fulfil certain formal requirements, e.g., regarding IT-format. Further, these nominations (and re-nominations) have to be made up and until exactly specified points in time. Also, when nominating quantities, allowance must be made for lead times for switching the operating mode of the storage facility (e.g., from injection to withdrawal);
- Minimum flow requirements
  In order to be able to guarantee the operability of the storage at each point in time, storage system operators might require their customers to maintain a minimum flow, for either the injection or the withdrawal, during defined periods. Either such a minimum flow requirement is an integral part of the contract and not reflected in the price, or the customer is explicitly incentivized to maintain a minimum flow by offering a discount on any of the price components;
- Maintenance
  In order to inspect or repair the storage facility, the storage system operator might shut down the facility partially or fully. Such maintenance is usually restricted regarding duration and timing.[34] Furthermore, such maintenance has to be announced in time by the storage system operator;
- Taxes and other public charges
  Storage fees are net amounts, i.e., value added tax or taxes due in connection with the operation of storages have to be added;
- Billing and payment
  Rules regarding invoices, payment dates, bank accounts, disputed invoices, and interest due for late payments have to be agreed;
- Credit issues
  In order to reduce credit risks, storage system operators undertake an assessment of the storage customer's solvency. For this, the storage customer has to disclose certain financial information. Furthermore, verifica-

[32]See Sect. 6.3.2.2.3; often indexed to electricity prices.

[33]See, e.g., Energy Charter Secretariat (2007), p. 35 et seq.

[34]The length of a maintenance period may be up to 30 d/a.

tions by rating agencies might be asked for by the storage system operator. In case the solvency assessment does not fulfil the storage system operator's requirements to a full extent, the storage system operator might ask for securities like bank guarantees, etc.;

- Confidentiality
  The parties shall treat the contents of the contract, as well as related information, as confidential;
- Force Majeure
  In cases of force majeure, the parties may be relieved from their obligations under the contract. 'Force Majeure' events are events beyond the control of the affected party;
- Liability
  The parties are mutually liable to each other for damages and violations of contractual obligations. Yet, usually, liabilities are limited;
- Term
  The start and end dates of the contract have to be stipulated in the contract. Typically, storage contracts are concluded for a term of one year. Longer and shorter tenures are also possible. Especially short-term storage contracts are also referred to as 'parking'.
  In Europe the start date of (one year) storage contracts is, typically, defined after the end of the winter, i.e., at the beginning of April or May;
- Applicable law and settlement of disputes
  The contract is governed by the laws of a specified country. Furthermore, the parties must agree upon whether disputes should be settled in front of an ordinary court or in in front of an arbitration tribunal and where such court or arbitration tribunal shall be located; and
- Other provisions
  Other clauses include rules regarding languages to be allowed under the contract, communication, severability clauses, etc.

Other clauses (or the lack thereof) are more specific to storage contracts.

- Penalties
  In case the storage customer overruns its booked capacities, penalties shall be paid. If e.g., the working gas volume is not at zero at the end of the contract period, a penalty is due to be paid by the customer to the storage system operator. To this end, the storage system operator might either ask for an additional payment or 'sell' the quantities left in storage on the wholesale market, reimbursing the customer usually only a fraction of the agreed price.
- Loyalty (or Hardship)
  An adjustment of contractual provisions (e.g., price) is only possible if unforeseen circumstances occur, having substantial impact on the storage contract that had not been considered when concluding the contract and if, consequently, any of the contractual provisions become unreasonable or unacceptable to a party.
  On the basis of such a loyalty or hardship clause, the revision of terms and conditions of the storage contract may be initiated. The legal obstacles to overcome, e.g., for a revision of the price under the loyalty clause, are higher than under a price review clause in a delivery contract and will have to be investigated in each case individually;
- Secondary market, transfer of rights and obligations
  Usually, the storage customer is entitled to market its booked capacities to third parties. Such secondary marketing does not relieve the customer from its obligations to the storage system operator. A full transfer of rights and obligations to a third party (assignment) is usually possible only with the prior written consent of the storage system operator; and
- Deprivation of storage capacities
  To prevent abusive hoarding of storage capacities, the storage system operator is entitled to take certain action in case the storage customer does not utilize the storage capacities during a certain period. Such action might, in extreme cases, even lead to the deprivation of capacities.

Remarkably, storage contracts, *inter alia*, do not contain the following stipulations

- Price review
  In contrast to many long-term GSAs or LNG SPAs, (long-term) storage contracts typically do not foresee price review clauses allowing the parties to renegotiate a storage price agreed upon or the price's indexation. However, there are exceptions, which were agreed in countries/regions where a (semi-) competitive storage market could be identified; and
- Clauses analogous to make-up or carry forward
  As the product sold under a storage contracts is capacity, there cannot be any rights granted to the customer for anticipation or deferral of volume injection or withdrawal.

8.6.4 Standard Contracts

Natural gas storages are typically not regarded as a natural monopoly. Nevertheless, some governments tend to regulate also the storage business, i.e., at a minimum they require the storage system operator to offer storage capacity on a non-discriminatory basis to other market participants. This leads necessarily to the development of standardized storage contracts. Such standard contracts are usually in the public domain and published by the storage system operators on their respective websites.

(Virtual) storage contracts are not standardized by statue, but recommendations for standardization exist in Europe.[35]

[35]As already described in Sect. 6.3.9, the European Federation of Energy Traders (EFET) introduced a so-called General Agreement Gas, which allows for appendices. One of these appendices reflects virtual gas storage transactions. See European Federation of Energy Traders (EFET) (2016b).

8.6.5 Categorization

Storage services and contracts can be categorized based on:

- the 'basis', i.e., if the facility in which the natural gas is stored is defined;
- the 'bundling', i.e., if bundled services with defined ratios of the three parameters injection rate, withdrawal rate, and working gas volume are offered;
- the 'commitment', i.e., if firm or interruptible services are offered;
- the 'services', i.e., if only storage services are offered or additional transportation services, too.

Of course, all these types of contracts can be combined, e.g., a physical, bundled, firm storage contract containing transportation (wheeling) services.

8.6.5.1 Basis: Physical and Virtual Storage Contracts

By virtual storage services, offered either by storage system operators or traders, the customer acquires the right (not the obligation) to 'inject', i.e., to deliver natural gas at a delivery point and to 'withdraw', i.e., to receive quantities not exceeding the 'injected' ones at a redelivery point identical to, or even different from, the delivery point. The sum of the quantities 'injected' and subsequently 'withdrawn' may not exceed the 'working gas capacity' reserved. Any company able and willing to receive and redeliver natural gas at its customers' request can offer such virtual storage contracts.

If such services are not physically backed by one clearly defined storage, it is irrelevant whether these services are backed by a portfolio of storages, i.e., several storages, or by a trader without assets, who is selling and purchasing quantities received and redelivered. As long as the contractual features correspond to the features of a storage contract, such a virtual storage contract will differ from a physical storage contract mainly, or only, by not defining the location where the natural gas is stored. This might have implications on the parties' risk exposure. For example, while a technical failure of the facilities of a storage might constitute force majeure under a physical storage contract, it is typically not the case under a virtual storage contract.

Consequently, a contract, according to which a storage system operator grants the (physical) storage of the natural gas delivered by its customer but does not specify the location where this natural gas is stored, must be categorized as virtual or as 'hybrid' storage contract. Again, implications to the risk exposure of the parties have to be defined.

Moreover, an even more restricted type of 'virtual' storage contract is a 'time swap'.[36]

8.6.5.2 Bundling: Bundled or Unbundled Services

In accordance with the pricing system and the product definition, contracts for bundled and unbundled services are offered.

For the storage system operator, offering bundled services, or bundled capacities, is advantageous insofar as this might enable him to sell the capacity following the technical parameters of the storage. Otherwise, the storage system operator is exposed to the risk of selling the maximum of one parameter and being left with idle capacity of another. For example, if sold as an unbundled service, the customers may require 100% of the available withdrawal rate by reserving only 50% of the available working gas capacity. This covers customers' demand but does not help the storage system operator maximize profits. Therefore, each storage pricing system will incentivize customers to purchase bundled products.

8.6.5.3 Commitment: Firm or Interruptible Contracts

Storage contracts are offered, similar to transportation and commodity sales contracts, as firm or interruptible contracts.[37] Here, the same requirements for definition of the parameters of interruptibility apply.

Yet, interruptible storage capacity will be booked primarily by customers not interested in the function of provision of security of supply.

8.6.5.4 Services: Storage or Combined Storage and Transportation Services

('Pure') storage contracts, with identical delivery and redelivery points, are used by customers to 'transfer' the natural gas in time (only), i.e., to inject and withdraw it at the same location at different points in time.

If, however, injection and withdrawal may be effected at different location and (of course) at different points in time, then such contracts are often also referred to as 'wheeling' contracts. A storage system operator must, in order to be able to offer such services, rent (or own) transportation capacity in addition to storage capacity. Only then can it transfer the natural gas between locations and time.

[36] See Sect. 5.5.4.2.3.1.

[37] See Sects. 6.3.10.4 and 7.6.4.1.2.

References

Energy Charter Secretariat (2007) Putting a price on energy—international pricing mechanisms for oil and gas, Brussels. Available at http://www.energycharter.org/fileadmin/DocumentsMedia/Thematic/Oil_and_Gas_Pricing_2007_en.pdf

Energy Information Administration (2004) The basics of underground natural gas storage. Available at http://www.eia.gov/pub/oil_gas/natural_gas/analysis_publications/storagebasics/storagebasics.html

Energy Information Administration (2011) Salt caverns account for 23% of U.S. underground natural gas storage daily deliverability

European Federation of Energy Traders (EFET) (2016b): Virtual Gas Storage Service Transaction - Individual Contract. Available at: http://www.efet.org/Standardisation/Legal-EFET-Standard-Contracts-and-Documentation/GasAndGasAnnexes

European Regulators' Group for Electricity and Gas (ERGEG) (2011) Guidelines for Good TPA Practice for Storage System Operators (GGPSSO) of 2 March 2005 as updated in July 2011, pp 1–20

Gas Infrastructure Europe (2016) GSE members. Available at http://www.gie.eu/index.php/about-us/gie-members/gse-members

Groupe International des Importateurs de Gaz Naturel Liquéfié (GIIGNL), The LNG Process Chain, (LNG Information Paper 2)

Johansson, J. (2003): High Pressure Storage of Gas in Lined Rock Caverns. Cavern Wall Design Principles. Royal Institute of Technology, Stockholm

Keyaertsa N, Hallack M, Glachant J-M, D'haeseleer W (2010) Gas balancing rules must take into account the trade-off between offering pipeline transport and pipeline flexibility in liberalized gas markets. Available at http://web.mit.edu/ceepr/www/publications/workingpapers/2010-014.pdf

Keyaertsa N, Hallack M, Glachant J-M, D'haeseleer W (2011) Gas market distorting effects of imbalanced gas balancing rules: inefficient regulation of pipeline flexibility. Energ Policy 39:865–876

Merriam-Webster (2016) Storage. Available at http://www.merriam-webster.com/dictionary/storage

Naturalgas.org (2013a) Storage of natural gas. Available at http://www.naturalgas.org/naturalgas/storage.asp

Oelizowski J, Walaszcyk J (2008) Long term stability evaluation of natural gas storage caverns. Available at https://meeri.eu/Wydawnictwa/GSM2441/slizowski-walaszczyk.pdf

Peebles MWH (1992) Natural Gas Fundamentals Shell International Gas Limited, London

Plaat H (2009) Underground gas storage: why and how, London

Portfolio Management 9

Summary

- Neither the subjects, nor the objectives nor the methods of portfolio management are universally shared within the natural gas industry.
- The application of methods to maximize future profits, while observing the risks and all given restrictions, is referred to as portfolio management.
- Portfolio management requires sophisticated methods for planning the deployment of all available assets.
- For traders in the natural gas industry, the assets are purchase contracts, sales contracts, transportation contracts, and storage contracts.
- Traditionally, linear, and later non-linear, optimization models were applied to calculate deterministically the maximum profit for future periods. Such calculations have to be based on forecasts of prices, quantities, and their influencing parameters, such as, e.g., temperatures.
- If hub-based trading dominates the markets, parameters, which were restrictions in the traditional market, turn into variables. The traditional optimization methods are replaced by tools applied in other commodity markets as well.
- *Inter alia*, to cope with the increased price and demand volatility and uncertainty, risk management constitutes an integral part of modern portfolio management.

As for many other terms, there is no generally accepted definition of portfolio management in the natural gas business. Even more, neither the subjects, nor the objectives nor the methods of portfolio management are universally shared within the natural gas industry.

Here, the outline of the subject (and object) shall be started with some general observations: A portfolio is commonly described as a collection of (financial) assets owned, or at the disposal of, an entity. Each asset is associated with risks, and each asset is expected to deliver a return. The asset holding entity, i.e., an entrepreneur, aims at making profit. Yet, no entrepreneurial activity is risk free. Enterprises, despite being prepared to take risks, try to restrict (not necessarily minimize) risk and maximize profits, while respecting legal, regulatory, technical, and commercial restrictions. This is generally seen as the objective of portfolio management.

The 'tools' to achieve these targets are supplied by portfolio theory. The basic idea is that a portfolio's total risk/return ratio does not only depend on the single assets' risk/return ratios but also on the correlation of these assets'

Subsequently only a very brief description of portfolio management as applied in mid- and downstream parts of the natural gas industry is provided. Any discussion of details is not intended, as it would exceed the scope of this textbook.

A. Pustišek and M. Karasz, *Natural Gas: A Commercial Perspective*,
DOI 10.1007/978-3-319-53249-3_9

parameters. The theory has been developed already in the 1950s[1] and applied to 'financial' assets, as, e.g., stocks and bonds, first.

Today, *inter alia*, as a consequence of the above-mentioned non-existence of a generally accepted view on portfolio management, a vast variety of methods is used for the management of portfolios. In addition to 'classic' optimization, these methods are, e.g., value at risk, constrained portfolio optimization, or even various forecasting techniques.[2]

In the natural gas industry, portfolio management is performed predominantly by mid- and downstream companies.[3] These companies purchase and sell natural gas and related services. They endeavor to maximize profit while trying to keep the risk below a defined level. To this end, the portfolio of assets is analyzed and subsequently optimized.

The assets of a natural gas (trading) company consist mainly of one or more of the following:

- purchase contracts,
- sales contracts,
- transportation capacity (contracts), and
- storage capacity (contracts).

The natural gas company's portfolio is the entirety of these assets. These assets have optionality and restrictions. Refining the definition above and focusing on the natural gas industry, portfolio management is understood as the utilization of this set of available assets with the objective of maximization of profits in future periods, while observing all legal, regulatory, technical, and commercial restrictions, among them a defined level of risk.[4]

Simplifying, maximizing future profits while observing the risks and all given restrictions requires sophisticated methods for planning the deployment of all available assets—the application of such methods is referred to as portfolio management.

9.1 Historical Development

In the traditional natural gas industry, portfolio management was not developed based on the theoretical framework mentioned above but with the objective of finding an optimum under constraints, i.e., solving an initially deterministic, linear optimization problem. Risk was not included in the restrictions. Largely, risk management was not performed. Such an approach was consistent and logical, as the assets' risks were not considered essential. The optimum, in which a traditional natural gas supply company was interested, was initially the minimum of supply costs and later, after sales became a variable, the maximization of profits, i.e., the difference between sales revenues and supply costs. The optimization, i.e., the maximization of profit and the 'actions' of natural gas purchasing, sales transportation, and storage utilization, was calculated for consecutive points in the future, while the 'traditional' portfolio theory sketched above referred to a single point in time only.

The steps of the historical development can be illustrated as follows:

- In the early days of the natural gas industry, the companies were vertically integrated, and networks did not branch out widely. When a midstream or downstream company commenced business, natural gas was delivered to a very limited number of customers.[5] For the delivery period, the midstream or downstream company purchased the natural gas to be delivered from one single supplier (in the simplest case, under one single contract). To this end, it had to transport and to store the

[1]See Markowitz (1952).

[2]See for further reference, also Mack (2014), Sect. 8.5.

[3]Upstream companies also use portfolio management techniques. However, these companies perform such activities in front of their overall (not necessarily related to natural gas) upstream-, midstream-, and downstream portfolios. As upstream companies are not in the focus of this textbook, the following sections refer to midstream and downstream companies only.

[4]Within this book, it is neglected that there is no scientific consensus for the quantification of risk.

[5]Only the case with one customer is selected here for illustration.

natural gas. Such a company did not require a sophisticated portfolio management or portfolio optimization, as it simply tried to deliver natural gas purchased at a positive margin to the customer, covering its total costs. The only optimization potential it had was to decide on the timely allocation of purchase quantities and the respective use of storage capacities.[6] Transportation and storage capacities had been constructed in accordance with the (anticipated) requirements (see Fig. 9.2a);

- This situation did not change when more customers were supplied. The demand had to be covered, supply was not a variable, and transportation and storage capacities owned by the vertically integrated companies were mere (technical) restrictions. The companies did not have any further options than described above and, therefore, could not optimize or manage a portfolio of assets. However, as pipeline systems evolved into networks, already at this stage technical optimization algorithms were implemented[7];
- Only after extending the purchases to multiple, flexible, differently priced supply contracts (see Fig. 9.2b), an optimization problem arose. Supply costs were variable. Yet, in the traditional market, sales were (still) considered a 'datum', i.e., a given. Consequently, the minimization of supply costs corresponded to the maximization of profits.[8]

  Maximization of profits was intended to be achieved during future periods. For example, the calculation of the optimization performed in December of a year aimed at fulfilling the objective function in each month of the following year. There were several, differently priced supply contracts offering flexibility, i.e., the quantities in each future period could be varied within limits. Therefore, when concluding supply (purchase) contracts, diversification of pricing (structures) was important. Optimization could be effected only by receiving physically or contractually diversified supplies or even supplies from one supplier, without physical diversification, containing different prices for different 'slices', i.e., parts of the total quantity.

  Transportation and storage contracts, as well as sales contracts, were (still) considered as restrictions of the optimization problem (i.e., the optimization model). Calculating future optima required that (sales and purchase) prices and sales quantities in all relevant future periods were assumed to be known. Necessarily, this applied to temperatures, economic development, and other factors influencing consumption, as well. These parameters were forecasted, and such a forecast was accepted as input of the optimization model.

  The result of the optimization revealed, for each relevant period in the future: the quantities to be off-taken from each of the different supply contracts, observing all restrictions; the quantities to be injected into or withdrawn from storages; and the quantities to be transported in each subsection of the network. Since only deterministic optimization[9] was applied, the practical significance was limited. Nonetheless, despite deficiencies, structuring of the problem already supported its solution. Optimization was not any more a technical problem (of flow optimization), but became a commercial problem considering technical restrictions.[10]
- When hubs emerged and, independently thereof, several interruptible sales contracts were introduced, sales quantities changed from being a 'datum' to a variable. Quantities could not only be sold to end users and

[6] The same applies to interruptible sales contracts.

[7] For a brief explanation of design optimization and operational optimization, see Sect. 7.1.

[8] As profit = (revenues − costs), minimization of costs equals maximization of profits under the assumption that revenues are given, i.e., not a variable.

[9] For example, purchase and sales prices in future periods and sales quantities were assumed to be given.

[10] This, *inter alia*, had organizational implications. Departments for technical optimization were merged with commercial optimization and planning.

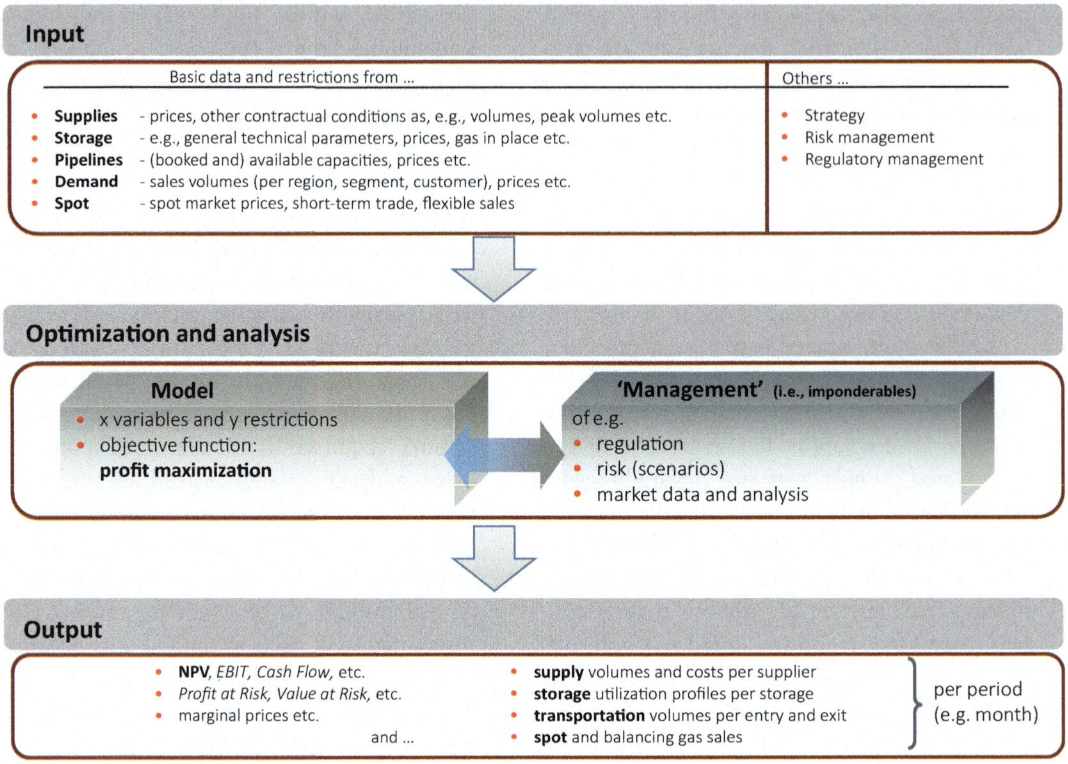

Fig. 9.1 Conceptual design of an integrated (deterministic) portfolio-optimization model

intermediaries as, e.g., utilities, but also sold and purchased at hubs. Customers could be interrupted, i.e., such contracts offered 'negative' flexibilities to the customers and 'positive' flexibilities to the supplier. *Ceteris paribus*, the objective function of the optimization problem had to change from cost minimization to profit maximization[11] (see Fig. 9.2c). The conceptual design of such an optimization model is depicted in Fig. 9.1.

With increasing complexity, the optimization models were not linear any more. Yet, it was still a deterministic optimization for future periods, i.e., forecasts of future prices and quantities were and had to be accepted, neglecting their inherent uncertainty. Consequently, the results calculated were accurate to the extent the forecasts were accurate. Some companies tried to mitigate the problem by calculating optima for future periods under different price- and quantity-scenarios. Hence, not one result of optimization but a set of results was calculated. Nevertheless, these scenarios were deterministic, and – implicitly – equal probabilities were assigned to each scenario. The selection of the relevant scenario and, thus, the acceptance of a result of optimization, including measures to be taken in future periods, were declared managerial tasks;

- When transportation and storage were unbundled and, as one of the consequences, transportation and storage capacities became tradable, they changed from a restriction to a variable, again increasing the complexity of the optimization models.[12]

[11]Once both sales and supply can be regarded as variables, profit maximization not cost minimization is the appropriate objective function of optimization.

[12]As a guideline, it can be assumed that at this stage even the deterministic optimization models of larger utilities had some 10^5 variables to be set up and 10^5 restrictions considered.

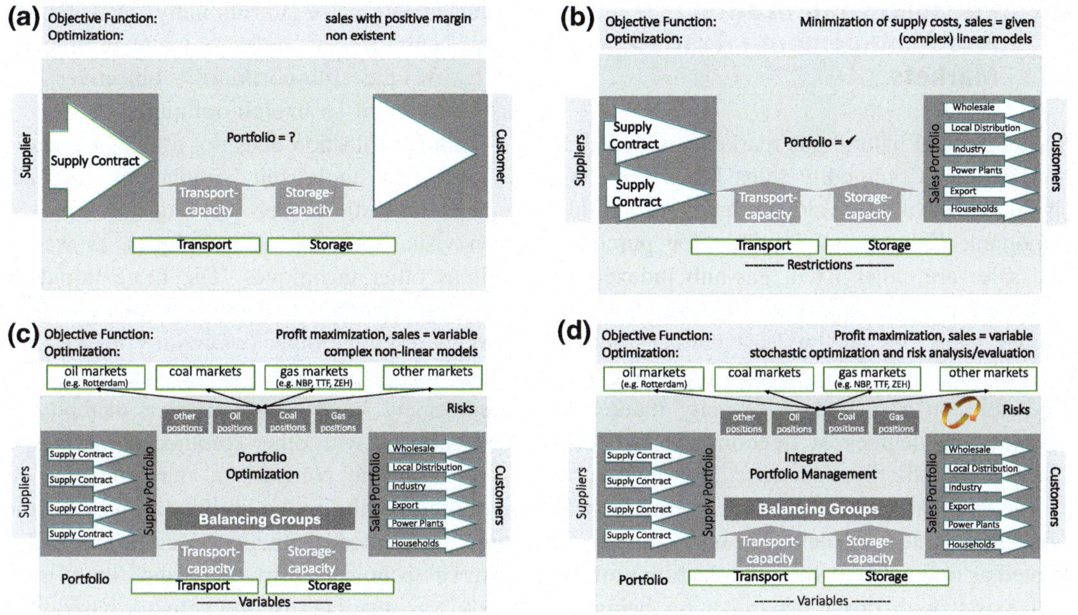

Fig. 9.2 Evolution of portfolio management

From this point onwards, optimization, along the lines described above, was primarily performed by (commodity) trading companies; and

- As traded markets became more liquid, *inter alia*, long-term supply contracts started to become less important, and commodity prices became more volatile. Mitigation of risks became an issue as, e.g., different oil product indexed prices and hub-based prices were used for purchasing and sales. Thus, the differences between purchase and sales prices, i.e., sales margins, became volatile. Simultaneously, sales became less predictable. Stochastic optimization, developed successfully in other branches of science and industry, was introduced into the natural gas industry (see Fig. 9.2d). Here, neither prices nor quantities were deterministically forecasted, but probability distributions for relevant restrictions of the optimization problem, different in different periods in the future, were assumed.[13]
  Hence, the result of optimization was not the optimal supply quantity, storage injection, or withdrawal, etc. during each future period, but a probability distribution of these parameters. In its simplest form, the standard deviation of such a probability distribution could be regarded as a measure of risk. Optimization and risk management were merged, i.e., the traditional natural gas industry's optimization approach was combined with risk management (at least the stage of risk identification) and became portfolio management.

[13]Of course, also the assumption of a certain probability distribution of a variable includes a quasi-deterministic forecast.

9.2 Portfolio Optimization and Management—Traded Markets

Today, several natural gas markets are based on hub trading. Commodity sales based on LTCs (including flexibility) already have become less important. Consequently, prices for purchases and sales are either fixed, gas-hub indexed, or (still) oil product indexed. Transportation and storage capacities are (at least partly) tradable. Even if reserved for longer periods, they can be (re-) sold on a secondary capacity market. In essence, the degrees of freedom increase. The number of restrictions decrease, and the number of variables increase. Effectively, former restrictions turned into variables, and such variables are treated as options. Yet, the optimization problem still exists, with profit maximization being the objective. However, the number of variables has increased in such a way that for companies the application of a holistic optimization does not appear to represent an 'optimal' solution to the optimization problem. Rather, portfolio management is established. Its focus is diverted from optimization (in its traditional form) to risk management and trading, seen as the enactment of short-term measures to maximize profit and reduce risk or to keep risk within prescribed limits. The company's total portfolio of assets is (re-)separated into sub-portfolios. Each sub-portfolio, also referred to as a 'book', is 'optimized', observing the individual sub-portfolio's legal, regulatory, commercial, and risk restrictions. The risk is prescribed and measured in term of limits, which are defined for each sub-portfolio and the company's total portfolio in terms of, e.g., maximum quantities to be traded with a single counterparty, maximum profit at risk, maximum value at risk, or any other measure. So-called 'open positions', i.e., trades that are not 'closed' with opposing trades, are essentially the objects or variables of trading (and optimization). Hedging such open positions reduces risk and chances but may safeguard a defined profit.

Despite being (re-)separated into sub-portfolios, the company's integral portfolio management is (organizationally and procedurally) centralized in order to maintain compatibility of the sub-portfolios' objectives and achievement of an overall optimum.

Such optimization does not utilize necessarily sophisticated tools but simulations, scenario models, heuristic models, scoring techniques, or even visual or mapping techniques, as well as well as other techniques. The more liquid the natural gas markets become, and the more these markets resemble other commodity markets, the closer portfolio management, including risk management[14] and optimization, is methodologically aligned with other commodity markets.

A simplistic comparison may illustrate the reasoning: While in the traditional market structure, the vertically integrated company had to observe technical restrictions and, thus, effectively was restricted by the network-based character of the industry, the trader in the current market does not have the necessity to observe the technical restrictions of the network. Transportation capacity is reserved based on an entry-exit capacity reservation system, natural gas is traded at (often virtual) hubs, and storage injections and withdrawals are sometimes delivered at hubs as well. In the abstract, trading natural gas does not need the 'view' on the network. Therefore, traders increasingly adopt the methodological approach from other commodity markets, rather than optimization models developed in the traditional network-based environment.

Such an approach is based on the following assumptions:

- markets are liquid,
- market prices of all assets are transparent and instantaneously available,
- all assets are tradable, and
- there is a defined market price for the commodity of natural gas.

There are, however, the following implications:

[14]Whereby risk management is to be clearly distinguished and organizationally separated from risk controlling.

- the general rule according to which the sum of optima of sub-portfolios can never exceed the optimum of the total portfolio is neglected, as the 'total' optimization problem is divided into parts;
- complexity is reduced, as optimization of a sub-portfolio, at each point in time and for each future period, can be achieved easier and without having to consider a vast variety of variables and restrictions;
- risk can be steered; by putting risk limits on each sub-portfolio and prescribing the organizational entities involved not to exceed such limits, the company's total money at risk can be set and distributed among the optimizing entities;
- the management and handling of sub-portfolios, as well as the company's total portfolio, has to be reflected organizationally; this implies further, that so-called front-, middle-, and back-office entities and their operational, procedural, and organizational interdependence, as well as the total portfolio management's interdependence to (former or still existing) purchase and sales entities, are clearly defined;
- companies will earn profits either by monetizing spreads speculatively on the time or locational value of natural gas or by selling additional services, such as, e.g., flexibility;
- risk management, i.e., the identification, analyzing, evaluating, 'treating', and monitoring of real or perceived risks, might become an integral part of portfolio management;
- trading is regarded as the portfolio management's operational entity;
- one of the results of portfolio management is the setting of internal transfer pricing mechanisms and rules. Acting (organizationally) between the traditional entities of (long-term) purchasing and sales, the company's alternative is at each point in time to purchase and sell on the traded market. Effectively, the trading entity operates as internal broker within the company;
- asset optimization is referring to the traded maximization of all asset's utilization, including re-sale of capacities; and diversification, an essential requirement of portfolio optimization in the traditional market with traditional contractual and pricing structures, becomes an issue of (national or regional) security of supply, but not portfolio optimization. Yet, diversifying counterparties is an objective because of limiting the company's credit exposures.

In addition to such short—and mid-term objectives, portfolio optimization is also used to support investment decisions,[15] support strategic planning, or, economically, may be used to support political decisions. Yet, such broad definitions and visions of portfolio management are, again, neither accepted nor applied universally.

References

Mack I (2014) Energy trading and risk management. Available at: http://www.scopus.com/inward/record.url?eid=2-s2.0-84927028907&partnerID=tZOtx3y1

Markowitz HM (1952) Portfolio selection. J Financ 7 (1):77–91

[15]See, e.g., Mack (2014), p. 191 et seq.

10 Epilogue

While, in general, conventional sources of energy are becoming less popular, natural gas—at least worldwide—is expected to continue being successful in terms of increasing quantities of production and consumption. Where implemented, even the 'German Energiewende' does not intend to abolish natural gas usages.

On the demand side, 'conventional' natural gas usages are forecasted to increase; new technologies satisfying new demand are expected to become technically and economically viable; and additional services, including flexibility, will be required. To cover such (increased) demand, additional sources of supply in remote regions or in different geological structures are anticipated. Yet, such projections have been discussed over recent years, if not decades. Revolutionary technological innovations were, even if invented, not economically successful. Often, regulation, legislation, and subsidization of competitive sources of energy have thwarted success.

A particularly European problem is growing dependency on imports. As indigenous reserves continue to decline, even stagnating natural gas consumption will require an increase of imports. This, in turn, initiated and initiates discussions about the general desirability of natural gas, because encouraging growth of natural gas consumption is equated to increasing import dependency, which, in turn, is alleged to decrease security of supply.

Worldwide, LNG has experienced some kind of 'hype' during the last years—both within the industry and governments. The recent developments in Australia might serve as an example. The reasons are numerous, among them decreasing fixed costs and (perceived) higher security of supply, largely independent of transit countries and diversification of sources. Consequently, various regasification terminals have been constructed worldwide. Yet, not only in Europe and in the US, these terminals are not constantly utilized, i.e., the respective load factors are partly very low. Nonetheless, this hype fostered the development of new reserves and further competition on the supply side. Therefore, it has to be assumed that globalization of the natural gas industry, traditionally restricted to the extent of pipeline networks, initiated by the growth of the LNG industry during the most recent two decades, will remain irreversible.

In addition, natural gas pipelines from the Middle East/Caspian region for the supply of Europe and Asia have been planned during the last three decades, endorsed by governments, primarily to promote diversification. These pipeline projects have not been realized so far. The reasons discussed are diverse and include questioning of the economic viability of such projects (promoted politically, but to be financed privately) and questioning the real increase in security for supply, as several countries, partly perceived as not stable, would have to be transited.

Downstream, upcoming digitalization, seen by some market participants as a threat rather than an opportunity, might have the potential to change not only the electricity industry but also

A. Pustišek and M. Karasz, *Natural Gas: A Commercial Perspective*,
DOI 10.1007/978-3-319-53249-3_10

the natural gas industry. In the electricity sector, customers are increasingly acquired online. New entrants, with high-end 'virtual economy' backgrounds, enter the business. As a reaction, many utilities have developed and launched applications for billing and payment, as well as the management of interruptions.[1] Natural gas companies at all levels will (probably) also have to cope with these challenges, primarily as such digitalization influences revenues, e.g., by expanding the customer base—regionally, nationally and internationally—and costs, e.g., by reduction of administrative costs. Given the experience in other sectors, e.g., banking and logistics, the expectations of potential benefits of digitalization in the natural gas industry are high.

Also in this context, block-chain technologies (or applications), enabling market participants to automate contracts and their management, processing larger quantities, increasing reliability, and decreasing costs, are an essential building block. It is expected that such automated trading applications might have a large impact primarily on contractual gas flows.

Companies preparing for the digitalization will have to identify the areas and challenges to be tackled first. This will be done based on revenue expectations and determine the strategy of digitalization, which might be discussed in parallel with a strategy of business transformation. Organizational implementation, be it structural or procedural, is not expected to constitute more than a 'usual' challenge to market participants. At all levels of the industry, participants are acquainted with re-organizations throughout the past years. Yet, the development of new digital products might be perceived differently, since this has not been the 'core' business of the industry. The development of innovative digital products might take time and effort and require different human resources. The business is becoming increasingly abstracted from the underlying technical reality.

Yet, such digitalization also is, and will be, responsible for a cultural shift within companies. While—e.g., in Europe—the industry has 'lost' its stable long-term perspective with individually negotiated terms and conditions of delivery during the last decade, digitalization might support even enhanced standardization of products—a tendency, which started to evolve during the most recent years. Customers do not receive individually shaped products, i.e., deliveries, but will have to assemble their very own bundles. Hence, some market participants expect the 'death' of the archetypical (gas) 'sales-man'—even in a prosperous future of the natural gas industry.

[1] Also referred to as 'Outage Management System' ('OMS').

Appendix A
Additional Recommended Literature

The following literature list is intended to provide interested readers an introductory, yet subjective, overview of the background of the topics discussed in this book. It is restricted to English-language literature.

Exploration and Production

Only literature providing a general overview is listed.

- Devold H (2013) Oil and gas production handbook—an introduction to oil and gas production, transport, refining and petrochemical industry, 3rd edn. ABB Oil and Gas. Available at: http://www04.abb.com/global/seitp/seitp202.nsf/0/f8414ee6c6813f5548257c14001f11f2/$file/Oil+and+gas+production+handbook.pdf
- Gluyas J, Swarbrick R (2008) Petroleum geoscience. Wiley-Blackwell
- Hyne NJ (2012) Nontechnical guide to petroleum geology, exploration, drilling and production, 3rd edn
- Sengbush RL (1986) Petroleum exploration: a quantitative introduction. Springer

Natural Gas Engineering

- Nasr GG, Connor NE (2014) Natural gas engineering and safety challenges, downstream process, analysis, utilization and safety. Springer

Economics

- Mankiw GN (2014) Principles of macroeconomics, 7th edn
- Pindyck R, Rubinfeld D (2012) Microeconomics, 8th edn. The Pearson series in economics

Physics

- Giancoli DC (2013) Physics: principles with applications, 7th edn
- Feynman RP, Leighton RB, Sands M (2011) The Feynman Lectures on physics, boxed set: the new millennium edition, Slp edn

Derivatives, Portfolio Management, and Risk Management

- Burger M, Graeber B, Schindlmayr G (2007) Managing energy risk—an integrated view on power and other energy markets. Wiley
- Fabozzi FJ, Pachamanova DA (2016) Portfolio construction and analytics (Frank J. Fabozzi Series), 1st edn. Wiley
- Hull JC (2014) Options, futures, and other derivatives, 9th edn
- Rachev ST, Stoyanov SV, Fabozzi FJ (2008) Advanced stochastic models, risk assessment, and portfolio optimization—the ideal risk, uncertainty and performance

A. Pustišek and M. Karasz, *Natural Gas: A Commercial Perspective*,
DOI 10.1007/978-3-319-53249-3

Regulation

- Decker C (2014) Modern economic regulation: an introduction to theory and practice.
- Knieps G (2015) Network economics, principles—strategies—competition policy. Springer

Some Regulatory Authorities

- Argentina: Ente Nacional Regulador del Gas

http://www.enargas.gov.ar
- Australia: Australian Energy Regulator

https://www.aer.gov.au/
- Austria: E-Control

https://www.e-control.at/
- Brazil: Agência Nacional do Petróleo, Gás Natural e Biocombustíveis

http://www.anp.gov.br
- Canada: National Energy Board, Office national de l'énergie

https://www.neb-one.gc.ca
- Czech Republic: Energetický Regulační Úřad (ERÚ)

http://www.eru.cz/en/
- European Union: Agency for the Cooperation of the Energy Regulators

http://www.acer.europa.eu/en/gas/Pages/default.aspx
- France: Commission de régulation de l'énergie

http://www.cre.fr/
- Germany: Bundesnetzagentur für Elektrizität, Gas, Telekommunikation, Post und Eisenbahnen

http://www.bundesnetzagentur.de/DE/Home/home_node.html
- Hungary: Magyar Energetikai és Közmű-szabályozási Hivata

http://www.mekh.hu/home
- India: Petroleum & Natural Gas Regulatory Board

http://www.pngrb.gov.in
- Italy: Autorità per l'energia elettrica il gas e il sistema idrico

http://www.autorita.energia.it
- Japan: Agency for Natural Resources and Energy

http://www.enecho.meti.go.jp/en/
- Netherlands: Autoriteit Consument & Markt (ACM)

https://www.acm.nl/en/
- Poland: Urząd Regulacji Energetyki

http://www.ure.gov.pl/en/
- Romania: Antoritatea Națională de Reglementare în domeniul Energiei

http://www.anre.ro/en
- Russia: Federal Tariff Service

http://www.fstrf.ru/eng
- South Africa: National Energy Regulator of South Africa

http://www.nersa.org.za
- Spain: Comisión Nacional de los Mercados y la Competencia

https://www.cnmc.es
- UK: Office of Gas and Electricity Markets, OFGEM

https://www.ofgem.gov.uk
- US: Federal Energy Regulatory Commission

http://www.ferc.gov/

Appendix B
Units and Prefixes

Prefixes

Tables B.1 and B.2[1]

Table B.1 Prefixes

Prefix	Abbreviation of prefix	Multiplication factor	'Conventional' notation
Yocto	y	10^{-24}	0.000 000 000 000 000 000 000 001
Zepto	z	10^{-21}	0.000 000 000 000 000 000 001
Atto	a	10^{-18}	0.000 000 000 000 000 001
Femto	f	10^{-15}	0.000 000 000 000 001
Pico	p	10^{-12}	0.000 000 000 001
Nano	n	10^{-9}	0.000 000 001
Micro	µ	10^{-6}	0.000 001
Milli	m	10^{-3}	0.001
Centi	c	10^{-2}	0.01
Deci	d	10^{-1}	0.1
		10^{0}	1
Deca	da	10^{1}	10
Hecto	h	10^{2}	100
Kilo	k	10^{3}	1000
Mega	M	10^{6}	1,000,000
Giga	G	10^{9}	1,000,000,000
Tera	T	10^{12}	1,000,000,000,000
Peta	P	10^{15}	1,000,000,000,000,000
Exa	E	10^{18}	1,000,000,000,000,000,000
Zetta	Z	10^{21}	1,000,000,000,000,000,000,000
Yotta	Y	10^{24}	1,000,000,000,000,000,000,000,000

[1]used alternatively to prefixes shown in Table B.1.

A. Pustišek and M. Karasz, *Natural Gas: A Commercial Perspective*,
DOI 10.1007/978-3-319-53249-3

Table B.2 Large number abbreviations[a]

Abbreviation	Name	Factor
m	Thousand	10^3
mm	Million	10^6
b	Billion	10^9
t	Trillion	10^{12}

[a]Used alternatively to prefixes shown in Table B.1

Volume

Table B.3

Energy

Table B.4

Remarks:

- The 'calorie' is a unit of energy. Yet, there are various definitions.
  - The '15° calorie' is the amount of energy (heat) required to raise the temperature of 1 g of water from 14.5 to 15.5 °C. This amount of energy equals 4.1858 Joule.
  - In Eastern Europe, particularly in Russia, the 20 °C-calorie is used (if at all). It is the amount of energy (heat) required to raise the temperature of 1 g of water from 19.5 to 20.5 °C. As the specific heat capacity of water is not a linear function of temperature, the energy required for heating is lower than at 15 °C, it equals 4.1819 Joule.
  - Another calorie is the International Table calorie (IT calorie), originally defined as 1/860 of a Watt-hour. It is equal to 4.1868 Joules.
- Only the 'Joule' and the 'Watt-hour' are consistent with the internationally accepted and applied 'système international d'unités' (SI-system; international system of units).
- Despite the fact that a tonne is a unit of mass, the tonne of coal equivalent (tce) is the amount of energy released by burning one tonne of coal. This analogously applies to the tonne of oil equivalent.
- It has to be noted: A 'ton' does not equal a 'tonne'. Both are units of mass.
  - A 'tonne' is a metric unit equaling 1000 kg. It is also referred to as 'metric ton'.
  - A 'ton' is a British unit equaling 2240 pounds or 1016.047 kg. It is also referred to as 'long ton', 'weight ton' or 'gross ton'.
  - In the US and Canada a 'ton' equals 2000 pounds or 907.1847 kg. It is also referred to as the 'short ton' or 'net ton'.

Pressure

Table B.5

Table B.3 Conversion of volume units

		Cubic meter	Liter	US gallon (liquid)	US petroleum barrel	Imperial gallon	Cubic foot
To convert from	Multiply by						
Cubic meter	m^3	1	1000	264.1721	6.2898	219.9692	35.3147
Liter	l	0.001	1	0.26417205	0.00628981	0.21996925	0.03531467
US gallon (liquid)	US liq. gal.	0.0037854	3.7854	1	0.02380952	0.83267418	0.13368056
US petroleum barrel	bbl	0.15898729	158.9873	42	1	34.9723	5.6146
Imperial gallon	Imp. gal	0.00454609	4.5461	1.2009	0.02859405	1	0.16054365
Cubic foot	ft^3	0.02831685	28.3168	7.4805	0.17810761	6.2288	1

Table B.4 Conversion of energy units

	Btu	Joule	Calorie (IT)	15° calorie	Therm US	Therm Europe	Tonne of coal equivalent	Tonne of oil equivalent	Kilowatt-hour
To convert from	Multiply by								
Btu	1	1.0551×10^{3}	2.52×10^{2}	2.5206×10^{2}	1.0002×10^{-5}	1×10^{-5}	3.5999×10^{-8}	2.52×10^{-8}	2.9307×10^{-4}
Joule	9.4782×10^{-4}	1	2.3885×10^{-1}	2.389×10^{-1}	9.4804×10^{-9}	9.4782×10^{-9}	3.4121×10^{-11}	2.3885×10^{-11}	2.7778×10^{-7}
Calorie (IT)	3.9683×10^{-3}	4.1868	1	1.0002	3.9693×10^{-8}	3.9683×10^{-8}	1.4286×10^{-10}	1×10^{-10}	1.163×10^{-6}
15° calorie	3.9674×10^{-3}	4.1858	9.9976×10^{-1}	1	3.9683×10^{-8}	3.9674×10^{-8}	1.4282×10^{-10}	9.9976×10^{-11}	1.1627×10^{-6}
Therm US	9.9976×10^{4}	1.0548×10^{8}	2.5194×10^{7}	2.521×10^{7}	1	9.9976×10^{-1}	3.5991×10^{-3}	2.5194×10^{-3}	2.93×10^{1}
Therm Europe	1×10^{5}	1.0551×10^{8}	2.521×10^{7}	2.5206×10^{7}	1.0002	1	3.5999×10^{-3}	2.52×10^{-3}	2.9307×10^{1}
Tonne of coal equivalent	2.7778×10^{7}	2.9308×10^{10}	7×10^{9}	7.0017×10^{9}	2.7785×10^{2}	2.7778×10^{2}	1	7×10^{-1}	8.141×10^{3}
Tonne of oil equivalent	3.9683×10^{7}	4.1868×10^{10}	1×10^{10}	1.0002×10^{10}	3.9693×10^{2}	3.9683×10^{2}	1.4286	1	1.163×10^{4}
Kilowatt-hour	3.4121×10^{3}	3.6×10^{6}	8.5985×10^{5}	8.6005×10^{5}	3.4121×10^{-2}	3.4121×10^{-2}	1.2284×10^{-4}	8.5985×10^{-5}	1

Table B.5 Conversion of pressure units

	Bar	Atmosphere technical	Atmospere standard	Centimetre of mercury	Pascal	Torr	Ounce/square inch	Pound/square inch
To convert from								
Bar	1	1.0197	0.9869	75.0062	100,000	50.0617	232.0604	14.5038
Atmosphere technical	0.980665	1	0.9678	73.5559	98,066.5	735.5592	227.5735	14.2233
Atmosphere standard	1.01325	1.033228	1	76	101,325.01	760.0001	235.1352	14.696
Centimeter of mercury	0.013332	0.013595	0.013158	1	1333.2239	10	3.0939	0.193368
Pascal	0.00001	0.00001	0.00001	0.00075	1	0.007501	0.002321	0.000145
Torr	0.001333	0.00136	0.001316	0.1	133.3224	1	0.309388	0.019337
Ounce/square inch	0.004309	0.004394	0.004253	0.323218	430.9223	3.232183	1	0.0625
Pound/square inch	0.068948	0.070307	0.068046	5.171492	6894.7573	51.7149	16	1

Units used in this book

°C	Degree Celsius
°F	Degree Fahrenheit
a	Year
atm	Atmosphere
bar a	Bar (absolute)
bar g	Bar (gauge)
bcm	Billion cubic meters
Btu	British thermal unit
D	Darcy
d	Day
EJ	Exajoule
ft	Foot (feet)
ft^3	Cubic foot (feet)
GJ	Gigajoule
h	Hour
J	Joule
K	Kelvin
kcal	Kilocalorie
kg	Kilogram
km^2	Square kilometer
kWh	Kilowatt hour
m	Meter
m^2	Square meter
m^3	Cubic meter
mD	Milli-Darcy
mg	Milligram
MJ	Megajoule
mmBtu	Million Btu
mmtpa	Million tonnes per year
MW	Megawatt
MWh	Megawatt hour
nD	Nano-Darcy
Nm^3	Normal cubic meter
psi	Pound(s) per square inch
Pa	Pascal
s	Second
scf	Standard cubic foot (feet)
tcm	Trillion cubic meters
TWh	Terawatt hour
µD	Micro-Darcy

Appendix C
Symbols and Abbreviations

Symbols

Symbol	Explanation	Dimension	Application in section
A	Cross sectional area of the formation	Area	Non-economic and non-commercial fundamentals
$A_{Storage}$	Total amount to be paid for the reservation of a storage product	Monetary unit	Storage
ACQ_{Max}	Maximum annual contract quantity	Quantity[a]	Sales along the value chain
ACQ_{Min}	Minimum annual contract quantity	Quantity[a]	Sales along the value chain
ACQ_{Norm}	Standard annual contract quantity	Quantity[a]	Sales along the value chain
b_{ij}	Conversion and weighting factors	Percentage	Sales along the value chain
$c_{Pipeline}$	Specific (pipeline-) length related costs	Monetary unit/length	Transportation
$c_{specific}$	Capacity related specific total cost	Monetary unit/capacity	Transportation
const	Constant	$Length^6/force^2atime^2$	Transportation
$C_{Pipeline}$	Fixed costs of pipeline construction	Monetary unit	Transportation
d_i	Relative weight of variable (e.g., $\Psi_i(t)$)	Percentage	Sales along the value chain, transportation, storage
$D_{Pipeline}$	Diameter of the pipeline	Length	Transportation
DCQ_{Max}	Maximum daily contract quantity	Quantity[a]	Sales along the value chain
DCQ_{Min}	Minimum daily contract quantity	Quantity[a]	Sales along the value chain
DCQ_{Norm}	Standard daily contract quantity	Quantity[a]	Sales along the value chain

(continued)

A. Pustišek and M. Karasz, *Natural Gas: A Commercial Perspective*,
DOI 10.1007/978-3-319-53249-3

(continued)

Symbol	Explanation	Dimension	Application in section
F_{Period}	Maximum/minimum boundary function for period	Quantity[a]	Appendix
GCV_{LNG}	Gross calorific value of LNG loaded or discharged	Energy/mass	Sales along the value chain
GCV_n	Gross calorific value under normal conditions	Energy/volume	Non-economic and non-commercial fundamentals
HCQ_{Max}	Maximum hourly contract quantity	Quantity[a]	Sales along the value chain
HCQ_{Min}	Minimum hourly contract quantity	Quantity[a]	Sales along the value chain
HCQ_{Norm}	Standard hourly contract quantity	Quantity[a]	Sales along the value chain
i, j, k, m, n	Integers to be defined		Sales along the value chain, storage, appendix
k_i^{i+n}	Number of periods i in period (i + n)		Sales along the value chain
$l_{Pipeline}$	Length of the pipeline	Length	Transportation
MCQ_{Max}	Maximum monthly contract quantity	Quantity[a]	Sales along the value chain
MCQ_{Min}	Minimum monthly contract quantity	Quantity[a]	Sales along the value chain
MCQ_{Norm}	Standard monthly contract quantity	Quantity[a]	Sales along the value chain
NCV_n	Net calorific value under normal conditions	Energy/volume	Non-economic and non-commercial fundamentals
N/G	Net to gross ratio	Percentage	Non-economic and non-commercial fundamentals
$\frac{\partial p}{\partial x}$	Fluid potential gradient	Pressure/length	Non-economic and non-commercial fundamentals
p_1	Pressure at the delivery (inlet) point	Pressure	Transportation
p_2	Pressure at the redelivery (outlet) point	Pressure	Transportation
p_{Amb}	ambient barometric pressure	pressure	Non-economic and non-commercial fundamentals
p_E	Measured pressure in pipeline	Pressure	Non-economic and non-commercial fundamentals
p_n	1.01325 bar normal pressure	Pressure	Non-economic and non-commercial fundamentals
p_S	Water vapor partial pressure	Pressure	Non-economic and non-commercial fundamentals
P_0	Base price (additive formula)	Monetary unit/quantity[a]	Sales along the value chain

(continued)

(continued)

Symbol	Explanation	Dimension	Application in section
$p_{Injection}$	Injection rate price	Monetary unit/capacity	Storage
P_M	Price (multiplicative formula)	Monetary unit/quantity[a]	Sales along the value chain
P_{M0}	Base price (multiplicative formula)	Monetary unit/quantity[a]	Sales along the value chain
P_{Market}	Commodity market price	Monetary unit/quantity[a]	Sales along the value chain
P_{Sales}	Commodity sales price	Monetary unit/quantity[a]	Economic and commercial fundamentals, storage
$P_{Storage}$	Storage price	Monetary unit/capacity or monetary unit/quantity[a]	Storage
$P_{Storage0}$	Base price for storage	Monetary unit/capacity or monetary unit/quantity[a]	Storage
P_{Supply}	Price from supplier	Monetary unit/quantity[a]	Economic and commercial fundamentals, storage
$P_{Transport}$	Transport price	Monetary unit/capacity	Transport
$P_{Transport0}$	Base price for transport	Monetary unit/capacity	Transport
$P_{Withdrawal}$	Withdrawal rate price	Monetary unit/capacity	Storage
$p_{Workinggas}$	Working gas capacity price	Monetary unit/quantity[a]	Storage
Q	quantity	quantity[a]	Sales along the value chain
Q(T)	Quantity as a function of temperature	Quantity[a]	Sales along the value chain
Q_0	Quantity calibration	Quantity[a]	Sales along the value chain
Q_{Gas}	Quantity of gas to ship to terminal	Quantity[a]	Sales along the value chain
Q_{i+n}^{Max}	Maximum quantity in period i +n	Quantity[a]	Sales along the value chain
Q_{i+n}^{Min}	Minimum quantity in period i+n	Quantity[a]	Sales along the value chain
Q_i^{Max}	Maximum quantity in period i	Quantity[a]	Sales along the value chain
Q_i^{Min}	Minimum quantity in period i	Quantity[a]	Sales along the value chain
QCQ_{Max}	Maximum quarterly contract quantity	Quantity[a]	Sales along the value chain
QCQ_{Min}	Minimum quarterly contract quantity	Quantity[a]	Sales along the value chain
QCQ_{Norm}	Standard quarterly contract quantity	Quantity[a]	Sales along the value chain
$R_{Remaining}$	Remaining reserves	Volume	Non-economic and non-commercial fundamentals
SCQ_{Max}	Maximum seasonal contract quantity	Quantity[a]	Sales along the value chain, appendix
SCQ_{Min}	Minimum seasonal contract quantity	Quantity[a]	Sales along the value chain, appendix

(continued)

(continued)

Symbol	Explanation	Dimension	Application in section
SCQ_{Norm}	Standard seasonal contract quantity	Quantity[a]	Sales along the value chain, appendix
t	Time of price (re-)calculation	Time	Sales along the value chain, transportation, storage
t^a	Number of periods of the term of contract duration		Sales along the value chain
t_{adj_i}	Adjustment period for the calculation of $\psi_i(t)$		Sales along the value chain
t_{lag_i}	Time lag applicable for the calculation of $\psi_i(t)$		Sales along the value chain
t_{ref_i}	Reference period for the calculation of $\psi_i(t)$		Sales along the value chain
T	Ambient temperature	Temperature	Non-economic and non-commercial fundamentals sales along the value chain
T_n	273.15 K norm temperature	Temperature	Non-economic and non-commercial fundamentals
TCQ	Total contract quantity	Quantity[a]	Sales along the value chain
V_{GIIP}	Gas initially in place	Volume	Non-economic and non-commercial fundamentals
V_{GRV}	Gross rock volume	Volume	Non-economic and non-commercial fundamentals
V_{LNG}	Volume of LNG loaded or discharged	Volume	Sales along the value chain
V_{nr}	Norm volume	Volume	Non-economic and non-commercial fundamentals
$V_{Produced}$	Volumes already produced	Volume	Non-economic and non-commercial fundamentals
V_r	Measured volume	Volume	Non-economic and non-commercial fundamentals
V_{UR}	Ultimate recovery	Volume	Non-economic and non-commercial fundamentals
$V_{Workinggas}$	Working gascapacity	Quantity[a]	Storage
$W_{G,n}$	Gross Wobbe index (or Wobbe number)	Energy/volume	Non-economic and non-commercial fundamentals
$W_{N,n}$	Net Wobbe index (or Wobbe number)	Energy/volume	Non-economic and non-commercial fundamentals
α_1, α_2	Parameters of the pipeline cost function	Monetary unit/length^2 and monetary unit/ length	Transportation

(continued)

(continued)

Symbol	Explanation	Dimension	Application in section
$\beta_1, \beta_2, \beta_3, \beta_4, \beta_5$	Parameters of 'sigmoid' function	Temperature, dimensionless	Sales along the value chain
γ	Coefficient (pipeline capacity)		Transportation
$\delta_H, \delta_D, \delta_M, \delta_Q, \delta_S, \delta_A$	Flexibility component per period (hour (H), day (D), month (M), quarter (Q), season (S) and year (A))	Percentage	Appendix
η	Permeability	Area	Non-economic and non-commercial fundamentals
ϑ^m, ϑ^l, ϑ^n	Total flexibility of a (contractual or physical) flexibility tool m, l, or n	Percentage	Appendix
$\kappa_{Injection}$	Injection rate	Capacity = quantity[a]/time	Storage
$\kappa_{Pipeline}$	Capacity of a pipeline	Capacity = quantity[a]/time	Transportation
$\kappa_{Withdrawal}$	Withdrawal rate	Capacity = quantity[a]/time	Storage
λ	Flow rate	Capacity = quantity[a]/time	Non-economic and non-commercial fundamentals
ν	Dynamic viscosity	Mass/(length [a] time)	Non-economic and non-commercial fundamentals
ρ_{Air}	Density of air	Mass/volume	Non-economic and non-commercial fundamentals
ρ_{Gas}	Density of natural gas	Mass/volume	Non-economic and non-commercial fundamentals
ρ_{LNG}	Density of LNG loaded or discharged	Mass/volume	Sales along the value chain
ρ_n	Relative density at normal conditions		Non-economic and non-commercial fundamentals
$\sigma_{Deviation}$	Gas law deviation factor [non dimensional]	Percentage	Non-economic and non-commercial fundamentals
$\sigma_{Formation}$	Formation volume factor	Percentage	Non-economic and non-commercial fundamentals
$\sigma_{Gas\ saturation}$	Gas saturation factor	Percentage	Non-economic and non-commercial fundamentals
$\sigma_{Recovery}$	Recovery factor	Percentage	Non-economic and non-commercial fundamentals
τ	Factor of proportionality		Transportation
φ	Flow rate	Capacity = quantity[a]/time	Transportation
χ	Relative gas humidity		Non-economic and non-commercial fundamentals

(continued)

(continued)

Symbol	Explanation	Dimension	Application in section
Δ	Price adjustment	Monetary unit/quantity[a]	Sales along the value chain
Θ	Quantity per period expressed as percentage of the norm quantity of such period	Percentage	Appendix
ϕ	Porosity	Percentage	Non-economic and non-commercial fundamentals
$\overline{\Psi}_i(t)$	Average value of the index i at time t	Monetary unit/quantity[a]	Sales along the value chain, storage
$\Psi_i(t))$	Variables of Indexed price formulae, e.g., oil product price quotations	Monetary unit/quantity[a]	Sales along the value chain, transportation, storage
Ψ_{io}	Base price for variable i	Monetary unit/quantity[a]	Sales along the value chain, transportation, storage

[a]Quantity: volume or energy

Abbreviations[2]

€	Euro(s); statutory means of payment in parts of Europe
€ct	Euro Cent(s), being 1/100 of a €
£	British Pound(s); statutory means of payment in the UK
ACER	Agency for the Cooperation of Energy Regulators
AIPN	Association of International Petroleum Negotiators
AT	Austria
AVO	Amplitude versus offset
B2C	'Business to consumer'
BE	Belgium
BIMCO	Baltic and International Maritime Council
CAPEX	Capital expenditures
CCGT	Combined cycle gas turbine
CFR	Cost and freight
CHP	Combined heat and power
cif	Cost, insurance, freight
CNG	Compressed natural gas
COMECON	Council for mutual economic assistance
CZ	Czech Republic
DAP	Delivered at place
DAT	Delivered at terminal
DE	Germany
DES	Delivered ex ship
DHI	Direct hydrocarbon indicators
DSO	Distribution system operator
DVGW	Deutscher Verband der Gas- und Wasserwirtschaft
e.g.	*exempli gratia*—for example
E&P	Exploration and production
EASEE	European Association for the Streamlining of Energy Exchange-gas
ECC	European Commodity Clearing
EPC	Engineering procurement construction
ERGEG	European Regulators' Group for Electricity and Gas
et al.	*et alii*—and others
et seq.	*et sequentes*—and that which follows
EU	European Union
FDP	Field development plan
FEED	Front end engineering design
FERC	Federal Energy Regulatory Commission
FID	Final investment decision

(continued)

[2]Units used in this book are listed in Appendix B.

(continued)

Fig.	Figure
FOB	Free on board
FPC	Federal Power Commission
GFU	Gassforhandlingsutvalget (Norwegian Gas Negotiation Committee)
GGPSSO	Guidelines for Good TPA Practice for Storage System Operators
GIE	Gas Infrastructure Europe
GIIGNL	Groupe International des Importateurs de Gaz Naturel Liquéfié
GIIP	Gas initially in place
GSA	(pipeline) Gas sales agreement
GSE	Gas Storage Europe
GTL	Gas to liquids
HOA	Heads of agreement
i.e.	*id est*—that is
IEA	International Energy Agency
ICC	International Chamber of Commerce
IMO	International Maritime Organization
IPE	International Petroleum Exchange
LNG	Liquefied natural gas
LNG SPA	LNG sales and purchase agreement
LOI	Letter of intent
LPG	Liquefied petroleum gas
LTC	Long-term contract
MCR	Maximum continuous rating
MOU	Memorandum of understanding
NAESB	North American Energy Standards Board
NDRC	National Development and Reform Commission
NEA	National Energy Act
NGL	Natural gas liquids
NGV	Natural gas vehicle(s)
NGPA	Natural Gas Policy Act
NGWDA	Natural gas wellhead decontrol act
NL	Netherlands
NOC	National oil company
NYMEX	New York Mercantile Exchange
OECD	Organisation for Economic Co-operation and Development
OMS	Outage management system
OPEC	Organization of Petroleum Exporting Countries

(continued)

(continued)

OTC	Over the counter
p.	Page
p	British pence, being 1/100 of a £
p.a.	Per annum—per year
PSA	Production sharing agreement
PSC	Production sharing contract
PXE	Power Exchange Central Europe
SBU	Standard bundled unit
SCADA	Supervisory control and data acquisition
SI	Système international d'unités—International System of Units
SK	Slovakia
SME	Small and medium enterprise(s)
SSO	Storage system operator
TSO	Transport(ation) system operator
UAE	United Arab Emirates
UK	United Kingdom of Great Britain and Northern Ireland
US	United States of America
US $	US Dollar(s); statutory means of payment in the US
WACOG	Weighted average cost of gas
WTI	(crude oil) West Texas Intermediate
ZBT	Zeebrugge hub natural gas trading

- For abbreviations in relation to companies, associations, hubs, and pipelines, reference is made to Appendix D.
- All units are referred to in Appendix B.
- All references to chemical elements are according to the rules of the International Union of Pure and Applied Chemistry (IUPAC).

Appendix D
Glossary

Note: *The following glossary does not intend to be a complete reference to all technical, commercial, and economic terms used in the natural gas industry. The (simple and main) purpose is to provide a comprehensive summary and definitions of essential terms used in this book. Terms, which may be used differently in other publications, are also included.*

General

- **AQUIFER STORAGE**: A sub-surface facility for storing natural gas constructed in (former) water-bearing layers topped by an impermeable cap rock.
- **AMERICAN OPTION**: Option that may be exercised at any time during its lifetime, in contrast to a European option.
- **ARBITRAGE**: The simultaneous purchase and sale of a commodity or derivative in two different markets (or, more generally, locations) or at two different points in time—done with the objective of exploiting price differentials.
- **ASK PRICE**: The price asked for by the seller, i.e., the lowest price a seller is prepared to accept for a good, in this case natural gas.
- **BACKWARDATION**: Situation where the price of a forward or futures contract is trading below the expected future spot price. The reverse situation is called contango.
- **BASELOAD**: The minimum continuous load or demand or off-take of natural gas over a given period.
- **BID PRICE**: The price offered by the buyer, i.e., the highest price a buyer is prepared to pay for a good, in this case natural gas.
- **BID-ASK SPREAD**: Difference between bid and ask prices. Also referred to as bid-offer spread.
- **CALORIFIC VALUE**: The calorific value of natural gas (in general, of a fuel) is the quantity of energy (equal to heat) produced by its combustion (i.e., 'contained' in a unit of volume) at defined (i.e., 'normal' or 'standard') conditions of pressure and temperature.

  As this combustion process generates water vapor, different calorific values are defined depending on the recovery of water vapor:
  - **GROSS CALORIFIC VALUE (GCV) (or Higher Calorific Value (HCV) or Higher Heating Value (HHV))**: The energy contained in the water vapor is recovered by condensing the water of combustion entirely;
  - **NET CALORIFIC VALUE (NCV) (or Lower Calorific Value (LCV) or Lower Heating Value (LHV))**: The heat in the water vapor is not recovered.
- **CAPACITY**: The (technical or contractually) maximum possible flow of natural gas, i.e., the potential throughput, of a pipeline or

A. Pustišek and M. Karasz, *Natural Gas: A Commercial Perspective*,
DOI 10.1007/978-3-319-53249-3

storage of a storage facility; measured as [volume/time] or, alternatively, [energy/time]. Capacity has to be distinguished from 'flow'.
  - **TECHNICAL CAPACITY**: The maximum capacity that the carrier, LNG liquefaction operator, LNG regasification operator, or storage operator can offer to customers, taking account the system integrity and the operational requirements of the system.
  - **FIRM CAPACITY**: Transportation, LNG liquefaction, LNG regasification, or storage capacity contractually and unconditionally guaranteed by the operator.
  - **INTERRUPTIBLE CAPACITY or NON-FIRM CAPACITY**[3]: Natural gas transportation, LNG liquefaction, LNG regasification, or storage capacity that can be interrupted by the operator according to the conditions stipulated in the respective contract.
  - **AVAILABLE FIRM CAPACITY**: The part of the technical capacity that is not allocated and is still available for reservation by customers.
- **CARRIER**: An entity, which operates a natural gas transportation system, often also referred to as a transmission system operator or transport(ation) system operator.
- **(SALT) CAVERN STORAGE**: A sub-surface facility for storing natural gas constructed in a hollowed-out salt dome.
- **CCGT**: See combined-cycle gas turbine.
- **CHP**: See Combined Heat and Power.
- **CIF**: Incoterms® 2010: Cost, Insurance, and Freight.
- **CITYGATE**: A point or measuring station at the entrance (or edge) of a distribution network.
- **CLEARING HOUSE**: Institution providing clearing and settlement services, *inter alia*, for exchanges. Used for reducing credit risks in exchange transactions.
- **CLEARING**: The process of matching and settling trades, often a service performed by exchanges.
- **COAL BED METHANE GAS**: Methane that is generated during coal formation and is contained in the coal microstructure.
- **COKE OVEN GAS**: The mixture of permanent gases produced by the carbonization of coal in a coke oven at temperatures in excess of 1000 °C.
- **COMBINED-CYCLE GAS TURBINE (CCGT)**: Gas-turbine systems where the first turbine produces electricity from fuel combustion. The produced hot gases are used in a boiler and a steam turbine to produce more electricity.
- **COMBINED HEAT AND POWER (CHP)**: Plant where a fuel is burned to produce both electricity and heat.
- **COMPRESSED NATURAL GAS (CNG)**: Natural gas that is compressed to a pressure at or above 165 bar, and stored in special high-pressure containers. It is used as a fuel for natural gas-powered vehicles.
- **CONTANGO**: Situation where the price of a forward or futures contract is trading above the expected future spot price. The reverse situation is called backwardation.
- **CONTRACT FOR DIFFERENCES (CFD)**: is a contract between two parties stipulating that party A will pay to party B or party B will pay to party A the difference (if positive) between the value of a commodity at a defined point in time and the settlement price at the time of conclusion of the contract.
- **CUSHION GAS (or BASE GAS)**: Volume of natural gas intended as permanent inventory in a storage required to maintain adequate pressure and deliverability rates throughout the withdrawal season.
- **CYCLING RATE**: 8760/(withdrawal time + injection time), expressed in cycles per year [1/a].
- **DAP**: Incoterms® 2010: Delivered at Place.
- **DARK SPREAD**: The difference between electricity prices and coal prices during a period in a region.

[3]Sometimes distinguished, these two terms shall be used synonymously.

- **DAT**: Incoterms® 2010: Delivered at Terminal.
- **DEGREE DAY**: is usually defined as the positive difference between a base temperature and the daily average outside temperature at a location (for heating degree-days) or vice versa (for cooling degree-days), putting negative values to zero. Thus, it is a measure of heating or cooling. The base temperatures used differ, typically in the range of 15–19 °C. If the difference between the base temperature and the average daily temperature is positive, the degree-day is referred to as heating degree-day; if the difference between the base temperature and the average daily temperature is negative, the positive value of the degree-day is referred to as cooling degree-day. The sum of degree-days is a measure for the heating or cooling requirements and, consequently, the natural gas consumption during a year or heating period at a given location.
- **DEMAND-SIDE MANAGEMENT (DSM)**: All activities undertaken by end users to manage and influence the utilization of natural gas in quantity and time.
- **DEMURRAGE**: Damages paid for a delay in loading or discharging cargo from a ship.
- **DEPLETED FIELD STORAGE**: A sub-surface facility for storing natural gas constructed in depleted natural gas or oil fields.
- **DES**: Incoterms® 2000: Delivered ex ship.
- **DISTRIBUTION SYSTEM**: A natural gas transportation system (or network), typically at low pressure and with small-diameter pipelines connecting the (national) transportation system (or network) to the end user.
- **DISTRIBUTION-SYSTEM OPERATOR (DSO)**: An entity that operates a natural gas distribution system.
- **END USER**: Final consumer of natural gas, including residential, commercial, industrial consumers, and electricity producers.
- **E&P**: Abbreviation for exploration and production.
- **EUROPEAN OPTION**: Option that may be exercised on its expiration date only, in contrast to an American option.
- **EXPORTS**: Natural gas deliveries to customers beyond the border of the country in which the natural gas is purchased or produced.
- **FEEDSTOCK**: Raw material used by any processing unit (e.g., in chemical industry).
- **FLARED NATURAL GAS**: Natural gas that is uneconomical for sale may be flared. The natural gas is disposed of by burning in flares usually at the production sites or at processing plants. For this purpose, the natural gas is transported to a(n) (often elevated) location at the facility and burned in an open flame.
- **FLOW**: The actual throughput of natural gas in a pipeline or a storage facility at any point in time; measured as [volume/time] or, alternatively, [energy/time]. Flow has to be distinguished from 'capacity'.
- **FOB**: Incoterms® 2010: Free on Board.
- **FORWARD CONTRACT**: A contract between a seller and a buyer to deliver and off-take a defined quantity of a commodity at a delivery point during a future period (or at a future point in time). The buyer will purchase and the seller will sell the specified commodity at a price determined at the time of entering into the (forward) contract.
- **FUEL GAS (or PIPELINE FUEL)**: Natural gas consumed for transportation, i.e., in the operation of pipelines, primarily as a fuel to run compressors.
- **FUTURE CONTRACT**: An exchange-traded standardized contract between a seller and a buyer to deliver and off-take a defined quantity of a commodity at a delivery point during a future period (or at a future point in time). The buyer will purchase and the seller will sell the specified commodity at a price determined at the time of entering into the (future) contract.
- **HEATING SEASON**: Typically begins (in the Northern Hemisphere) on October 1st of a

year and ends on March 31st of the following calendar year.

- **HENRY HUB**: A physical point located in Erath, Louisiana. Here nine interstate and four intrastate pipelines are interconnected.
- **HUB**: A hub describes a (physical or even virtual) 'location', at which natural gas is traded, i.e., sold and purchased.
- **IMPORTS**: Natural gas deliveries from suppliers beyond the border of the country in which the natural gas is consumed.
- **INJECTION RATE**: Amount of natural gas that can be injected into a storage on an hourly or daily basis, usually expressed in, e.g., [m^3/h] or [m^3/d]. Some storage system operators, however, express and offer injection rates in terms of energy per unit of time, e.g., [kWh/h = kW] or [kWh/d] or [MJ/h] etc.
- **INJECTION TIME**: Time needed to inject natural gas up to the working gas capacity into a natural gas storage, expressed in [h].
- **INTERCONNECTION**: Facilities that connect two natural gas systems (or networks).
- **INTERRUPTIBLE DEMAND**: The amount of natural gas demand of a customer that, in accordance with the terms and conditions of a delivery or transportation contract, can be interrupted by the supplier or carrier and, if required, replaced by other fuels by the customer.
- **LINE PACK (or LINEPACK)**: The amount of natural gas contained in a network at a given point in time.
- **LIQUEFIED NATURAL GAS (LNG)**: Natural gas that has been liquefied by reducing its temperature to −161.5 °C at atmospheric pressure.
- **LOAD**: Realized demand for natural gas during a period.
- **LOCAL DISTRIBUTION COMPANY (LDC)**: A company engaged primarily in the retail sale of natural gas to end users using a distribution system.
- **MARGINS**: Cash (or securities) that must be deposited with a clearinghouse for security against losses.
  Note: not to be confused with the 'sales margin'.
- **MARKETER**: A company that arranges purchases and sales of natural gas. That is, a marketer does not own physical assets commonly used in the supply of natural gas (term primarily used in the US).
- **NATURAL GAS LIQUIDS (NGL)**: Hydrocarbons including ethane, propane, and higher hydrocarbons. Extracted as a by-product of natural gas or crude oil production.
- **NETWORK**: An interconnected system of natural gas pipelines, compressors, measurement stations, etc.
- **NETWORK CODE**: Rules governing the utilization of networks. They regulated the conditions and subjects of network use. Network codes might be applied to national or cross-border transactions.
- **OFFER PRICE**: See ask-price.
- **ON-THE-DAY COMMODITY MARKET (OCM)**: The market used to balance the NBP gas market. Also described as the successor to the flexibility mechanism, being a trading system managed by ICE Endex to effectuate trades at the NBP.
- **OPTION**: The right (but not the obligation) to purchase (call-option) or sell (put-option) a commodity at a delivery point during a future period (or at a future point in time) at a strike (or exercise) price determined at the time of entering into the (forward) contract.
- **OVER-THE-COUNTER (OTC)**: Private contracts negotiated between two parties. In contrast to standardized contracts traded on exchanges.
- **PEAK LOAD**: The maximum quantity per hour or in another period, e.g., during a day or a year.
- **REFILL SEASON**: When natural gas has been withdrawn from the storage and no requirements for immediate additional withdrawals are anticipated, the period in which natural gas is re-injected into the storage begins. This refill season (in the Northern Hemisphere) typically starts at the beginning of April and lasts through the end of September.

- **REPRESSURING**: The (re-)injection of natural gas into reservoirs to enhance the recovery rate.
- **RESERVE**: A resource that is discovered, recoverable, economically viable, and sustainable.
- **RESERVOIR**: A porous, permeable, and delimited rock or underground formation containing natural gas and/or oil, confined by impermeable rock or water barriers.
- **RESERVOIR ROCK**: A sedimentary rock in which natural gas can be stored. This is usually a porous and permeable rock.
- **RESOURCE**: Concentrations of natural gas (or, generally, other materials) in such form that economic extraction is—at least—potentially feasible.
- **RESIDENTIAL CONSUMPTION**: Natural gas used in private households for heating, air-conditioning, cooking, water heating, and other domestic uses.
- **SETTLEMENT PRICE**: A price established by an exchange for natural gas deliveries at hubs at the close of a trading day that is used to calculate the settlement of futures contracts.
- **SHALE GAS**: Shale gas is natural gas found in shale formations. It is natural gas produced from wells that are open to shale or similar fine-grained rocks, i.e., not yet trapped in the reservoir formations occupied by so-called conventional gas but in bedrocks.
- **SPARK SPREAD**: The difference between electricity prices and natural gas prices during a period in a region.
- **SPOT PRICE**: Price paid for natural gas in the short-term physical market.
- **SPREAD**: The differential between two prices for deliveries at hubs. Spreads can refer to different futures periods (e.g., summer-winter spread), different locations (location spread) or commodities (dark spread) , etc.
- **STORAGE INJECTIONS (also called: ADDITIONS)**: Quantities of natural gas injected into underground natural gas storages or liquefied natural gas storages.
- **STORAGE SYSTEM OPERATOR (SSO)**: An entity, which operates a natural gas storage system.
- **STORAGE WITHDRAWALS**: Volume of natural gas withdrawn from an underground storage or from a liquefied natural gas storage.
- **TARIFF**: Prices or rates for transportation or storage of natural gas that a regulated entity will charge to its customers as well as the terms and conditions for providing such services.
- **THIRD-PARTY ACCESS (TPA)**: The right of a shipper, not being the owner or affiliated to an owner of a (pipeline) system, i.e., a 'third party', to reserve transportation capacity and receive related transportation services of a pipeline company for a charge. Analogously applied to storage capacity reservations and related storage services.
- **TOTAL GAS IN STORAGE**: Volume of natural gas remaining in the storage at a particular time.
- **TOTAL GAS STORAGE CAPACITY**: Maximum volume of natural gas that can be stored in a certain storage facility. As such, it is the sum of cushion gas and working gas capacity.
- **TRANSIT**: Transportation of natural gas across a country, i.e., the entry point being one border and the exit point being another border of a country.
- **TRANSMISSION**: Transportation of natural gas.
- **TRANSPORT(ATION) SYSTEM OPERATOR (TSO)**: See carrier.
- **VENTED GAS**: Natural gas that is released into the air, mostly on the production site or at processing plants.
- **VOLATILITY**: (usually defined as) The normalized standard deviation of the ratio of logarithmic values of a parameter to its

predecessor in equidistant, successive periods.
For (n + 1) values S_i, S_i being the time series to be investigated, e.g., prices of a commodity, applicable in periods i = 0, …, n:

$$u_i = \ln\left(\frac{S_i}{S_{i-1}}\right)$$

an estimate of the mean of this time series u_i is given by:

$$\bar{u} = \frac{1}{(n)}\sum\nolimits_{i=1}^{n}(u_i)$$

and an estimate of the standard deviation of this time series u_i by:

$$s = \sqrt{\frac{1}{(n-1)}\sum\nolimits_{i=1}^{n}(u_i - \bar{u})^2}$$

Then the volatility is defined by $\sigma = \frac{s}{\sqrt{\tau}}$, with τ being the reciprocal length of the time interval T investigated, i.e., $\tau = \frac{1}{T}$.[4]

- **WET NATURAL GAS**: Natural gas existing in the gaseous phase or in solution with crude oil in porous rock formations at reservoir conditions.
- **WITHDRAWAL RATE (or DELIVERABILITY)**: Quantity of natural gas that can be withdrawn from a storage during a short period, e.g., on an hourly or daily basis, usually expressed in [m^3/h] or [m^3/d]. Some storage system operators, however, express and offer withdrawal rates in terms of energy per unit of time, i.e., [kWh/h=kW] or [kWh/d] or [MJ/h], etc.
- **WITHDRAWAL TIME**: Time needed to withdraw natural gas up to the working gascapacity from a natural gas storage, expressed in [h].
- **WORKING GAS CAPACITY**: Capacity that can be used in the storage to inject and withdraw natural gas.
- **WORKING GAS VOLUME**: Volume of natural gas in the storage exceeding cushion gas. The working gas volume can be injected and withdrawn. It is at the disposal of the users.

Selected Examples of Companies, Associations, Hubs, and Pipelines

- **ALMACENAMIENTO OPERATIVO COMERCIAL (AOC)**: A notional/virtual point within the Spanish natural gas pipeline network. Started in 2004, replaced by Punto Virtual de Balance (PVB).
- **CENTRAL EUROPEAN GAS HUB (CEGH)**: Started as a physical trading point in Baumgarten, Austria, acting as an exchange offering natural gas contracts at VTP and VOB, Austria.
- **CENTRAL EASTERN EUROPEAN GAS EXCHANGE (CEEGEX)**: Trading platform for Central Eastern Europe.
- **CME GROUP**: Energy exchange providing natural gas contracts, e.g., in North America, *inter alia*, Nymex Henry Hub.
- **EUROPEAN ENERGY EXCHANGE (EEX)**: Energy exchange in Germany.
- **EUROPEAN FEDERATION OF ENERGY TRADERS (EFET)**: EFET was founded in 1999. It is an association of nearly 100 energy-trading companies. EFET focuses on market design, policy and regulation, standard contracts, and data exchange standards.
- **EXCHANGE TRANSFER FACILITY (ETF)**: A notional/virtual point within the Danish natural gas pipeline network.
- **GASPOOL (GPL)**: A notional/virtual point within the German natural gas pipeline network. Started in 2009.
- **GESTORE MERCATI ENERGETICI (GME)**: Exchange providing, *inter alia*,

[4]Note that other definitions of the elements of the time series are possible and common. Instead of assuming $u_i = \ln\left(\frac{S_i}{S_{i-1}}\right)$ one could also assume e.g. $u_i = \frac{S_i - S_{i-1}}{S_{i-1}}$.

natural gas contracts at PSV and storage capacity, Italy.

- **INTERCONTINENTAL EXCHANGE (ICE-Endex)**: Exchange providing, *inter alia*, natural gas contracts at Henry Hub, TTF, OCM and ZTP, and natural gas storage services in Germany and the Netherlands, UK.
- **NATIONAL BALANCING POINT (NBP)**: A notional/virtual point within the British natural gas pipeline network. Started in 1996.
- **NET CONNECT GERMANY (NCG)**: A notional/virtual point within the German natural gas pipeline network. Started in 2008.
- **PEGAS**: Joint venture of EUROPEAN ENERGY EXCHANGE (EEX) and POWERNEXT. Exchange providing natural gas contracts in Europe, France.
- **POINT D'ÉCHANGE DE GAZ NORD (PEG (N))**: A notional/virtual point within the French natural gas pipeline network (GRTgaz). Started in 2004.
- **POINT D'ÉCHANGE DE GAZ SUD (PEG (S))**: A notional/virtual point within the French natural gas pipeline network (GRTgaz). Started in 2004, merged in 2015.
- **POINT D'ÉCHANGE DE GAZ SUD-OUEST (PEG (S-O))**: A notional/virtual point within the French natural gas pipeline network (TIGF). Started in 2004, merged in 2015.
- **POWERNEXT**: Energy exchange, France.
- **PUNTO DI SCAMBIO VIRTUALE (PSV)**: A notional/virtual point within the Italian natural gas pipeline network. Started in 2003.
- **PUNTO VIRTUAL DE BALANCE (PVB)**: A notional/virtual point within the Spanish natural gas pipeline network. Started in 2016, replaced Almacenamiento Operativo Comercial (AOC).
- **TITLE TRANSFER FACILITY (TTF)**: A notional/virtual point within the Dutch natural gas pipeline network.
- **TRADING REGION SOUTH (TRS)**: PEG (S) and PEG (S-O) merged, forming a virtual point within the French natural gas pipeline system (April 1, 2015).
- **VIRTUAL POINT GAZ-SYSTEM (VPGS)**: A notional/virtual point within the Polish natural gas pipeline network. Started in 2014.
- **VIRTUAL TRADING POINT (VTP)**: A notional/virtual point within the Austrian natural gas pipeline network. Started in 2013.
- **VIRTUÁLNÍ OBCHODNÍ BOD (VOB)**: A notional/virtual point within the Czech natural gas pipeline network. Started in 2011.
- **ZEEBRUGGE HUB (ZEE)**: A physical point at the Zeebrugge terminal, Belgium. Started in 2000.
- **ZEEBRUGGE TRADING POINT (ZTP)**: A notional/virtual point within the Belgian natural gas pipeline network. Started in 2012.

Appendix E
Flexibility Definition

Prerequisites

In addition to the prerequisites of a volume flexibility definition[5], the following conditions that are more formal, shall be fulfilled:

For ϑ^{m}, being the total flexibility (to be defined below) of a (contractual or physical) flexibility tool m, the following conditions shall be fulfilled:

- Positivity

$$\vartheta^{m} \geq 0$$

i.e., the total flexibility shall always have a positive value.

- Transitivity

$$\vartheta^{l} \geq \vartheta^{m} \text{ and } \vartheta^{m} \geq \vartheta^{n}, \text{ then } \vartheta^{l} \geq \vartheta^{n}$$

i.e., if the total flexibility of a contractual or physical flexibility tool m is smaller than the total flexibility of another contractual or physical flexibility tool l and larger than the total flexibility of a third contractual or physical flexibility tool n, then the total flexibility of tool l has to be larger than the total flexibility of tool n, as well.

- Inverse sub-additivity or portfolio reflectiveness

$$\vartheta^{m+n} \geq \vartheta^{m} + \vartheta^{n}$$

i.e., the aggregated total flexibility of two contractual or physical flexibility tools m and n shall be at least equal to the sum of flexibilities of each single contractual or physical flexibility tool m and tool n. This requirement reflects the (potential) existence of portfolio effects.

Flexibility Components and Flexibility Index

All restrictions on single flexibility components shall be used to define contractual or physical flexibility. Thus, a single number will be assigned to each flexibility tool. The first steps towards a calculation of such a flexibility number are descriptions of single-flexibility components' boundary functions.

To this end, the periods used shall be: hour (H), day (D), month (M), quarter (Q), season (S), and year (A)[6]. In general, the flexibility components shall be defined as follows: The hourly flexibility component (δ_H) is the difference between the maximum hourly quantity Θ_{HMax} and the minimum hourly quantity Θ_{HMin}, i.e., $\delta_H = \Theta_{HMax} - \Theta_{HMin}$, with Θ_{HMax} and Θ_{HMin} expressed as percentage of the hourly contract quantity, HCQ_{Norm}, i.e., $\Theta_{HMax} = \frac{HCQ_{Max}}{HCQ_{Norm}}$, etc. Analogously, the daily flexibility component (δ_D); the monthly flexibility component (δ_M); the quarterly flexibility component (δ_Q); the seasonal flexibility component (δ_S) and the

[5]See Sect. 6.3.1.1.4.

[6]For the definition, see Footnote 23 of Sect. 6.3.1.1.2.

A. Pustišek and M. Karasz, *Natural Gas: A Commercial Perspective*,
DOI 10.1007/978-3-319-53249-3

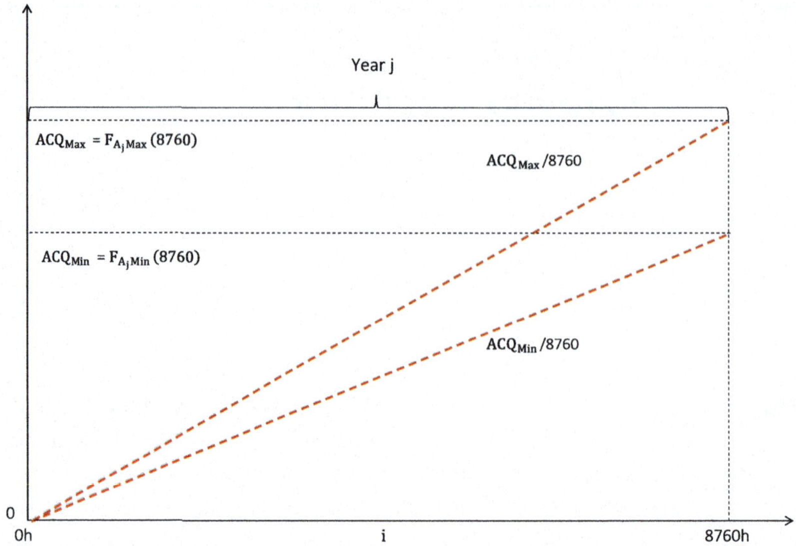

Fig. E.1 Illustration of annual components

annual (yearly) flexibility component (δ_A) shall be defined.

Annual Components' Boundary Function

The annual flexibility component has been defined by $\delta_A = \Theta_{AMax} - \Theta_{AMin}$. With an annual contract quantity, ACQ_{Norm}, the maximal annual quantity, $ACQ_{Max} = ACQ_{Norm} * \Theta_{AMax}$ and the minimal annual quantity $ACQ_{Min} = ACQ_{Norm} * \Theta_{AMin}$, the annual boundary functions for any year j, $F_{A_jMax}(i_{Year})$ and $F_{A_jMin}(i_{Year})$, with $i_{Year} = 1, \ldots, 8760 \in \mathbb{N}$, describing the maximal and minimal off-take during each hour of the year required to fulfil the absolute annual restrictions ACQ_{Max} and ACQ_{Min}, respectively are given by:

$$F_{A_jMax}(i_{Year}) = \sum_{k=1}^{i_{Year}} \frac{ACQ_{Max}}{8760} = i_{Year}\frac{ACQ_{Max}}{8760}$$

$$F_{A_jMin}(i_{Year}) = \sum_{k=1}^{i_{Year}} \frac{ACQ_{Min}}{8760} = i_{Year}\frac{ACQ_{Min}}{8760}$$

Consequently, $F_{A_jMax}(8760) = ACQ_{Max}$ and $F_{A_jMin}(8760) = ACQ_{Min}$.

This is illustrated in Fig. E.1.

Seasonal Components' Boundary Function

Assuming that the seasonal quantity equals $SCQ_{Norm} = \frac{ACQ}{2} = \frac{ACQ_{Norm}}{2}$, the maximal seasonal quantity, SCQ_{Max}, and the minimal seasonal quantity, SCQ_{Min}, are:

$$SCQ_{Max} = SCQ_{Norm} * \Theta_{SMax}$$

and

$$SCQ_{Min} = SCQ_{Norm} * \Theta_{SMin}$$

Evidently, $\Theta_{SMin} \le \Theta_{AMin}$ and $\Theta_{SMax} \ge \Theta_{AMax}$, as otherwise the annual flexibility component could not be fully utilized[7]. The seasonal components, however, have to fulfil the restrictions imposed by the annual components, as well. Hence, as shown subsequently, they are not independent.

- Minimal seasonal quantities

[7]Note that the minimum and maximum seasonal boundary functions, $F_{S_jMin}(i_{Season})$ and $F_{S_jMax}(i_{Season})$, are described subsequently for the year $j \in \mathbb{N}$.

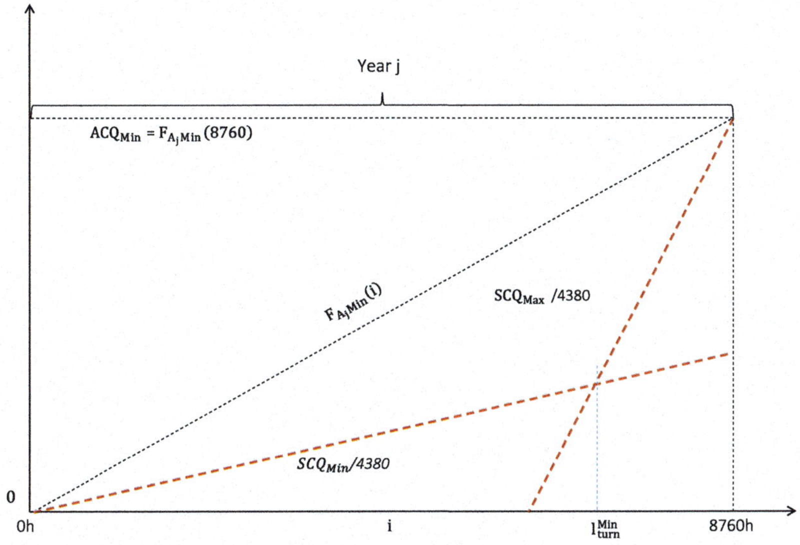

Fig. E.2 Minimal seasonal off-take pattern

As sketched in Fig. E.2, the sum of minimal seasonal quantities has to equal the minimal annual quantity, ACQ_{Min} by the end of the year, i.e., by the end of the 8760th hour. That is, starting at 0, the buyer may take each hour the minimal hourly quantity required to fulfil the seasonal minimal restriction, i.e., $\frac{SCQ_{Min}}{4380}$ until the l_{turn}^{Minth} hour, from which it has to take the maximum hourly quantity permitted according to the seasonal restriction, i.e., $\frac{SCQ_{Max}}{4380}$, in order to reach the required annual minimum quantity during that year, i.e., ACQ_{Min}.

The minimal seasonal off-take shall be defined by the seasonal boundary functions, being discrete functions $F_{S_jMin}(i_{Season})$ each $\mathbb{R} \rightarrow \mathbb{R}$ with $i_{Season} = 1, \ldots, 8760 \in \mathbb{N}$, for the seasons during each year $j \in \mathbb{N}$ by:

$$F_{S_jMin}(i_{Season}) = \begin{cases} i_{Season} * \frac{SCQ_{Min}}{4380} & \text{for } 0 < i_{Season} \leq l_{turn}^{Min} \\ ACQ_{Min} - (8760 - i_{Season}) * \frac{SCQ_{Max}}{4380} & \text{for } l_{turn}^{Min} < i_{Season} \leq 8760 \end{cases}$$

The hour of intersection, i.e., l_{turn}^{Min} can be calculated by:

$$l_{turn}^{Min} = \frac{ACQ_{Min} - 2SCQ_{Max}}{(SCQ_{Min} - SCQ_{Max})}4380$$

Hence the boundary function's value for this hour, l_{turn}^{Min}, is:

$$F_{S_jMin}(l_{turn}^{Min}) = \frac{(ACQ_{Min} - 2SCQ_{Max})}{(SCQ_{Min} - SCQ_{Max})}SCQ_{Min}$$

- Maximal seasonal quantities

Analogously the seasonal maximal quantities are calculated: (Fig. E.3)

The maximal seasonal off-take shall be defined by the seasonal boundary functions, being discrete functions $F_{S_jMax}(i_{Season})$ each $\mathbb{R} \rightarrow \mathbb{R}$ with $i_{Season} = 1, \ldots, 8760 \in \mathbb{N}$, for the seasons during each year $j \in \mathbb{N}$ by:

$$F_{S_jMin}(i_{Season}) = \begin{cases} i_{Season} * \frac{SCQ_{Min}}{4380} & \text{for } 0 < i_{Season} \leq l_{turn}^{Min} \\ ACQ_{Min} - (8760 - i_{Season}) * \frac{SCQ_{Max}}{4380} & \text{for } l_{turn}^{Min} < i_{Season} \leq 8760 \end{cases}$$

The hour of intersection, i.e., l_{turn}^{Max}, can be calculated by:

$$l_{turn}^{Max} = \frac{ACQ_{Max} - 2SCQ_{Min}}{(SCQ_{Max} - SCQ_{Min})}4380$$

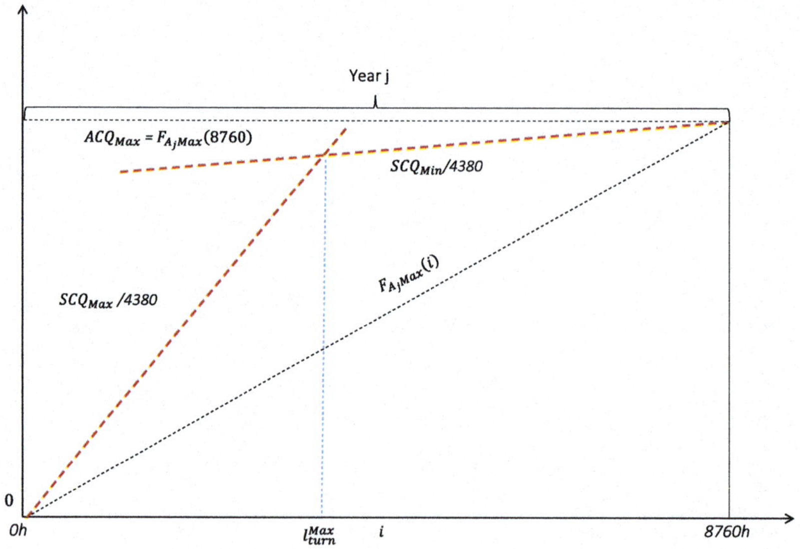

Fig. E.3 Maximal seasonal off-take pattern

Hence the boundary function's value for this hour, l^{Max}_{turn}, is:

$$F_{S_jMin}\left(l^{Max}_{turn}\right) = \frac{(ACQ_{Max} - 2SCQ_{Min})}{(SCQ_{Max} - SCQ_{Min})}SCQ_{Max}$$

Other Components

Other flexibility components' boundary functions are calculated analogously[8]. The results are summarized in Table E.1. The calculation reveals that both minimal and maximal quantities to be taken during each subordinate period have to fulfil the respective restrictions of the superordinate period. For example, hourly quantities cannot be taken without observing the daily restrictions, which, in turn, have to fulfil monthly restrictions etc.

Flexibility Index

Consequently, the boundary function for the hour, be it the minimum or the maximum boundary function, is a (nested) function of the respective boundary functions of all superordinate periods, i.e., for the minimum boundary function:

$$F_{H_jMin}(i) = F_{H_jMin}\Big(i_{Hour}, F_{D_jMin}\Big(i_{Day}, F_{M_jMin}\Big(i_{Month}, F_{Q_jMin}\Big(i_{Quarter}, F_{S_jMin}\big(i_{Season}, F_{A_jMin}(i_{Year,})\big)\Big)\Big)\Big)\Big)$$

This analogously applies to the maximum boundary function, $F_{H_jMax}(i)$.

Using this, flexibility, ϑ^m, of a contract or physical source or tool m can formally be expressed as:

$$\vartheta^m = \frac{1}{ACQ}\sum_{j=1}^{365}\sum_{i=1}^{24}\left[F_{H_jMax}(i) - F_{H_jMin}(i)\right]$$

Thus, it is a function of all components concerned.

[8]Evidently $\Theta_{\text{Subordinate Period Minimum}} \leq \Theta_{\text{Superordinate Period Minimum}}$ and $\Theta_{\text{Subordinate Period Maximum}} \geq \Theta_{\text{Superordinate Period Maximum}}$ as otherwise the flexibility component of the subordinate period could not be fully utilized.

Table E.1 Summary of flexibility components' boundary functions

	Maximum boundary function	Minimum boundary function
Hour	$F_{H_j Max}(i_{Hour}) = \begin{cases} F_{D_{(j-1)} Max}(24) + i_{Hour} * HCQ_{Max} & \text{for } 0 < i_{Hour} \leq 1_{turn}^{Max} \\ F_{D_j Max}(24) - (24 - i_{Hour}) * HCQ_{Min} & \text{for } 1_{turn}^{Max} < i_{Hour} \leq 24 \end{cases}$	$F_{H_j Min}(i_{Hour}) = \begin{cases} F_{D_{(j-1)} Min}(24) + i_{Hour} * H_{CQMin} & \text{for } 0 < i_{Hour} \leq 1_{turn}^{Min} \\ F_{D_j Min}(24) - (24 - i_{Hour}) * HCQ_{Max} & \text{for } 1_{turn}^{Min} < i_{Hour} \leq 24 \end{cases}$
Day	$F_{D_j Max}(i_{Day}) = \begin{cases} F_{M_{(j-1)} Max}(720) + i_{Day} * \frac{DCQ_{Max}}{24} & \text{for } 0 < i_{Day} \leq 1_{turn}^{Max} \\ F_{M_j Max}(720) - (720 - i_{Day}) * \frac{DCQ_{Min}}{24} & \text{for } 1_{turn}^{Max} < i_{Day} \leq 720 \end{cases}$	$F_{D_j Min}(i_{Day}) = \begin{cases} F_{M_{(j-1)} Min}(720) + i_{Day} * \frac{DCQ_{Min}}{24} & \text{for } 0 < i_{Day} \leq 1_{turn}^{Min} \\ F_{M_j Min}(720) - (720 - i_{Day}) * \frac{DCQ_{Max}}{24} & \text{for } 1_{turn}^{Min} < i_{Day} \leq 720 \end{cases}$
Month	$F_{M_j Max}(i_{Month}) = \begin{cases} F_{Q_{(j-1)} Max}(2190) + i_{Month} * \frac{MCQ_{Max}}{720} & \text{for } 0 < i_{Month} \leq 1_{turn}^{Max} \\ F_{Q_j Max}(2190) - (2190 - i_{Month}) * \frac{MCQ_{Min}}{720} & \text{for } 1_{turn}^{Max} < i_{Month} \leq 2190 \end{cases}$	$F_{M_j Min}(i_{Month}) = \begin{cases} F_{Q_{(j-1)} Min}(2190) + i_{Month} * \frac{MCQ_{Min}}{720} & \text{for } 0 < i_{Month} \leq 1_{turn}^{Min} \\ F_{Q_j Min}(2190) - (2190 - i_{Month}) * \frac{MCQ_{Max}}{720} & \text{for } 1_{turn}^{Min} < i_{Month} \leq 2190 \end{cases}$
Quarter	$F_{Q_j Max} F_{Q_j Max}(i_{Quarter}) = \begin{cases} F_{S_{(j-1)} Max}(4380) + i_{Quarter} * \frac{QCQ_{Max}}{2190} & \text{for } 0 < i_{Quarter} \leq 1_{turn}^{Max} \\ F_{S_j Max}(4380) - (4380 - i_{Quarter}) * \frac{QCQ_{Min}}{2190} & \text{for } 1_{turn}^{Max} < i_{Quarter} \leq 4380 \end{cases}$	$F_{Q_j Min}(i_{Quarter}) = \begin{cases} F_{S_{(j-1)} Min}(4380) + i_{Quarter} * \frac{QCQ_{Min}}{2190} & \text{for } 0 < i_{Quarter} \leq 1_{turn}^{Min} \\ F_{S_j Min}(4380) - (4380 - i_{Quarter}) * \frac{QCQ_{Max}}{2190} & \text{for } 1_{turn}^{Min} < i_{Quarter} \leq 4380 \end{cases}$
Season	$F_{S_j Max}(i_{Season}) = \begin{cases} i_{Season} * \frac{SCQ_{Max}}{4380} & \text{for } 0 < i_{Season} \leq 1_{turn}^{Max} \\ ACQ_{Max} - (8760 - i_{Season}) * \frac{SCQ_{Min}}{4380} & \text{for } 1_{turn}^{Max} < i_{Season} \leq 8760 \end{cases}$	$F_{S_j Min}(i_{Season}) = \begin{cases} i_{Season} * \frac{SCQ_{Min}}{4380} & \text{for } 0 < i_{Season} \leq 1_{turn}^{Min} \\ ACQ_{Min} - (8760 - i_{Season}) * \frac{SCQ_{Max}}{4380} & \text{for } 1_{turn}^{Min} < i_{Season} \leq 8760 \end{cases}$
Year	$F_{A_j Max}(i_{Year}) = \sum_{k=1}^{i_{Year}} \frac{Y_{Max}}{8760} = i_{Year} \frac{ACQ_{Max}}{8760}$	$F_{A_j Min}(i_{Year}) = \sum_{k=1}^{i_{Year}} \frac{ACQ_{Min}}{8760} = i_{Year} \frac{ACQ_{Min}}{8760}$

	Indices to the above mentioned boundary functions
Hour	$j = 1, \ldots, 365 \in \mathbb{N}$ $i_{Hour} = 1, \ldots, 24 \in \mathbb{N}$
Day	$j = 1, \ldots, 12 \in \mathbb{N}$ $i_{Day} = 1, \ldots, 720 \in \mathbb{N}$
Month	$j = 1, \ldots, 4 \in \mathbb{N}$. $i_{Month} = 1, \ldots, 2190 \in \mathbb{N}$.
Quarter	$j = 1, \ldots, 2 \in \mathbb{N}$ $i_{Quarter} = 1, \ldots, 4380 \in \mathbb{N}$
Season	$j \in \mathbb{N}$ $i_{Season} = 1, \ldots, 8760 \in \mathbb{N}$
Year	$j \in \mathbb{N}$ $i_{Year} = 1, \ldots, 8760 \in \mathbb{N}$

Appendix F
Equivalence of Additive and Multiplicative Formulae

Both additive and multiplicative price formulae are (multi-dimensional) linear equations. Under certain assumptions, a multiplicative formula could be transformed into an additive one and vice versa.

Such equivalence shall be shown for simplified formulae with n = 3 variables. In addition, without loss of generality, it is assumed that t_{ref_i}, t_{lag_i}, and t_{adj_i} shall be identical for the multiplicative and the additive formula.

Let the additive formula be given by:

$$\begin{aligned} P(t, \Psi_1, \Psi_2, \Psi_3) = P_0 + b_{11} * b_{12} * b_{13} * (\Psi_1 - \Psi_{10}) \\ + b_{21} * b_{22} * b_{23} * (\Psi_2 - \Psi_{20}) \\ + b_{31} * b_{32} * b_{33} * (\Psi_3 - \Psi_{30}) \end{aligned}$$

This shall be transformed to a multiplicative formula of the form:

$$P_M(t, \Psi_1, \Psi_2, \Psi_3) = P_{M0} * \left(d_1 * \frac{\Psi_1}{\Psi_{10}} + d_2 * \frac{\Psi_2}{\Psi_{20}} + d_3 * \frac{\Psi_3}{\Psi_{30}} \right)$$

To this the equation $P(t, \Psi_1, \Psi_2, \Psi_3) = P_M(t, \Psi_1, \Psi_2, \Psi_3)$, i.e.,

$$\begin{aligned} P_0 + b_{11} * b_{12} * b_{13} * (\Psi_1 - \Psi_{10}) + b_{21} * b_{22} * b_{23} * (\Psi_2 - \Psi_{20}) \\ + b_{31} * b_{32} * b_{33} * (\Psi_3 - \Psi_{30}) = P_{M0} * \left(d_1 * \frac{\Psi_1}{\Psi_{10}} + d_2 * \frac{\Psi_2}{\Psi_{20}} + d_3 * \frac{\Psi_3}{\Psi_{30}} \right) \end{aligned}$$

is solved for, e.g., d_1, assuming that d_2, and d_3 are given. Then, with

$$\begin{aligned} d_1 = -\frac{1}{P_{M0}\Psi_1\Psi_{20}\Psi_{30}} \Psi_{10}(d_3 P_{M0}\Psi_{20}\Psi_3 + d_2 P_{M0}\Psi_2\Psi_{30} \\ - P_0\Psi_{20}\Psi_{30} - b_{11}b_{12}b_{13}\Psi_1\Psi_{20}\Psi_{30} \\ + b_{11}b_{12}b_{13}\Psi_{10}\Psi_{20}\Psi_{30} - b_{21}b_{22}b_{23}\Psi_2\Psi_{20}\Psi_{30} \\ + b_{21}b_{22}b_{23}\Psi_{20}^2\Psi_{30} - b_{31}b_{32}b_{33}\Psi_{20}\Psi_3\Psi_{30} \\ + b_{31}b_{32}b_{33}\Psi_{20}\Psi_{30}^2) \end{aligned}$$

The additive and the multiplicative price formula will be equivalent, i.e.:

$$P(t, \Psi_1, \Psi_2, \Psi_3) = P_M(t, \Psi_1, \Psi_2, \Psi_3).$$

Another approach to derive a multiplicative formula from a given additive formula is:

- to start with the additive price formula and its variables Ψ_1, Ψ_2, and Ψ_3,
- to summarize the constants to deliver an adjusted base price, P_0^{Total},
- to set this adjusted base price, P_0^{Total}, to zero, and
- to choose the constants of the multiplicative formula (which is to be determined) to equal the remaining coefficients of the rearranged additive formula.

A. Pustišek and M. Karasz, *Natural Gas: A Commercial Perspective*,
DOI 10.1007/978-3-319-53249-3

This is shown next.

$$
\begin{aligned}
&P_0 + b_{11}{*}b_{12}{*}b_{13}{*}(\Psi_1 - \Psi_{10}) + b_{21}{*}b_{22}{*}b_{23}{*}(\Psi_2 - \Psi_{20}) \\
&\quad + b_{31}{*}b_{32}{*}b_{33}{*}(\Psi_3 - \Psi_{30}) \\
&\quad = P_0 + b_{11}{*}b_{12}{*}b_{13}{*}\Psi_1 - b_{11}{*}b_{12}{*}b_{13}{*}\Psi_{10} + b_{21}{*}b_{22}{*}b_{23}{*}\Psi_2 \\
&\qquad - b_{21}{*}b_{22}{*}b_{23}{*}\Psi_{20} + b_{31}{*}b_{32}{*}b_{33}{*}\Psi_3 - b_{31}{*}b_{32}{*}b_{33}{*}\Psi_{30} \\
&\quad = (P_0 - b_{11}{*}b_{12}{*}b_{13}{*}\Psi_{10} - b_{21}{*}b_{22}{*}b_{23}{*}\Psi_{20} - b_{31}{*}b_{32}{*}b_{33}{*}\Psi_{30}) \\
&\qquad + b_{11}{*}b_{12}{*}b_{13}{*}\Psi_1 + b_{21}{*}b_{22}{*}b_{23}{*}\Psi_2 + b_{31}{*}b_{32}{*}b_{33}{*}\Psi_3 \\
&\quad = P_0^{\mathrm{Total}} + b_{11}{*}b_{12}{*}b_{13}{*}\Psi_1 + b_{21}{*}b_{22}{*}b_{23}{*}\Psi_2 + b_{31}{*}b_{32}{*}b_{33}{*}\Psi_3
\end{aligned}
$$

With all the constants summarized as:

$$
\begin{aligned}
P_0^{\mathrm{Total}} = (&P_0 - b_{11} * b_{12} * b_{13} * \Psi_{10} - b_{21} * b_{22} \\
&* b_{23} * \Psi_{20} - b_{31} * b_{32} * b_{33} * \Psi_{30})
\end{aligned}
$$

Then[9], if, $P_0^{\mathrm{Total}} = 0$ and

$$
\begin{aligned}
b_{11}{*}b_{12}{*}b_{13} &= P_{M0} * \frac{d_1}{\Psi_{10}}, \\
b_{21}{*}b_{22}{*}b_{23} &= P_{M0} * \frac{d_2}{\Psi_{20}}, \text{ and} \\
b_{31}{*}b_{32}{*}b_{33} &= P_{M0} * \frac{d_3}{\Psi_{30}},
\end{aligned}
$$

the additive and multiplicative formulae are equivalent.

[9]Note that this applies for the sum only, and does not imply $P_0 = 0$.

Index

A. Pustišek and M. Karasz, *Natural Gas: A Commercial Perspective*,
DOI 10.1007/978-3-319-53249-3

The manufacturer's authorised representative in the EU is Springer
Nature Customer Service Centre GmbH, Europaplatz 3, 69115 Heidelberg,
Germany. If you have any concerns regarding our products, please
contact ProductSafety@springernature.com

Printed and bound by CPI Group (UK) Ltd, Croydon, CR0 4YY
01/10/2026
02225870-0020